# Wavelets

# Wavelets
# A Concise Guide

Amir-Homayoon Najmi

The Johns Hopkins University Press
Baltimore

The Johns Hopkins University Press

2715 North Charles Street

Baltimore, Maryland 21218-4363

www.press.jhu.edu


ISBN 13: 978-1-4214-0495-0 (hc)
ISBN 13: 978-1-4214-0496-7 (pbk)
ISBN 10: 1-4214-0495-8 (hc)
ISBN 10: 1-4214-0496-6 (pbk)


Library of Congress Control Number: 2011936858


A catalog record for this book is available from the British Library.


*Special discounts are available for bulk purchases of this
book. For more information, please contact Special
Sales at 410-516-6936 or specialsales@press.jhu.edu.*


The Johns Hopkins University Press uses environmentally
friendly book materials, including recycled text paper that is
composed of at least 30 percent post-consumer waste,
whenever possible.

To

*Linda, Amir Christopher,*
*Sasan Jonathan, and Jeffrey Darius*

**In memoriam**

*Mahinjoon*

# Contents

List of Tables     xi

List of Figures     xiii

List of Acronyms     xix

Preface     xxi

Acknowledgments     xxix

| 1 | Analysis in Vector and Function Spaces | 1 |
|---|---|---|
| 1.1 | Introduction | 1 |
| 1.2 | The Lebesgue Integral | 3 |
| 1.3 | Discrete Time Signals | 5 |
| 1.4 | Vector Spaces | 5 |
| 1.5 | Linear Independence | 6 |
| 1.6 | Bases and Basis Vectors | 7 |
| 1.7 | Normed Vector Spaces | 8 |
| 1.8 | Inner Product | 9 |
| 1.9 | Banach and Hilbert Spaces | 12 |
| 1.10 | Linear Operators, Operator Norm, the Adjoint Operator | 13 |
| 1.11 | Reproducing Kernel Hilbert Space | 15 |
| 1.12 | The Dirac Delta Distribution | 18 |
| 1.13 | Orthonormal Vectors | 20 |
| 1.14 | Orthogonal Projections | 21 |
| 1.15 | Multi-Resolution Analysis Subspaces | 22 |
| 1.16 | Complete and Orthonormal Bases in $L_2(\mathbb{R})$ | 25 |
| 1.17 | The Dirac Notation | 28 |
| 1.18 | The Fourier Transform | 31 |
| 1.19 | The Fourier Series Expansion | 34 |
| 1.20 | The Discrete Time Fourier Transform | 36 |
| 1.21 | The Discrete Fourier Transform | 37 |
| 1.22 | Band-Limited Functions and the Sampling Theorem | 38 |

1.23  The Basis Operator in $L_2(\mathbb{R})$                                    41
1.24  Biorthogonal Bases and Representations in $L_2(\mathbb{R})$                43
1.25  Frames in a Finite Dimensional Vector Space                              45
1.26  Frames in $L_2(\mathbb{R})$                                               50
1.27  Dual Frame Construction Algorithm                                        54
1.28  Exercises                                                                56

**2  Linear Time-Invariant Systems                                          59**
2.1   Introduction                                                            59
2.2   Convolution in Continuous Time                                          59
2.3   Convolution in Discrete Time                                            60
2.4   Convolution of Finite Length Sequences                                  61
2.5   Linear Time-Invariant Systems and the **Z** Transform                   63
2.6   Spectral Factorization for Finite Length Sequences                      66
2.7   Perfect Reconstruction Quadrature Mirror Filters                        68
2.8   Exercises                                                               73

**3  Time, Frequency, and Scale Localizing Transforms                       75**
3.1   Introduction                                                            75
3.2   The Windowed Fourier Transform                                          79
3.3   The Windowed Fourier Transform Inverse                                  81
3.4   The Range Space of the Windowed Fourier Transform                       81
3.5   The Discretized Windowed Fourier Transform                              83
3.6   Time-Frequency Resolution of the Windowed Fourier Transform            88
3.7   The Continuous Wavelet Transform                                        90
3.8   The Continuous Wavelet Transform Inverse                                93
3.9   The Range Space of the Continuous Wavelet Transform                     95
3.10  The Morlet, the Mexican Hat, and the Haar Wavelets                      96
3.11  Discretizing the Continuous Wavelet Transform                          101
3.12  Algorithm A' Trous                                                     104
3.13  The Morlet Scalogram                                                   107
3.14  Exercises                                                              110

**4  The Haar and Shannon Wavelets                                         111**
4.1   Introduction                                                           111
4.2   Haar Multi-Resolution Analysis Subspaces                               112
4.3   Summary and Generalization of Results                                  119
4.4   The Spectra of the Haar Filter Coefficients                            122
4.5   Half-Band Finite Impulse Response Filters                              124
4.6   The Shannon Scaling Function                                           125

| | | |
|---|---|---:|
| 4.7 | The Spectrum of the Shannon Filter Coefficients | 130 |
| 4.8 | Meyer's Wavelet | 131 |
| 4.9 | Exercises | 133 |

**5 General Properties of Scaling and Wavelet Functions** — **135**

| | | |
|---|---|---:|
| 5.1 | Introduction | 135 |
| 5.2 | Multi-Resolution Analysis Spaces | 135 |
| 5.3 | The Inverse Relations | 140 |
| 5.4 | The Shift-Invariant Discrete Wavelet Transform | 143 |
| 5.5 | Time Domain Properties | 145 |
| 5.6 | Examples of Finite Length Filter Coefficients | 149 |
| 5.7 | Frequency Domain Relations | 150 |
| 5.8 | Orthogonalization of a Basis Set: $b_1$ Spline Wavelet | 157 |
| 5.9 | The Cascade Algorithm | 159 |
| 5.10 | Biorthogonal Wavelets | 163 |
| 5.11 | Multi-Resolution Analysis Using Biorthogonal Wavelets | 167 |
| 5.12 | Exercises | 170 |

**6 Discrete Wavelet Transform of Discrete Time Signals** — **173**

| | | |
|---|---|---:|
| 6.1 | Introduction | 173 |
| 6.2 | Discrete Time Data and Scaling Function Expansions | 174 |
| 6.3 | Implementing the DWT for Even Length $h_0$ Filters | 179 |
| 6.4 | Denoising and Thresholding | 185 |
| 6.5 | Biorthogonal Wavelets of Compact Support | 187 |
| 6.6 | The Lazy Filters | 191 |
| 6.7 | Exercises | 191 |

**7 Wavelet Regularity and Daubechies Solutions** — **193**

| | | |
|---|---|---:|
| 7.1 | Introduction | 193 |
| 7.2 | Zero Moments of the Mother Wavelet | 194 |
| 7.3 | The Form of $H_0(z)$ and the Decay Rate of $\Phi(\omega)$ | 199 |
| 7.4 | Daubechies Orthogonal Wavelets of Compact Support | 200 |
| 7.5 | Wavelet and Scaling Function Vanishing Moments | 204 |
| 7.6 | Biorthogonal Wavelets of Compact Support | 207 |
| 7.7 | Biorthogonal Spline Wavelets | 211 |
| 7.8 | The Lifting Scheme | 215 |
| 7.9 | Exercises | 217 |

**8   Orthogonal Wavelet Packets**                                       **221**
   8.1   Introduction                                             221
   8.2   Review of the Orthogonal Wavelet Transform               221
   8.3   Packet Functions for Orthonormal Wavelets                224
   8.4   Discrete Orthogonal Packet Transform of Finite Length Sequences                                                231
   8.5   The Best Basis Algorithm                                 236
   8.6   Exercises                                                239

**9   Wavelet Transform in Two Dimensions**                              **241**
   9.1   Introduction                                             241
   9.2   The Forward Transform                                    242
   9.3   The Inverse Transform                                    247
   9.4   Implementing the Two-Dimensional Wavelet Transform       248
   9.5   Application to Image Compression                         249
   9.6   Image Fusion                                             257
   9.7   Wavelet Descendants                                      258
   9.8   Exercises                                                259

**Bibliography**                                                         **261**

**Index**                                                                **267**

# Tables

2.1    Properties of Z transform.    65

2.2    Relations between Fourier and **Z** transforms.    65

3.1    Properties of the continuous wavelet transform (CWT).    92

4.1    Polynomial solutions used in the Fourier transform of the Meyer scaling and wavelet functions.    132

5.1    $h_0$ coefficients for DAUB-4, DAUB-8, DAUB-12, and DAUB-16 Daubechies compactly supported wavelets of orders 4, 8, 12, and 16.    162

6.1    Support of compact biorthogonal scaling and wavelet functions in terms of support of corresponding filter sequences.    189

9.1    9/7 biorthogonal wavelet transform of $512 \times 512$ image (figure 9.3): fraction of total transform energy for the final LL portion as a function of the number of transform levels 1–5.    256

# Figures

P.1   Quartet No. 4 in C Major, K.157, Violin 1 by Mozart.     xxi

P.2   A spectrogram and a continuous Morlet wavelet transform of a bowhead whale sound.     xxiii

P.3   One stage forward and inverse wavelet transform filter bank     xxiv

P.4   The two-dimensional DWT: Resolution level 4 DWT coefficients (LL, LH, HL, and HH) from resolution level 3 approximation (LL).     xxv

P.5   Multi-resolution property of the DWT: approximations at resolution levels 1–5.     xxvi

1.1   Approximation of a vector using nested subspaces and their orthogonal complements.     23

1.2   Three typical functions that are constant on intervals of length 1/2, 1, and 2.     24

1.3   Two biorthogonal pairs of vectors.     43

2.1   Linear time-invariant (LTI) system convolution: continuous and discrete time.     64

2.2   Finite length discrete sequence $x[n]$, $n = -(M-1), \ldots, N-1$, and its time-reversed form $x[-n]$.     65

2.3   Linear time-invariant (LTI) system correlation: continuous and discrete time.     66

2.4   PR-QMF analysis and synthesis filter banks.     69

3.1   Bowhead whale sound and its Fourier spectrum.     75

3.2   Three spectrograms of the bowhead whale sound of figure 3.1.     76

3.3   Basis functions $\xi_{\tau\omega}(t)$ for a triangular window, $\tau = 0$ and $\omega = 0.1\pi$, $0.2\pi$, and $0.3\pi$.     80

3.4   The range space of the windowed Fourier transform.     83

3.5   Trapezoidal window and its dual.     86

3.6   Resolution cells for the windowed Fourier transform.     89

3.7   Unit-norm basis functions $\psi_{\tau\eta}(t)$ for a generic $\psi(t)$, $\tau = 0$ and $\eta = 1$, $0.5$, and $2$.     91

3.8   Resolution cells for the continuous wavelet transform (CWT).   92

3.9   The range space of the continuous wavelet transform (CWT).   96

3.10  Real and imaginary parts of the Morlet mother wavelet ($f_s =$ 1.25 Hz and $\omega_0 = 0.1\pi$, $\sigma = 0.8$ seconds) and its spectrum.   97

3.11  The Morlet constant-$Q$ filter bank: $f_0 = 0.4$ Hz, $\eta = 1, 2, 4, 8, 16$.   98

3.12  The Mexican hat mother wavelet.   99

3.13  The Mexican hat mother wavelet filter bank members: $\eta = 1, 2, 2^2, 2^3$.   99

3.14  Spectrum of the Haar wavelet.   101

3.15  The A' Trous filter for $m = 1$.   105

3.16  XMorlet: an IDL widget to design Morlet wavelets and compute CWT coefficients of arbitrary signals.   109

3.17  CWT coefficients of bat chirp data on the time-frequency plane. 109

4.1   Four examples of the Haar wavelet functions.   112

4.2   The Haar function $\phi_{mn}(t)$.   112

4.3   The Haar scaling function.   114

4.4   The Haar scaling function approximations to $t \exp(-t^2)$, $t \geq 0$ at levels $m = -1$ and $m = -3$.   116

4.5   Down-sampling (decimation) by a factor of 2.   121

4.6   Discrete convolution interpretation for the scaling and wavelet coefficients.   122

4.7   Spectra of the Haar filter coefficients.   122

4.8   Frequency band implication of figure 4.6.   123

4.9   Normalized spectra of the Haar scaling and wavelet functions. 124

4.10  Support of the spectra for functions in $\mathcal{V}_{-1}, \mathcal{V}_0, \mathcal{V}_0^\perp$.   126

4.11  The Shannon scaling and mother wavelet functions.   128

4.12  The Shannon filter coefficients $h_0[n]$ and $h_1[n]$.   129

4.13  Spectra of the Shannon filter coefficients $H_0(\omega)$ and $H_1(\omega)$.   131

4.14  Spectra of Meyer and Shannon scaling functions.   133

4.15  The triangle function and its spectrum.   134

5.1   Discrete convolution interpretation for the inverse scaling and wavelet coefficients.   142

5.2   Discrete convolution interpretation for computing the discrete wavelet coefficients.   142

5.3   The shift-invariant wavelet transform coefficients' recursion with A' Trous interpolated filter sequences.   144

5.4   The shift-invariant wavelet transform inverse (even and odd indices).   144

5.5  $b_1$ spline orthogonal scaling and wavelet functions.  159

5.6  Daubechies scaling and wavelet functions of compact support for $M = 4, 8$ after eight iterations.  161

5.7  Daubechies scaling and wavelet functions of compact support for $M = 12, 16$ after eight iterations.  162

5.8  Single-stage biorthogonal wavelet analysis and synthesis.  168

6.1  Three-stage finite resolution wavelet transform and its inverse. 178

6.2  Linearly chirped signal and its DWT coefficients (six stages).  182

6.3  Signal and reconstruction error (six-stage DAUB-4 transform). 183

6.4  Signal and reconstruction (six-stage DAUB-4 transform) using direct convolution method.  183

6.5  Binary signal: original and noise corrupted.  186

6.6  Six-level Haar transform of signals in figure 6.5.  186

6.7  Binary signal: original and denoised.  186

6.8  Hard and soft thresholding: threshold $= \gamma$.  187

6.9  10/6 biorthogonal wavelet: forward transform matrix.  190

6.10 10/6 biorthogonal wavelet: transpose of inverse transform matrix.  190

6.11 The lazy wavelet filter bank.  191

7.1  A function consisting of a quadratic and a cubic with a very small jump discontinuity at $t = 0$.  193

7.2  Four levels of DWT coefficients using DAUB-16.  194

7.3  Coiflet scaling and wavelet functions of order 6 with two vanishing scaling function moments and one vanishing wavelet function moment.  207

7.4  The 8/8 biorthogonal pair scaling and wavelet functions.  210

7.5  The wavelet function corresponding to the Haar scaling function $b_0(t)$.  212

7.6  The scaling and wavelet functions biorthogonal to $b_0(t)$ and $\psi(t)$ of figure 7.5.  212

7.7  The scaling and wavelet functions corresponding to the spline $b_1(t)$.  214

7.8  The scaling and wavelet functions biorthogonal to the functions in figure 7.7.  214

7.9  The Deslauriers-Dubuc interpolating function $\phi(t)$ and the associated biorthogonal wavelet system.  217

7.10  Coiflet scaling and wavelet functions of order 12 with four vanishing scaling function moments and three vanishing wavelet function moments (excluding the zeroth moment).                    218

7.11  Coiflet scaling and wavelet functions of order 18 with six vanishing scaling function moments and five vanishing wavelet function moments (excluding the zeroth moment).                    219

7.12  The 9/7 biorthogonal pair scaling and wavelet functions.      219

7.13  The 10/6 biorthogonal pair scaling and wavelet functions.     220

8.1   Three-level orthogonal wavelet transform.                     222

8.2   Two decomposition levels in the orthogonal wavelet transform in the frequency domain (positive frequencies are shown).      223

8.3   Time and frequency plane division of a three-level orthogonal wavelet transform of a 16-point data set.                     223

8.4   Orthogonal wavelet packet three-level decomposition (see figure 8.1 for the meaning of the arrows).                         224

8.5   General structure of packet coefficients: stage $m-1$ to stage $m$ (see figure 8.1 for the meaning of the arrows).              225

8.6   Packet functions, their translated and scaled versions, and calculation of packet coefficients.                             226

8.7   DAUB-4 packet function pairs 1 through 4.                      230

8.8   DAUB-2 (Haar) packet function pairs — Walsh functions.        231

8.9   Two-stage orthogonal packet transform of 16-point data set.   232

8.10  Signal reconstruction from packet coefficients: level $m$ from level $m-1$, $0 \leq j \leq m-1$.                               232

8.11  Four-level discrete wavelet and packet transforms for a linearly chirped signal using DAUB-4 orthogonal wavelets of compact support.                                             233

8.12  The orthogonal wavelet packet table for four levels.          235

8.13  The orthogonal wavelet basis location within the orthogonal wavelet packet table.                                            236

8.14  An orthogonal basis, different from the wavelet basis, chosen from the orthogonal wavelet packet table.                       236

8.15  Magnitude squared coefficient entropy function of a unit vector in $\mathbb{R}^3$.                                           237

8.16  Bowhead whale sound and its best packet basis (black) using a six-level DAUB-8 wavelet packet table (the orthogonal wavelet basis is shown in gray).                             239

9.1   Recursive computation of the $\mathbf{cc}_m$ coefficients (the letter above the operation of taking the even numbered samples indicates the index over which the operation is performed).   243

9.2   The forward two-dimensional orthogonal wavelet transform coefficients.   245

9.3   Example of a one-stage forward two-dimensional orthogonal wavelet transform: $\mathbf{cc}_{onk}$ is a $512 \times 512$ image decomposed into four orthogonal components of size $256 \times 256$ each.   246

9.4   Three-stage forward Haar transform of the leftmost image.   246

9.5   The forward two-dimensional orthogonal wavelet transform coefficients.   248

9.6   Data compression using transform coding.   250

9.7   A simple four-level quantizer.   251

9.8   JPEG2000 fundamental building blocks.   252

9.9   DCT: energy versus % number of coefficients sorted on magnitude (decreasing left to right).   253

9.10  9/7 biorthogonal DWT: energy versus % number of coefficients sorted on magnitude (decreasing left to right).   254

9.11  DCT reconstructed images using: (**a**) top 0.1%, (**b**) 1%, and (**c**) 2% of the coefficients with largest magnitude.   254

9.12  9/7 biorthogonal 3-level DWT reconstructed images using: (**a**) top 0.1%, (**b**) 1%, and (**c**) 2% of the coefficients with largest magnitude — map of used coefficients, with a 60 dB range, is shown above each image.   255

9.13  9/7 biorthogonal 5-level DWT reconstructed images using: (**a**) top 0.1%, (**b**) 1%, and (**c**) 2% of the coefficients with largest magnitude — map of used coefficients, with a 60 dB range, is shown above each image.   255

9.14  PSNR values for 5-level 9/7 biorthogonal DWT and DCT— $1024 \times 1024$ version of image shown in figure 9.3.   256

9.15  PSNR curves for JPEG2000 test images: bike, cafe, and woman.   257

# Acronyms

**CWT** Continuous Wavelet Transform

**DC** Direct Current - zero frequency in this text

**DCT** Discrete Cosine Transform

**DFT** Discrete Fourier Transform

**DWT** Discrete Wavelet Transform

**FFT** Fast Fourier Transform

**FIR** Finite Impulse Response

**IIR** Infinite Impulse Response

**JPEG** Joint Photographic Experts Group

**LTI** Linear Time-Invariant

**MRA** Multi-Resolution Analysis

**MRI** Magnetic Resonance Imaging

**RDWT** Redundant Discrete Wavelet Transform

**SIDWT** Shift-Invariant Discrete Wavelet Transform

**SVD** Singular Value Decomposition

**UDWT** Undecimated Discrete Wavelet Transform

**WFT** Windowed Fourier Transform

# Preface

I began studying wavelets when I worked at the Shell Oil Bellaire Geophysical Research Center in Houston in the mid to late 1980s. I then moved to the Applied Physics Laboratory of the Johns Hopkins University where I began using continuous wavelets to analyze electromagnetic and underwater acoustic data, as well as human heart sounds. In 1996, this work led me to produce a wavelet-based heart murmur detector. In the same year, I taught an internal training class on time-frequency localization techniques including applications of the continuous wavelet transform.

In 2000, I began teaching a graduate course in wavelets for the continuing education master's degree program in Applied Sciences and Engineering of the Whiting School of the Johns Hopkins University. With 15 programs, the Whiting School is the largest master's degree program for professionals in the country, with a very diverse student population. The graduate students who have taken the wavelets course since its inception have had degrees in electrical engineering, applied mathematics, physics, and computer science. Thus, I have successfully met the challenge of teaching students with a variety of backgrounds and interests. The present text is, therefore, primarily directed to a varied student population, with an emphasis on both the mathematical and theoretical foundations as well as the implementations of the relevant algorithms.

The study of wavelets can begin with considering the problem of localization of a signal simultaneously in time and frequency. This is the reverse of what normally takes place in a musical composition. Figure P.1 shows a short piece of music in which every note and its duration are precisely defined. Playing the notes according to the score produces the music that plays the role of a signal. Thus, localization of a signal in time and frequency is analogous to producing the score from the music.

Figure P.1: Quartet No. 4 in C Major, K.157, Violin 1 by Mozart.

A basic problem of signal analysis is representing a signal in terms of a collection of continuously parametrized basis functions with useful properties, e.g., the Fourier transform basis functions $e^{i\omega t}$. Although the Fourier transform can determine the frequency components of a signal, it fails to show the times and durations of those components since the basis functions are not localized in time. A time-dependent form of the Fourier transform, the windowed Fourier transform (WFT), uses analyzing functions of the form $w(t - \tau)e^{i\omega t}$ with a localized window function centered at $\tau$. The analyzing functions are constructed by performing two operations on a single localized window function $w(t)$: time-shift by $\tau$ and frequency modulation by $\omega$. Useful digital implementation requires discretization of the time-shift and modulating frequency parameters. Orthonormal discrete WFT analyzing functions are easy to construct on a regular grid, easy to implement, and for decades have been the main tool for investigating and characterizing signals in the time-frequency plane.

A straightforward extension of the windowed Fourier transform to two spatial dimensions and using real valued cosine functions based on localized $8 \times 8$ blocks is known as the discrete cosine transform (DCT) which has provided a low-complexity and memory-efficient backbone of the JPEG standard for image compression. DCT representation of images have good energy concentration properties. Although much of the transform energy is packed into a few coefficients, the locations of the coefficients in the two-dimensional spatial frequency space (the transform domain) are spread throughout that space and reconstructions at low compression suffer from blocking artifacts.

Wavelets[1] provide another way of constructing continuously parametrized basis functions by performing operations of time dilation (scaling) and shift on a localized mother function $\psi(t)$. Thus, wavelet basis functions $\psi\left(\frac{t-\tau}{\eta}\right)$ are compressed or expanded versions of the mother wavelet. The mother wavelet must satisfy only one condition (the admissibility condition): its integral over the real line must vanish. The expansion of a time function in terms of wavelet basis functions is known as the continuous wavelet transform (CWT), which is used to study properties of signals in the time-scale domain.

Figure P.2 shows a comparison between the WFT spectrogram and a Morlet CWT (defined in section 3.10) scalogram (with an equivalent frequency axis instead of scale) for a 15-minute section of a bowhead whale

---

[1] From the French ondelette meaning little wave. Although in the context of a transform the word wavelet was introduced by Morlet and Grossman in 1985 [1], the Ricker wavelet, has been used by exploration geophysicists since 1953 [2].

sound with a nominal sampling frequency of 1 Hz. The CWT image clearly shows five nonlinear chirps that are less well resolved in the WFT spectrogram. The main difference between the two methods is that the WFT analyzing functions have fixed resolutions in the time-frequency plane with the time and frequency resolutions inversely proportional to each other. Whereas the CWT analyzing functions have variable resolutions in the time-scale plane: they are spread out at higher scales (low frequencies) and compressed at lower scales (high frequencies).

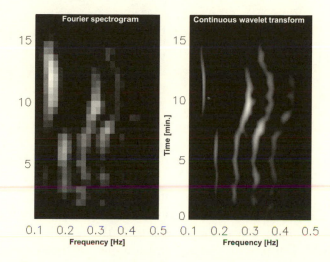
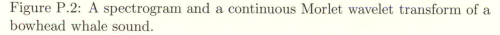

Figure P.2: A spectrogram and a continuous Morlet wavelet transform of a bowhead whale sound.

The discretization of the continuous parameters $\tau$ and $\eta$, in general, leads to an overcomplete and nonorthogonal basis set known as a frame. Although frames provide a powerful method of signal representation because of their redundancy, orthogonal or biorthogonal bases are much easier to implement in both the forward and the inverse directions. These transforms are known as the discrete wavelet transform (DWT). Through the work of Daubechies on the families of orthogonal and biorthogonal wavelets, and the work of Mallat on multi-resolution decomposition, the DWT has become a very valuable tool in signal and image analysis and compression. Mallat's algorithm radically changed the issue of the implementation of the discrete wavelet transform so that only a collection of finite impulse response filter coefficients and an ability to perform linear correlation and convolution is required.

A single stage of the DWT and its inverse is illustrated in figure P.3 in terms of its equivalent bank of analysis and synthesis filters. In the forward transform (the left half of figure P.3) data is filtered (convolved) with the time-reversed versions of the pair of filters $h'_0$ and $h'_1$ (equivalently, correlated with the pair $h'_0$ and $h'_1$), decimated (down-sampled) by a factor of 2. The two outputs are the approximation to the signal at a coarser scale (lower resolution) and the detail lost in this approximation, respectively.

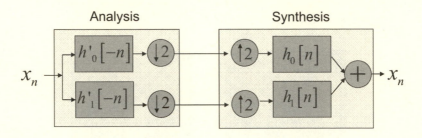

Figure P.3: One stage forward and inverse wavelet transform filter bank

The inverse is performed by first inserting zeros between samples (up-sampling by a factor of 2), filtering them with the pair $h_0$ and $h_1$, and finally summing them to reproduce the original signal. The filter pairs depicted here are biorthogonal, which implies definite relations between the forward and inverse filter pairs. For orthogonal filters, the initial and final pairs are identical. The two filters on the top are low-pass filters, while the two on the bottom are high-pass filters.

Figure P.3 is simple to understand and fast to implement, and is formally identical to the filter bank structures of multirate signal analysis that has been enormously enriched by the wide variety of filter functions related to families of orthogonal and biorthogonal wavelets. Thus, each stage of the one-dimensional DWT produces two data sets at half the length of the input data: one is the coarse approximation to the previous data set (the half-band low-pass filtered data that is down-sampled by 2), and the other is the detail (the half-band high-pass filtered data that is down-sampled by 2). Both are represented by the two arrows in the middle of figure P.3.

The structure shown in figure P.3 is easily extended to two dimensions by performing the operations first on the rows and then on the columns. The output of one stage of the two-dimensional DWT (two one-dimensional DWT applications in the two dimensions) consists of four sub-images (four ways of combining the low-pass and high-pass filters): the low-low (LL) is the approximation to the original at the present level, while the low-high (LH), high-low (HL), and high-high (HH) are the detail DWT coefficients.

This procedure is illustrated in figure P.4 in which the approximation at resolution level 3 produces four sub-images (LL, LH, HL, and HH) at the next level 4.

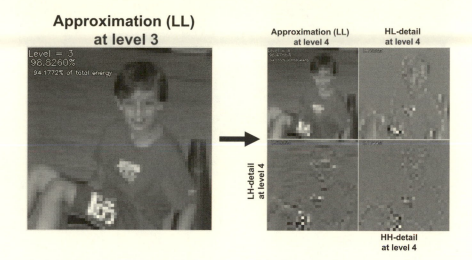

Figure P.4: The two-dimensional DWT: Resolution level 4 DWT coefficients (LL, LH, HL, and HH) from resolution level 3 approximation (LL).

The low-low sub-image is then transformed in the same manner and the transformation process ends after a number of iterations dictated by the size of the initial data. The final set of DWT coefficients consists of the last level's approximation together with all the detail coefficients at each stage. Figure P.5 shows five intervening approximations at five resolution levels 1–5. The size of each approximation is one-half the size of the previous level approximation. Note that the DWT coefficients include the last image (numbered 5) together with the intervening detail images (one set of LH, HL, and HH for each level) that are not shown here.

The inverse to this process (the right hand side of figure P.3) in two dimensions reproduces a given low-low (LL) approximation at resolution level $m$ by using the right hand side of figure P.3 to combine the coarser LL approximation at level $m+1$ together with that level's DWT detail coefficients (low-high LH, high-low HL, and high-high HH). This procedure can be seen in figure P.4 by reversing the arrow: the level-4 LL approximation together with the three sets of level 4 DWT coefficients (HL, LH, and HH) are filtered and combined (the right hand half of figure P.3 to produce the LL approximation at level 3. Figures P.4 and P.5 illustrate the multi-resolution

Figure P.5: Multi-resolution property of the DWT: approximations at resolution levels 1–5.

property of the discrete wavelet transform.

The fast recursive algorithm for computing the discrete wavelet transform, together with an excellent energy compaction property (almost all the energy resides in a small percentage of the coefficients), led to the adoption of the DWT as the backbone of the JPEG2000 standard for image compression. Other benefits of the DWT in image compression include its abilities to decorrelate an image at larger scales and to eliminate blocking artifacts at high compression ratios (a major issue for the DCT and the older JPEG standard).

The two-dimensional wavelet transform has spawned many new and improved image decompositions. For instance, the curvelet transform represents an image in multiple scales and multiple angles. Other areas of applied mathematics have found new wavelet-related methods that solve existing problems more efficiently and faster. For example, wavelets are used to find more accurate and faster methods of solving the wave and the diffusion equations. Simultaneous localization in space and scale dimensions of wavelets, when extended to the surface of a sphere, has found applications to cosmology. Wavelet analysis of the cosmic microwave background radiation (CMBR) data from the WMAP (Wilkinson Microwave Anisotropy Probe) satellite has been successfully used to constrain the parameters of the present $\Lambda CDM$ (cosmological constant, cold dark matter) model of the universe [3].

A full understanding of wavelets and their extensions and applications, however, requires a thorough review of important concepts in functional analysis and linear system theory, which form the basis for this text. Chapter 1, the longest chapter, is devoted to functional analysis and the theory of frame decomposition. Chapter 2 is an introduction to linear system theory, convolution, and the **Z** transform: essential for understanding the Daubechies solutions and for computer imlementation of the Mallat algorithm. Chapter 3 introduces time-frequency and time-scale transforms: the windowed Fourier transform (WFT) and the continuous wavelet transform (CWT). The Mallat multi-resolution analysis interpretation of the discrete wavelet transform and the underlying scaling and wavelet functions are introduced in Chapter 4 through the two well-known examples of the Haar and the Shannon functions. Chapter 5 describes general properties of orthogonal and biorthogonal wavelet systems in both time and frequency domains. Simple implementation and applications of the discrete wavelet transforms of compact support to discrete and finite length data are discussed in Chapter 6. Concepts of regularity and vanishing moments and the Daubechies method of constructing orthogonal wavelets, Coiflets, and biorthogonal wavelets are described in detail in Chapter 7. Chapter 8 describes the more general concept of wavelet packets, of which the standard wavelet transform is a subset, and the best basis algorithm using an entropy cost function. Chapter 9 is a straightforward extension of the one-dimensional discrete wavelet transform to two-dimensional data (images). Each chapter includes a set of exercises.

I have made every effort to produce a self-contained text using accurate, consistent, and comprehensible mathematical notation that should be comfortable for both physicists and applied mathematicians, as well as useful diagrams with which electrical engineering students should be familiar. Ultimately, however, a textbook of this kind reflects the author's path in exploring a specific field. I could not hope to cover all topics in the subject but I have included those I consider necessary for a comprehensive understanding of this fascinating and elegant field.

# Acknowledgments

I am grateful to the Whiting School of the Johns Hopkins University for providing me the opportunity to teach in the professional master's degree program since 1992, and wish to thank Dr. Dexter Smith, currently the Associate Dean of the Whiting School, for encouraging me to develop and teach a course on wavelets.

This textbook could not have been completed without the generous Stuart Janney Fellowship awarded to me by the Johns Hopkins University Applied Physics Laboratory (JHU-APL) and guidance and encouragement from the editors at the Johns Hopkins University Press: Dr. Trevor Lipscombe and Dr. Vincent Burke.

The particular style and emphasis in this text are the result of years of interaction with more than 120 graduate students who have taken this class since its inception. Jon Zombek, William Sweeney, and Christie O'Hara have been particularly helpful. Dr. Rick Chapman, Dr. Don Thompson, Dr. Bill Peter, Dr. Shawn Johnson, Mr. Frank Monaldo, and Mr. Scott Peacock from JHU-APL, Dr. Majid Rabbani, Director of Kodak Image Sciences, Dr. Christopher Burrows, and Dr. William Franklin have made constructive comments on portions of the manuscript.

All technical figures in this book were produced by programs written in the scientific computing language IDL from ITT Visual Systems (formerly RSI), who have provided me a free personal license. I thank Mr. Ray Sterner of the Space Department of JHU-APL for teaching me IDL and providing an unmatched IDL subroutine library.

# Wavelets

# CHAPTER 1

# Analysis in Vector and Function Spaces

## 1.1 Introduction

Independent variables are either continuous or discrete. If we denote by $t$ a continuous variable, $-\infty < t < \infty$, then the range of $t$ defines the real line $\mathbb{R}$. If we denote a discrete variable by $n$, then the range $n = \ldots, -2, -1, 0, 1, 2, \ldots$ defines the set of integers $\mathbb{Z}$. Thus, we write $t \in \mathbb{R}$ and $n \in \mathbb{Z}$ introducing the symbol $\in$ to mean "belongs to" or "is a member of." We define the range of ordered pairs $(x, y)$, $x, y \in \mathbb{R}$, to be the real plane and we denote it by $\mathbb{R}^2$. The set of all "grid" points $(m, n)$, $m, n \in \mathbb{Z}$, is denoted by $\mathbb{Z}^2$. There are times when we need to refer to the positive or negative "half" of the real line or the set of integers in which case we modify the notation in an obvious way: $\mathbb{R}^+$ denotes the positive real line (including 0) and $\mathbb{Z}^+$ denotes the non-negative integers. The set of complex numbers $\mathbb{C}$ is defined as the range space of the complex variable $z = x + iy$, $x, y \in \mathbb{R}$, and $i \equiv \sqrt{-1}$.[1] Complex conjugation is denoted by superscripted $*$: the complex conjugate of $z = x + iy$ is $z^* = x - iy$. In addition, we use the symbol $\forall$ to mean for all, $\exists$ meaning there exists, $\Leftrightarrow$ meaning if and only if, and set theoretic symbols $\cup$ for union of sets and $\cap$ for their intersection.

We think of all signals as real or complex valued functions of either a continuous time variable $t \in \mathbb{R}$, e.g., $x(t)$, or a discrete time variable $n \in \mathbb{Z}$, e.g., $x[n]$. Although many signals of interest are real quantities, some circumstances necessarily produce complex signals. For instance, the

---

[1]The same notation $\mathbb{C}$ is often used to denote the complex plane which is the space $\mathbb{R}^2$ of ordered pairs $(x, y)$, $x, y \in \mathbb{R}$, together with addition and multiplication defined by $(x_1, y_1) + (x_2, y_2) \equiv (x_1 + x_2, y_1 + y_2)$ and $(x_1, y_1)(x_2, y_2) \equiv (x_1 x_2 - y_1 y_2, x_1 y_2 + x_2 y_1)$, respectively. Defining the complex number $z \equiv x + iy$ the latter rules lead to the usual complex number algebra with $(1, 0)$ as the identity and $1/z = (x/r^2, -y/r^2)$ with $r^2 \equiv x^2 + y^2$. The symbol $\equiv$ is used to mean "defined" or "is equivalent to" and is distinct from the equality sign used to mean "is equal to."

Fourier transform (see section 1.18 of chapter 1) of a real signal is a complex signal. A second example is a signal-processing operation known as complex demodulation whereby a real signal whose spectrum is limited to a small band of frequencies around a very high carrier frequency $\omega_c$ is shifted down in frequency, retaining the original shape of the spectrum but now centered around zero frequency (also known as DC). The process of shifting down of the frequencies involves a multiplication of the original real signal with the complex exponential $\exp(i\omega_c t)$, thus producing a complex signal. A third example of complex signals is provided by magnetic resonance imaging (MRI) in which the circularly polarized changing magnetic field, the source of the MRI signal, results in two data streams with a $90°$ phase difference that are combined together to form a single complex sequence of numbers. Whether real or complex, we think of the space of all signals as a space of functions with properties that we will discuss in this chapter. Two spaces of particular interest are those in the following two definitions.

**Definition 1.** The space of continuous time signals $x(t)$ with finite energy, $\int_{-\infty}^{\infty} |x(t)|^2 dt < \infty$, is denoted by $L_2(\mathbb{R})$. This is also known as the space of square integrable functions.

**Definition 2.** The space of continuous time signals $x(t)$ that are absolutely integrable, i.e., functions $x(t)$ for which $\int_{-\infty}^{\infty} |x(t)| dt < \infty$, is denoted by $L_1(\mathbb{R})$.

In the next section we describe the meaning of the integrals in the above definitions. It is important to note that in signal analysis there are two broad categories of functions of interest: functions with finite energy (definition 1), and functions such as thermal noise at a receiver that do not possess finite energy but have finite power (energy per unit time) defined below

$$\lim_{T \to +\infty} \frac{1}{2T} \int_{-T}^{T} |x(t)|^2 \, dt.$$

Although signals with finite energy are of interest, all functions with finite power (such as thermal noise) could be included if we assume that a window function of finite duration is applied: that is, the function is set to zero outside a finite time range.

## 1.2   The Lebesgue Integral

All integrals in this text are Lebesgue integrals which are based on the concept of the Lebesgue measure[4]. The Lebesgue integral is a more general concept than that of the Riemann integral which is based on the concept of "the limit of a sum." A difficulty with the Riemann integral can be illustrated with the following example. Consider the function $x(t)$ defined on the closed interval $[0, 1]$ by the rule that it is equal to 1 when $t$ is a rational number, and it is equal to 0 when $t$ is irrational. Proceeding with the concept of a Riemann integral we divide the interval into $N$ subintervals $[t_n, t_{n+1}]$ of equal length $\Delta t$, $t_0 = 0$ and $t_N = 1$, and consider the sum

$$\sum_{n=0}^{N} x\left(\bar{t}_n\right) \Delta t,$$

where $\bar{t}_n$ is any point in the interval $[t_n, t_{n+1}]$. The limit of this sum as $N \to \infty$, if it exists, is the Riemann integral of our example function. The result, of course, must be independent of the choice of evaluation point $\bar{t}_n$. It is clear, however, that no matter how large $N$ is the corresponding intervals contain at least one rational and one irrational number and the result of the sum can be made equal to 1 if we always choose a rational number for all the points $\bar{t}_n$, or zero if we choose irrationals! Thus, the limit does not exist and the function is not integrable in the Riemann sense.

The Lebesgue integral, on the other hand, is based on the concept of the Lebesgue measurable sets. The simplest example of a measurable set is the closed interval $[T_1, T_2]$ in $\mathbb{R}$ whose measure is the length of the interval $T_2 - T_1$. In fact, the Lebesgue measures of the semi-open sets $(T_1, T_2]$ and $[T_1, T_2)$ and that of the open set $(T_1, T_2)$ are all given by the same length. Thus, the Lebesgue measure of one or two points is zero, as is the measure of a countable set of points, where countable means that the set can be put into a one-to-one correspondence with the integers.[2] For instance, consider the set of all positive rational numbers $\mathbb{Q}^+$ whose members are of the form $n/m$

---

[2]Although we take the measure of a countable set to be zero, countability is not a necessary condition for a set to have zero measure. The best known example of an uncountable set with zero measure is the Cantor set that is created by repeatedly deleting the open middle thirds of a set of line segments. Starting with the closed interval $[0, 1]$ we delete the open middle third $(1/3, 2/3)$, leaving the set $[0, 1/3] \cup [2/3, 1]$, and continue by removing the middle thirds of each of the remaining intervals. The final removed set is then $\bigcup_{j=1}^{\infty} \bigcup_{k=0}^{(3^{j-1}-1)} 3^{-j}(1 + 3k, 2 + 3k)$ which can be shown to have measure 1. The Cantor set is the uncountable closed set of remaining points and has measure 0.

($n, m$ nonzero positive integers). We can "count" them by writing them in the form of a matrix whose row is labeled by $n$ and whose column is labeled by $m$ and then mapping the $(1, 1)$ element to 1, elements $(2, 1), (1, 2)$ to $2, 3$, elements $(3, 1), (2, 2), (1, 3)$ to $4, 5, 6$, and so on. Consequently the set $\mathbb{Q}^+$ is countable and has measure zero.[3]

Now consider the function $x(t)$ defined on the interval $[t_a, t_b]$ whose range is $[x_a, x_b]$. In general, $x(t)$ is a many-to-one map between the two intervals in the sense that there may be many $t$ values that will give the same $x$ value. The Riemann integral is defined by dividing the domain of the map $[t_a, t_b]$ into a number of equal length subintervals as was discussed in the last paragraph. To define the Lebesgue integral we begin by dividing the range of the map into $N$ equal length subintervals $[x_k, x_{k+1}]$, $k = 0, \ldots, N$, $x_0 = x_a$, $x_N = x_b$. Denoting the Lebesgue measure of the set $x^{-1}([x_k, x_{k+1}])$ (i.e., the set whose image under the function $x$ is the interval $[x_k, x_{k+1}]$) by $M_k$, the Lebesgue integral of the function $x(t)$ on the interval $[t_a, t_b]$ is then defined by

$$\lim_{N \to +\infty} \sum_{k=0}^{N} M_k x_k,$$

provided that the limit exists. It is important to note, however, that if the Riemann integral of a given function exists, then it is equal to the Lebesgue integral and there is no distinction between the two in that case. Later in this chapter, when we define inner products and norms of functions, we will have to make no distinction between functions in $L_2(\mathbb{R})$ that are the same everywhere except on a set of measure zero. For instance, the function defined earlier on $[0, 1]$ that took the value 1 on all the rationals and 0 on all the irrationals will be considered to be the same as the function 0 defined on the same interval, and so its Lebesgue integral is 0. Thus, we are implicitly considering equivalence classes of functions[4] instead of individual functions: functions belonging to the same class differ from each other only on a set of measure zero. In other words, finite or countable numbers of points (as well

---

[3]The definition of the Lebesgue measure includes many properties[4]. The one property of the Lebesgue measure on $\mathbb{R}$ used here is that if a subset $S$ of $\mathbb{R}$ is a disjoint union of countably many disjoint Lebesgue measurable subsets of $\mathbb{R}$, then $S$ is itself Lebesgue measurable and its measure is equal to the sum of the measures of the underlying measurable sets.

[4]Equivalence class structures arise from equivalence relations. A binary relation $R$ on a set $\{x, y, z, \ldots\}$, written $x \, R \, y$ to denote the relation between $x$ and $y$, is said to be an equivalence relation if it is symmetric, i.e., : $x \, R \, y \Rightarrow y \, R \, x$, and it is reflexive, i.e., : $x \, R \, x$, and finally it is transitive, i.e., : $x \, R \, y$, $y \, R \, z \Rightarrow x \, R \, z$. The equivalence class $[x]$ then contains all elements of the set that are related (equivalent) to $x$.

as any other set of measure zero) do not count!

**Definition 3.** Two functions $x(t)$ and $y(t)$ in $L_2(\mathbb{R})$ are said to be the same almost everywhere if they differ from each other only on a set of measure zero, i.e., the measure of the set $\{t \in \mathbb{R} : x(t) \neq y(t)\}$ is zero.

Hence, an equivalence class structure in $L_2(\mathbb{R})$ can be defined where all functions that are the same almost everywhere belong to the same class. The equivalence class structure of $L_2(\mathbb{R})$ will allow us in later chapters to define inner products and norms unambiguously.

## 1.3   Discrete Time Signals

In most modern applications a continuous time signal $x(t)$ is converted to a discrete time series denoted by $x[n]$ representing the signal $x(n\Delta t)$, where $\Delta t$ is the sampling interval of the analog-to-digital converter and is related to the sampling frequency $f_s$ by $\Delta t = 1/f_s$. When $\Delta t$ is measured in seconds the sampling frequency is measured in Hertz (Hz). The space of finite energy discrete time signals is defined as follows.

**Definition 4.** The space of discrete time signals of finite energy $x[n]$, $\sum_{n=-\infty}^{\infty} |x[n]|^2 < \infty$, is denoted by $l_2(\mathbb{Z})$.

## 1.4   Vector Spaces

The most basic algebraic structure of interest is the vector space [5]. In its most concise form it defines the space of all functions of interest to be such that if $x_1(t)$ and $x_2(t)$ are two signals in the space, then for all complex numbers $\alpha$ and $\beta$, the linear combination $\alpha x_1(t) + \beta x_2(t)$ also belongs to the space. Thus, the space is closed under the rules of addition of functions and multiplication of functions by complex numbers. The sets of real numbers $\mathbb{R}$, complex numbers $\mathbb{C}$, and rationals $\mathbb{Q}$ are examples of a mathematical entity known as a field which is required in the definition of a vector space. An abstract field is a set with notions of addition and multiplication and their corresponding inverses, namely, subtraction and division.

**Definition 5.** A vector space $\mathcal{V}$ defined over a field $\mathbb{F}$ is a collection of objects **x** called vectors together with a binary operation called addition, and a scalar multiplication by the members of the field satisfying the following axioms.

- To every pair $\mathbf{x}$ and $\mathbf{y}$ of vectors corresponds a vector $\mathbf{x} + \mathbf{y}$ that is called their sum. Addition of vectors is commutative and associative.

- There is a unique vector $\mathbf{0}$ such that $\mathbf{x} + \mathbf{0} = \mathbf{x}$ for all $\mathbf{x} \in \mathscr{V}$.

- For every vector $\mathbf{x} \in \mathscr{V}$ there is a unique vector $-\mathbf{x}$ such that $\mathbf{x} + (-\mathbf{x}) = \mathbf{0}$.

- For every vector $\mathbf{x} \in \mathscr{V}$ and every scalar $\alpha \in \mathbb{F}$ there corresponds a vector $\alpha\mathbf{x}$, the product of $\alpha$ and $\mathbf{x}$, and multiplication by scalars is associative.

- The product $1\mathbf{x}$ is equal to $\mathbf{x}$ for every vector $\mathbf{x} \in \mathscr{V}$.

- Multiplication by scalars is distributive with respect to vector addition, i.e., $\alpha(\mathbf{x} + \mathbf{y}) = \alpha\mathbf{x} + \alpha\mathbf{y}$.

- Multiplication by vectors is distributive with respect to scalar addition, i.e., $(\alpha + \beta)\mathbf{x} = \alpha\mathbf{x} + \beta\mathbf{x}$.

If the field is the set of real numbers $\mathbb{R}$ then $\mathscr{V}$ is called a real vector space, and if the field is $\mathbb{C}$ then $\mathscr{V}$ is a complex vector space. These are the only two fields of interest to us in this book.

## 1.5   Linear Independence

A linear combination of vectors denoted by $\sum_{j=1}^{N} c_j\mathbf{x}_j$ is clearly a member of the vector space.

**Definition 6.** A set of vectors $\{\mathbf{x}_j : j = 1, \ldots, N\}$ is linearly independent provided

$$\sum_{j=1}^{N} c_j\mathbf{x}_j = 0 \Leftrightarrow c_j = 0 \quad \forall j. \tag{1.1}$$

**Theorem 1.** *The set of nonzero vectors $\{\mathbf{x}_j : j = 1, \ldots, N\}$ is linearly dependent if and only if some $\mathbf{x}_k$, $k = 1, \ldots, N$, is a linear combination of the preceding vectors.*

## 1.6   Bases and Basis Vectors

**Definition 7.** A basis in a vector space consists of a set of linearly independent vectors, called basis vectors, such that every vector in the given space is a linear combination of the basis vectors. A vector space is finite dimensional if it has a finite basis.

The basis is not unique, although once chosen, the representation of any vector in terms of the basis (the "components") is unique. Every linearly independent set can be extended to a basis by the following theorem.

**Theorem 2.** *The set of linearly independent vectors $\{\mathbf{x}_j : \ j = 1, \ldots, M\}$ in a finite dimensional vector space either forms a basis or can be extended to a set $\{\mathbf{x}_j : \ j = 1, \ldots, N\}$, where $N > M$ which is a basis.*

**Theorem 3.** *The number of elements in any basis of a finite dimensional vector space is the same as in any other basis. This number $N$ is the dimension of the corresponding vector space. In addition, every set of $N + 1$ vectors in an $N$ dimensional vector space is linearly dependent.*

**Definition 8.** Given a set of vectors $\{\mathbf{x}_j : \ j = 1, \ldots, N\}$ which are not necessarily linearly independent, and a field $\mathbb{F}$, the span of this set is defined by

$$\text{span}\,\{\mathbf{x}_j : \ j = 1, \ldots, N\} = \left\{ \sum_{j=1}^{N} c_j \mathbf{x}_j, \ \forall c_j \in \mathbb{F} \right\}. \tag{1.2}$$

Note that the span of a set of vectors satisfies all the conditions required for a vector space. If the span of a set of vectors is not the entire space, then it is a proper subspace.

**Definition 9.** An infinite-dimensional vector space is a vector space in which no linear span of the form (1.2) is the entire space, for any integer $N$.

In other words, for any given integer $N$, however large, and any set of vectors $\mathbf{x}_j, \ j = 1, \ldots, N$, there is at least one vector $\mathbf{y}$ that is not in the linear span of those $N$ vectors.

We continue to denote a vector in a finite dimensional vector space by bold letters, e.g., $\mathbf{x}$. An infinite dimensional vector space in this text is a space of functions of a real variable, e.g., $x(t)$, which we often abbreviate to $x$, dropping the dependence on the real variable $t$. Although many definitions and properties in this chapter use bold letters that refer to vectors in finite dimensional vector spaces, most are equally applicable to infinite dimensional space of functions. The two notations are for the most part interchangeable unless otherwise stated.

**Definition 10.** If $\mathbb{V}$ and $\mathbb{W}$ are subspaces of the vector space $\mathscr{V}$, the inner sum, denoted by $\mathbb{V} + \mathbb{W}$, is a subspace defined by

$$\mathbb{V} + \mathbb{W} \equiv \{\mathbf{u} \in \mathscr{V} : \; \mathbf{u} = \mathbf{v} + \mathbf{w}; \; \mathbf{v} \in \mathbb{V}, \mathbf{w} \in \mathbb{W}\},$$

while the direct sum of the two subspaces, denoted by $\mathbb{V} \oplus \mathbb{W}$, is defined by

$$\mathbb{V} \oplus \mathbb{W} \equiv \{(\mathbf{v}, \mathbf{w}) : \; \mathbf{v} \in \mathbb{V}, \mathbf{w} \in \mathbb{W}\}.$$

If $\mathbb{V}$ and $\mathbb{W}$ are disjoint, i.e., $\mathbb{V} \cap \mathbb{W} = \{\mathbf{0}\}$, then $\mathbb{V} + \mathbb{W}$ and $\mathbb{V} \oplus \mathbb{W}$ are isomorphic, i.e., they have identical algebraic structure.

In order to do analysis we must be able to tell how close vectors are to each other, and so we introduce the concept of normed vector spaces in the next section [6].

## 1.7   Normed Vector Spaces

**Definition 11.** A normed vector space is equipped with a non-negative real valued function (the norm), $\|\cdot\| : \mathscr{V} \to \mathbb{R}$, such that

$$\|\mathbf{x}\| \geq 0 \; \forall x \in \mathscr{V}, \tag{1.3a}$$

$$\|\mathbf{x}\| = 0 \Leftrightarrow \mathbf{x} = \mathbf{0}, \tag{1.3b}$$

$$\|a\mathbf{x}\| = |a| \, \|\mathbf{x}\|, \tag{1.3c}$$

$$\|\mathbf{x} + \mathbf{y}\| \leq \|\mathbf{x}\| + \|\mathbf{y}\|, \quad \forall \mathbf{x}, \mathbf{y} \in \mathscr{V}. \tag{1.3d}$$

Condition (1.3d) is known as the triangle inequality. For instance, if $\mathscr{V}$ is the infinite dimensional vector space of all square-integrable functions defined on the interval $[a, b] \in \mathbb{R}$, the $L_2$ norm of a function $x(t)$ is defined by

$$\|x(t)\|_2 = \left[\int_a^b |x(t)|^2 \, dt\right]^{1/2}.$$

This can be generalized to an $L_p$ norm where $1 \leq p < \infty$ and $\mathscr{V}$ is the vector space of all functions whose magnitude raised to the power $p$ is integrable

$$\|x(t)\|_p = \left[\int_a^b |x(t)|^p \, dt\right]^{1/p}.$$

The norm enables us to define convergence in a normed vector space. In the infinite dimensional space $L_2(\mathbb{R})$, the space of functions with finite $L_2$ norm, three types of convergence can be defined.

**Definition 12.** A sequence $x_n(t)$, $n = 0, 1, 2, \ldots$, converges in the mean to the function $x(t)$ if $\|x_n - x\| \to 0$ for $n \to \infty$.

Unless otherwise stated, this is the definition for convergence that is used throughout this book. Here we will state the other two possible definitions for convergence.

**Definition 13.** A sequence $x_n(t)$, $n = 0, 1, 2, \ldots$, converges pointwise to the function $x(t)$ if, given the point $t$ and a small positive number $\varepsilon$, an integer $N_t$ can be found for which $|x_n(t) - x(t)| < \varepsilon$ for all $n \geq N_t$.

**Definition 14.** A sequence $x_n(t)$, $n = 0, 1, 2, \ldots$, converges uniformly to the function $x(t)$ if, for any small positive number $\varepsilon$, an integer $N$ can be found that is independent of $t$ and for which $|x_n(t) - x(t)| < \varepsilon$ for all $n \geq N$.

Given a sequence of functions $x_n(t)$ converging in the mean to the function $x(t)$ and using the triangle inequality we have

$$\|x_n - x_m\| = \|x_n - x + x - x_m\| \leq \|x_n - x\| + \|x_m - x\|,$$

which in view of the convergence property, can be made as small as we please by taking the integers $n$ and $m$ sufficiently large. This motivates the following definition.

**Definition 15.** A sequence $x_n(t)$, $n = 0, 1, 2, \ldots$, is called a Cauchy sequence if, for any $\varepsilon > 0$ there is a positive integer $N$ (which could depend on $\varepsilon$) such that $\|x_n - x_m\| < \varepsilon$ for $n, m \geq N$.

All convergent sequences are Cauchy sequences, but the converse is not true.

**Definition 16.** A subspace of a normed vector space is closed if all convergent sequences in that subspace converge to points within that subspace.

## 1.8 Inner Product

An inner product can be used to project vectors in arbitrary directions, in addition to introducing the concept of orthogonality between vectors.

**Definition 17.** The inner product is a complex function $\langle \cdot, \cdot \rangle : \mathscr{V} \times \mathscr{V} \to \mathbb{C}$ with the following properties[5]:

$$\langle \mathbf{x}, \mathbf{y} \rangle = \langle \mathbf{y}, \mathbf{x} \rangle^*, \tag{1.4a}$$

$$\langle \mathbf{x}, a\,\mathbf{y} \rangle = a \langle \mathbf{x}, \mathbf{y} \rangle, \tag{1.4b}$$

$$\langle \mathbf{x} + \mathbf{y}, \mathbf{z} \rangle = \langle \mathbf{x}, \mathbf{z} \rangle + \langle \mathbf{y}, \mathbf{z} \rangle, \tag{1.4c}$$

$$\langle \mathbf{x}, \mathbf{x} \rangle > 0, \quad \forall \mathbf{x} \neq \mathbf{0}, \tag{1.4d}$$

$$\langle \mathbf{x}, \mathbf{x} \rangle = 0 \Leftrightarrow \mathbf{x} = \mathbf{0}. \tag{1.4e}$$

For example, if $\mathscr{V} = \mathbb{R}^n$ and $\mathbf{x} = [x_1, x_2, \ldots, x_n]^T$, where $x_j \in \mathbb{R}$ and superscript $T$ denotes transposition, then

$$\langle \mathbf{x}, \mathbf{y} \rangle = \mathbf{x} \cdot \mathbf{y} = \mathbf{x}^T \mathbf{y} = \sum_{j=1}^{n} x_j y_j. \tag{1.5}$$

If, on the other hand, $\mathscr{V} = \mathbb{C}^n$ and $x_j \in \mathbb{C}$, then

$$\langle \mathbf{x}, \mathbf{y} \rangle = \mathbf{x}^* \cdot \mathbf{y} = \mathbf{x}^H \mathbf{y} = \sum_{j=1}^{n} x_j^* y_j, \tag{1.6}$$

where the superscript $H$ denotes the Hermitian conjugate, $\mathbf{x}^H \equiv [x_1^*, \ldots, x_n^*]$. Note that the relation

$$\langle a\mathbf{x}, \mathbf{y} \rangle = a^* \langle \mathbf{x}, \mathbf{y} \rangle$$

is a consequence of the inner product defining properties (1.4a) and (1.4b). The usual example of an inner product on a function space is

$$\langle x, y \rangle = \int\limits_{-\infty}^{\infty} x^*(t) y(t)\, dt$$

where and $x$ and $y$ are square-integrable complex valued functions of the continuous variable $t \in \mathbb{R}$.

    An inner product space has a natural norm, called the induced norm, associated with the inner product,

$$\|\mathbf{x}\| \equiv \sqrt{\langle \mathbf{x}, \mathbf{x} \rangle}.$$

There is a converse to the above relation in the sense that the knowledge of the norm in an inner product space is sufficient to recover the associated inner product, as shown in the following theorem known as the polarization identity.

---

[5]The space $\mathscr{V} \times \mathscr{V}$ denotes the set of all ordered pairs $(\mathbf{x}, \mathbf{y})$ where $\mathbf{x}, \mathbf{y} \in \mathscr{V}$.

**Theorem 4.** *Let $\mathscr{V}$ be an inner product space and let $\mathbf{x}$ and $\mathbf{y} \in \mathscr{V}$. Then, if $\mathscr{V}$ is a real space*

$$\langle \mathbf{x}, \mathbf{y} \rangle \equiv \frac{1}{4} \|\mathbf{x} + \mathbf{y}\|^2 - \frac{1}{4} \|\mathbf{x} - \mathbf{y}\|^2,$$

*and if $\mathscr{V}$ is a complex space*

$$\langle \mathbf{x}, \mathbf{y} \rangle \equiv \frac{1}{4} \|\mathbf{x} + \mathbf{y}\|^2 - \frac{1}{4} \|\mathbf{x} - \mathbf{y}\|^2 + \frac{i}{4} \|\mathbf{x} + i\mathbf{y}\|^2 - \frac{i}{4} \|\mathbf{x} - i\mathbf{y}\|^2.$$

The induced norm satisfies the Cauchy-Schwartz inequality stated in the following theorem.

**Theorem 5.** *Given an inner product space $\mathscr{V}$ and two elements $\mathbf{x}$ and $\mathbf{y}$, then*

$$|\langle \mathbf{x}, \mathbf{y} \rangle| \le \|\mathbf{x}\| \|\mathbf{y}\| \tag{1.7}$$

*with equality if, and only if, $\mathbf{y} = a\mathbf{x}$ for some (complex) constant $a$.*

Two vectors are orthogonal if their inner product is zero: $\mathbf{x} \perp \mathbf{y} \Leftrightarrow \langle \mathbf{x}, \mathbf{y} \rangle = 0$. One can, using the Cauchy-Schwartz inequality, define an angle $\theta$ between two vectors by

$$cos\,(\theta) \equiv \frac{|\langle \mathbf{x}, \mathbf{y} \rangle|}{\|\mathbf{x}\| \|\mathbf{y}\|}, \quad 0 \le \theta \le \frac{\pi}{2}. \tag{1.8}$$

The concept of orthogonality can be extended to subspaces of an inner product vector space.

**Definition 18.** Given a subspace $\mathbb{V}$ of an inner product space $\mathscr{V}$, the space $\mathbb{V}^{\perp}$ is defined to contain all vectors that are orthogonal to every vector in $\mathbb{V}$. That is, $\langle \mathbf{v}, \mathbf{w} \rangle = 0$ whenever $\mathbf{v} \in \mathbb{V}$ and $\mathbf{w} \in \mathbb{V}^{\perp}$.

**Theorem 6.** $\mathbb{V}^{\perp}$ *is a subspace of $\mathscr{V}$ and $\mathbb{V} \subset \mathbb{V}^{\perp\perp}$ and $\mathbb{V}^{\perp} = \mathbb{V}^{\perp\perp\perp}$.*

In transform theory a function $x(t)$ in $L_2(\mathbb{R})$ is mapped to a set of transform coefficients by calculating the inner products of the given function with a set of basis function $\xi_a(t)$. Using theorem 5, the magnitude squared of the inner product $\langle \xi_a, x \rangle$ satisfies the inequality

$$|\langle \xi_a, x \rangle|^2 \le \|\xi_a\|^2 \|x\|^2$$

and so the quantity $|\langle \xi_a, x \rangle|^2$ can be used to find the matching value $a$ of the basis function that $x(t)$ may be known to be proportional to. However, if the function of interest $x$ is not proportional to any of the basis functions, then the best match is found by minimizing the magnitude squared of the sum or the difference $x \pm \xi_a$. Thus, the minima of the quantities

$$\|x \pm \xi_a\|^2 = \|x\|^2 + \|\xi_a\|^2 \pm 2\mathrm{Re}\,\langle \xi_a, x \rangle, \qquad (1.9)$$

occur at the maximum of $(\mathrm{Re}\,\langle \xi_a, x \rangle)^2$.

## 1.9   Banach and Hilbert Spaces

Although in a normed space we have the means for doing analysis and studying convergence, we are not guaranteed that converging sequences of vectors actually have a limit inside the space itself [6]. A Banach space is a normed space in which every convergent Cauchy sequence (see definition 15) of vectors converges to a vector inside the space. We take this as our working definition of completeness.[6]

**Definition 19.** A Banach space is a complete normed vector space.

**Definition 20.** A Hilbert space $\mathscr{H}$ is a complete inner product space, or equivalently a Banach space whose norm is induced by an existing inner product, $\|x\| = \sqrt{\langle x, x \rangle}$.

Hilbert spaces commonly occur in mathematics and physics as function spaces. Their study is, therefore, contained in a branch of mathematics known as functional analysis. Hilbert spaces generalize the familiar Euclidean three-dimensional space $\mathbb{R}^3$: most important geometrical results in the latter have direct generalizations to Hilbert spaces. For instance, an orthogonal projection from the tip of a standard three dimensional vector onto an arbitrary plane through the origin has its analog in Hilbert spaces in the form of an orthogonal projection onto a linear subspace and is at the foundation of linear optimal filter theory. Orthogonal projections are discussed in section 1.14.

An indispensable property of Hilbert spaces in almost all of physics and applied mathematics is that of separability: separable Hilbert spaces possess a countable orthonormal basis (see section 1.16 for orthonormal bases). The function space of interest to us is $L_2(\mathbb{R})$ which is a separable Hilbert space.

---

[6]A more general definition in a metric space relies on first defining the diameter of closed subspaces and then requiring that every decreasing sequence of nonempty closed subsets of the corresponding space whose diameters tend to 0 must have a nonempty intersection, i.e., a point that is common to all of them.

One such basis can be found starting with the windowed (or short-time) Fourier transform and its discretization in both time and frequency. The resulting basis, although localized, has fixed time and frequency resolutions that are inversely proportional to each other. The continuous wavelet transform is another approach to arrive at variable resolution functions that are localized in time and scale. Both methods are discussed in chapter 3.

## 1.10 Linear Operators, Operator Norm, the Adjoint Operator

**Definition 21.** Given two vector spaces $\mathscr{V}_1$ and $\mathscr{V}_2$, a transformation (operator) $T$ between the two denoted by $T : \mathscr{V}_1 \to \mathscr{V}_2$ is said to be linear if, for every $\mathbf{x}, \mathbf{y} \in \mathscr{V}_1$ and all $\alpha, \beta \in \mathbb{C}$, $T(\alpha\mathbf{x} + \beta\mathbf{y}) = \alpha T(\mathbf{x}) + \beta T(\mathbf{y})$.

For instance, if the two vector spaces are $\mathbb{C}^n$ and $\mathbb{C}^m$, then $T$ is an $m \times n$ matrix of complex numbers. When the underlying spaces are both copies of $L_2(\mathbb{R})$, a linear transformation takes the form of an integral operator, e.g., $X(\omega) = \int_{-\infty}^{\infty} T(\omega, t) x(t) \, dt$. An example of such an operator is the Fourier transform (see equations (1.57) and (1.58)).

In order to ensure that the image of a linear operator is in fact a member of $L_2(\mathbb{R})$ we must extend the notion of norm to linear operators. An operator norm must have all the required properties of a norm shown in equation (1.3). The $p$-norm of a linear operator $T : \mathscr{V}_1 \to \mathscr{V}_2$ is defined by

$$\begin{aligned} \|T\|_p &= \left\{ \max \|Tx\|_p ;\ x \in \mathscr{V}_1 \text{ and } \|x\|_p = 1 \right\} \\ &\equiv \left\{ \max \left( \|Tx\|_p \Big/ \|x\|_p \right);\ x \in \mathscr{V}_1 \text{ and } \|x\|_p \neq 0 \right\} \end{aligned} \tag{1.10}$$

where $\|x\|_p$ is the $L_p$ norm defined previously. The $p$-norm of an operator satisfies the following inequalities

$$\|Tx\|_p \leq \|T\|_p \|x\|_p \quad \text{and} \quad \|T_1 T_2\|_p \leq \|T_1\|_p \|T_2\|_p . \tag{1.11}$$

When the operator is defined as $T : \mathscr{V} \to \mathscr{V}$ and it has an inverse we can show that

$$\left\|T^{-1}\right\|_p = \frac{1}{\left\{ \min \|Tx\|_p ;\ \|x\|_p = 1,\ x \in \mathscr{V} \right\}} . \tag{1.12}$$

**Definition 22.** A bounded operator is a linear operator with finite norm.

**Theorem 7.** *A linear operator is bounded if, and only if, it is continuous.*

The following theorem describes a useful formula, known as the Neumann expansion, for bounded operators.

**Theorem 8.** *For a bounded operator $T$ with unity bound, i.e., $\|T\| < 1$, where the norm satisfies the inequalities (1.11), we have the following formula [7]:*

$$(1 - T)^{-1} = \sum_{k=0}^{\infty} T^k. \tag{1.13}$$

Every bounded linear operator $T : \mathscr{H}_1 \to \mathscr{H}_2$ between two Hilbert spaces $\mathscr{H}_1$ and $\mathscr{H}_2$, has an associated adjoint $T^+ : \mathscr{H}_2 \to \mathscr{H}_1$.

**Definition 23.** The adjoint of a bounded linear operator $T : \mathscr{H}_1 \to \mathscr{H}_2$ is a linear operator $T^+ : \mathscr{H}_2 \to \mathscr{H}_1$ defined by the equation

$$\langle s_2, T s_1 \rangle \equiv \langle T^+ s_2, s_1 \rangle \tag{1.14}$$

for every $s_1 \in \mathscr{H}_1$ and $s_2 \in \mathscr{H}_2$. When the two Hilbert spaces are the same, $\mathscr{H}_1 = \mathscr{H}_2 = \mathscr{H}$, a linear bounded operator is self-adjoint if $T = T^+$.

To every linear operator $T : \mathscr{H}_1 \to \mathscr{H}_2$ there correspond two fundamental linear subspaces, namely, the range space $\mathscr{R}_T$ and the null space $\mathscr{N}_T$.

**Definition 24.** The range space of a linear operator $T : \mathscr{H}_1 \to \mathscr{H}_2$ is the image of the linear map and it is a linear subspace of $\mathscr{H}_2$:

$$\mathscr{R}_T = T\left(\mathscr{H}_1\right) = \{\mathbf{y} \in \mathscr{H}_2 : \ \mathbf{y} = T\mathbf{x}, \text{ for some } \mathbf{x} \in \mathscr{H}_1\}.$$

The null space is a linear subspace of $\mathscr{H}_1$ and contains all elements of $\mathscr{H}_1$ that map to the zero of $\mathscr{H}_2$:

$$\mathscr{N}_T = \{\mathbf{x} \in \mathscr{H}_1 : \ T\mathbf{x} = \mathbf{0}\}.$$

The range and null spaces of a linear operator and its adjoint are related through the following theorem.

**Theorem 9.** *For a bounded linear operator $T : \mathscr{H}_1 \to \mathscr{H}_2$ whose range space $\mathscr{R}_T$ is closed, and whose adjoint range space $\mathscr{R}_{T^+}$ is also closed,[7] the following relations hold:*

$$\mathscr{R}_T{}^{\perp} = \mathscr{N}_{T^+}, \mathscr{N}_T{}^{\perp} = \mathscr{R}_{T^+}, \tag{1.15a}$$

$$\mathscr{R}_{T^+}{}^{\perp} = \mathscr{N}_T, \mathscr{N}_{T^+}{}^{\perp} = \mathscr{R}_T, \tag{1.15b}$$

---

[7]Although all bounded linear operators between finite dimensional spaces have closed ranges, this is not necessarily the case in infinite dimensional spaces. The theorem still holds, however, if we replace the corresponding range spaces with their closures.

*where the subspaces denoted by the symbol $\perp$ are defined in definition 18. In addition,*

$$\mathcal{H}_1 = \mathcal{R}_{T^+} \oplus \mathcal{N}_T, \ \mathcal{H}_2 = \mathcal{R}_T \oplus \mathcal{N}_{T^+},$$

*where the direct sum $\oplus$ is defined in definition 10.*

Given a self-adjoint (and bounded) positive definite operator $T$, and denoting its smallest and largest eigenvalues[8] by $\lambda_{\min}$ and $\lambda_{\max}$, where $0 < \lambda_{\min} < \lambda_{\max}$, the spectral radius of $T$ is defined to be its largest eigenvalue $\rho_T \equiv \lambda_{\max}$. Following this, if $T$ is not a self-adjoint operator we define the $l_2$ norm of $T$ by the following equation

$$\|T\|_2 = \sqrt{\rho_{T^+T}}. \tag{1.16}$$

This definition relies on the fact that the operator $T^+T$ is self-adjoint and positive definite for any operator $T$. It is self-adjoint because $(T^+T)^+ = T^+T^{++} = T^+T$, and positive definite since $\langle x, T^+Tx \rangle = \langle Tx, Tx \rangle$. The latter is real and positive by the postulates of an inner product shown in equation (1.4). When $T$ is a self-adjoint operator, i.e., $T^+ = T$, it follows that $\|T\|_2 = \rho_T$.

**Definition 25.** A unitary operator is one that satisfies the equation $T^+T = 1$, i.e., the operator has a left inverse which is the same as its adjoint.

## 1.11  Reproducing Kernel Hilbert Space

A special case of linear operators is a linear functional as defined below [6].

**Definition 26.** A linear functional on a Hilbert space $\mathcal{H}$ is a linear map $f : \mathcal{H} \to \mathbb{C}$.

For instance, suppose that $\phi \in \mathcal{H}$ is fixed. Then

$$f_\phi [x] = \langle \phi, x \rangle = \int_{-\infty}^{\infty} \phi^*(t)x(t)dt, \ \ \forall x \in \mathcal{H},$$

is a linear functional. An important example is the evaluation functional $\mathcal{E}_t$.

---

[8]For a positive definite operator $T$, the matrix element $\langle x, Tx \rangle$ is real and positive for all $x \in \mathcal{V}$. The eigenvalues and the eigenfunctions of the operator $T$ are defined by the equation $Tx = \lambda x$, where $x$ is an eigenvector (eigenfunction) corresponding to the eigenvalue $\lambda$.

**Definition 27.** The evaluation functional $\mathscr{E}_t$ is a linear map between a Hilbert space of functions $\mathscr{H}$ and the complex numbers $\mathbb{C}$

$$\mathscr{E}_t : \mathscr{H} \to \mathbb{C} , \quad \mathscr{E}_t[x] = x\,(t) , \tag{1.17}$$

where the functions are assumed to have been defined for every point $t \in \mathbb{R}$.

**Definition 28.** A linear functional $f$ on a Hilbert space is bounded if there exists a constant $C$, such that $|f\,[x]| \le C\,\|x\|$, $\forall x$ in the Hilbert space. Note that $C$ is independent of $x$.

An example is the Hilbert space $L_2\,(\mathbb{R})$, and the linear functional

$$f_w\,[x] \equiv \int\limits_{-\infty}^{\infty} w^*\,(t)\,x\,(t)\,dt.$$

Then by the Cauchy-Schwartz inequality (1.7) we have

$$|f_w\,[x]| \le \|w\|\,\|x\| ,$$

and so we may take $C = \|w\|$.

**Definition 29.** A linear functional $f$ is continuous if, given that $x_n\,(t) \to x\,(t)$ as $n \to \infty$, then $f\,[x_n] \to f\,[x]$.

**Theorem 10.** *A linear functional $f$ defined on a Hilbert space is bounded if, and only if, it is continuous.*

This theorem, to be used below, can be proven as follows. Given that $x_n \to x$ and if $f$ is bounded, we use linearity to write

$$|f\,(x_n) - f\,(x)| = |f\,(x_n - x)| \le C\,\|x_n - x\| ,$$

which proves the continuity of the functional. Conversely, if $f$ is continuous but unbounded then there must exist a sequence $x_n$ such that $f\,[x_n] > n\,\|x_n\|$. Now let $x'_n = x_n/(n\,\|x_n\|)$. Clearly $\|x'_n\| = n^{-1}$ and so $x'_n \to 0$. Then $|f\,[x'_n]| = f\,[x_n]/(n\,\|x_n\|) > 1$ which shows that $f\,[x'_n]$ will not tend to zero, even though $x'_n \to 0$, thus contradicting the continuity assumption. Consequently if $f$ is continuous it must be bounded.

In a Hilbert space, bounded linear functionals have a particularly simple representation, embodied in the Frechet-Riesz theorem.

**Theorem 11.** *(Frechet-Riesz) If $f$ is a bounded linear functional on a Hilbert space $\mathscr{H}$, then there exists a unique element $\phi \in \mathscr{H}$ such that $f\,[x] = \langle \phi, x \rangle$ for all $x \in \mathscr{H}$.*

**Definition 30.** The evaluation functional $\mathscr{E}_t$ of definition 27 is said to be bounded if, given $x(t) \in \mathscr{H}$, defined for every value of the independent variable $t$, there exist constants $C_t$ (that depend on $t$ but not on $x$) for which the following inequality holds for all $x$:

$$|x(t)| \leq C_t \|x\|. \tag{1.18}$$

If the evaluation functional is bounded then theorem 11 shows that $\mathscr{E}_t[x]$ can be written as the inner product between a fixed element $e_t \in \mathscr{H}$, and $x$, i.e.,

$$x(t) = \langle e_t, x \rangle = \int_{-\infty}^{\infty} e_t^*(t') \, x(t') \, dt'. \tag{1.19}$$

**Definition 31.** A Hilbert space with a bounded evaluation functional is a reproducing kernel Hilbert space with the reproducing kernel defined by [8]

$$K(t, t') \equiv e_t(t') = \langle e_{t'}, e_t \rangle. \tag{1.20}$$

Equation (1.19) can now be written in terms of the reproducing kernel

$$x(t) = \int_{-\infty}^{\infty} K^*(t, t') x(t') \, dt' \equiv \langle K, x \rangle_t, \tag{1.21}$$

where the subscript $t$ indicates that in the corresponding inner product the integration is over the variable $t'$ and that $t$ is held fixed. This is the reproducing kernel property of any Hilbert space in which the evaluation functional is a bounded (equivalently, continuous) linear map. Thus, we have the following theorem.

**Theorem 12.** *(Aronszajn) A necessary and sufficient condition that a Hilbert space possesses a reproducing kernel is that the evaluation functional be bounded.*

**Theorem 13.** *If a Hilbert space possesses a reproducing kernel then the kernel is unique.*

For if we have two reproducing kernels $K_1(t, \tau)$ and $K_2(t, \tau)$, then

$$\|K_1(t, \tau) - K_2(t, \tau)\|_t^2 = \langle K_1(t, \tau) - K_2(t, \tau), K_1(t, \tau) - K_2(t, \tau) \rangle_t$$
$$= \langle K_1(t, \tau) - K_2(t, \tau), K_1(t, \tau) \rangle_t - \langle K_1(t, \tau) - K_2(t, \tau), K_2(t, \tau) \rangle_t$$
$$= K_1(t, t) - K_2(t, t) - K_1(t, t) + K_2(t, t) = 0.$$

The final result, of course, holds for all $t$ and so the reproducing kernel is unique.

## 1.12    The Dirac Delta Distribution

In most Hilbert spaces the evaluation functional is unbounded. For instance, functions in $L_2(\mathbb{R})$ could be unbounded on sets of measure zero and so a reproducing kernel cannot be defined. Examples of reproducing kernel Hilbert spaces include the space of band-limited functions (a subspace of $L_2(\mathbb{R})$ defined in section 1.22) as well as the range spaces of both the windowed Fourier transform and the continuous wavelet transform (subspaces of $L_2(\mathbb{R}^2)$), as will be shown later.

The Dirac delta distribution[9] [9] $\delta(t - t')$ is a singular form of a reproducing kernel existing in a space far larger than $L_2(\mathbb{R})$: the dual space of a space of "sufficiently well behaved (good)" functions.

**Definition 32.** Consider the space of functions $\mathscr{S}$ whose elements (in general, complex functions) are absolutely integrable on the real line $\mathbb{R}$, and are infinitely many times differentiable at every $t \in \mathbb{R}$, and, together with any of their derivatives, decay to zero faster than any negative power of $t$ as $|t| \to \infty$ [10]. We will call this the space of test functions of rapid decay.[10]

An example of the sort of function contained in $\mathscr{S}$ is $\exp(-t^2)$.

**Definition 33.** The space of all linear functionals defined on $\mathscr{S}$, whose range is either $\mathbb{R}$ or $\mathbb{C}$, is known as the dual space of $\mathscr{S}$ and is denoted by $\mathscr{S}'$.

Thus, if $\phi_1(t) \in \mathscr{S}'$, then $\int\limits_{-\infty}^{\infty} \phi_1^*(t)\,\phi(t)\,dt$ is a number (in general, complex).

The dual space $\mathscr{S}'$ includes all functions in $\mathscr{S}$ as well as all functions in $L_2(\mathbb{R})$, and thus it includes regular functions, since

$$\left| \int\limits_{-\infty}^{\infty} x^*(t)\phi(t)dt \right| < \infty, \ x \in L_2(\mathbb{R}) \text{ or } x \in \mathscr{S}, \ \phi \in \mathscr{S}'.$$

But there is more: $\mathscr{S}'$ includes objects that are not ordinary functions. For instance, consider the linear functional that maps every member of $\mathscr{S}$ to

---

[9]The Dirac delta is popularly known as the Dirac delta function. The usual definition "zero everywhere except at one point," however, clearly does not satisfy the basic requirements for a function.

[10]Technically, the space of test functions $\mathscr{K}$, contained within $\mathscr{S}$, is defined to include only functions with continuous derivatives of all order and with compact support. Although, $\mathscr{K}$ is used to define generalized functions, in practice $\mathscr{S}$ is often used and so we take that as our function space for the definition of the Dirac delta and other generalized functions.

the value of that member at a fixed point $t'$. This linear map, a member of $\mathscr{S}'$, is the Dirac delta generalized function (or distribution). Thus, the defining relation for the Dirac delta is

$$\int_{-\infty}^{\infty} \delta\left(t - t'\right) \phi\left(t'\right) dt' = \phi\left(t\right). \tag{1.22}$$

Comparison of the above with the reproducing kernel property (1.21) suggests that the Dirac delta, which is defined only on functions in $\mathscr{S}$, is a generalized reproducing kernel: it is commonly used with members of $L_2(\mathbb{R})$ which do not often even belong to $\mathscr{S}$! Although we can still use equation (1.22) for functions in $L_2(\mathbb{R})$ at points $t$ for which the function is defined, there is no meaning to the relation at points $t$ where the function is not defined (comprising a set of measure zero). Thus, the Dirac delta may be considered to be a singular (or generalized) reproducing kernel in $L_2(\mathbb{R})$.

Another set of useful functions that are members of $\mathscr{S}'$, but not $L_2(\mathbb{R})$, are the complex exponentials $e_\omega(t) \equiv (2\pi)^{-1/2} \exp(i\omega t)$, $\omega \in \mathbb{R}$; they are used to construct the Fourier transforms of members of $L_2(\mathbb{R})$. The variable $\omega$ (also known as the frequency) is used as a continuous index to distinguish different exponential functions. The inverse Fourier transform (the reconstruction formula) is based on the relation

$$\int_{-\infty}^{\infty} e_\omega^*\left(t\right) e_\omega\left(t'\right) d\omega = \delta\left(t - t'\right). \tag{1.23}$$

As we shall see later, when defining the Fourier transform and its inverse, the above equation shows that any member of $L_2(\mathbb{R})$ can be expanded in terms of $e_\omega(t)$,

$$x(t) = \int_{-\infty}^{\infty} e_\omega(t) X(\omega) d\omega, \quad x(t) \in L_2(\mathbb{R}), \tag{1.24}$$

where the expansion coefficients $X(\omega)$ are

$$X(\omega) = \int_{-\infty}^{\infty} e_\omega^*(t) x(t) dt. \tag{1.25}$$

Thus, the functions $e_\omega(t)$ are a continuously labeled basis for members of $L_2(\mathbb{R})$ within the space $\mathscr{S}'$. Equation (1.23) is a resolution of identity and often referred to as the completeness relation for the continuously indexed basis functions $e_\omega(t)$. However, there is no discretization of the frequency variable $\omega$ of the form $\omega_n = n\Delta\omega$, $n \in \mathbb{Z}$, for which the corresponding functions $e_{\omega_n}(t)$ satisfy the completeness relation.

As is shown later in section 1.22, if we limit the frequency band of the function space $L_2(\mathbb{R})$ to the interval $[-\Omega, \Omega]$, $\Omega > 0$, the resulting Hilbert space (space of band-limited functions that are square integrable, a subspace of $L_2(\mathbb{R})$) will admit a regular reproducing kernel that is the band-limited form of the Dirac delta, namely, $\sin[\Omega(t - t')]/\pi(t - t')$, in the sense that

$$\sin\left[\Omega\left(t - t'\right)\right] / \left[\pi\left(t - t'\right)\right] \to \delta(t - t'), \quad \text{as } \Omega \to \infty.$$

This reproducing kernel is a regular function (not a distribution) and is often used as an example of a Dirac delta convergent sequence. We will return to the Dirac delta and reproducing kernels later in the context of frames and orthonormal bases.

## 1.13   Orthonormal Vectors

Orthogonal sets of vectors are particularly important in most applications and if, in addition, they are of unit norm then they are referred to as orthonormal sets [5].

**Definition 34.** A set of vectors $\{\mathbf{x}_j\}$ is said to be orthonormal provided $\langle \mathbf{x}_j, \mathbf{x}_k \rangle = \delta_{jk}$, where $\delta_{jk}$ is the Kroenecker delta

$$\delta_{jk} = \begin{cases} 1, & \text{if } j = k, \\ 0, & \text{if } j \neq k. \end{cases} \tag{1.26}$$

Given any set of vectors $\{\mathbf{x}_1, \ldots, \mathbf{x}_n\}$ the Gramm-Schmidt method [7] is used to obtain an orthonormal set $\{\mathbf{e}_1, \ldots, \mathbf{e}_m\}$ where $m \leq n$ and both sets of vectors span the same vector space. The method is to choose $\mathbf{e}_1 = \mathbf{x}_1/\|\mathbf{x}_1\|$ and

$$\mathbf{e}'_k = \mathbf{x}_k - \sum_{l=1}^{k-1} \langle \mathbf{x}_k, \mathbf{e}_l \rangle \mathbf{e}_l, \quad \mathbf{e}_k = \left\|\mathbf{e}'_k\right\|^{-1} \mathbf{e}'_k, \ k \geq 2. \tag{1.27}$$

For example, if we use the space of square integrable real functions on $[-1, 1]$ with the inner product $\langle f, g \rangle = \int\limits_{-1}^{1} f(t)g(t)\,dt$, starting with the vectors $\{1, t, t^2, \ldots, t^n\}$, the Gramm-Schmidt method will produce the Legendre polynomials up to degree $n$. If, on the other hand, we use the weighted inner product $\langle f, g \rangle = \int\limits_{-1}^{1} f(t) (1 - t^2)^{-1/2} g(t)\, dt$, the Gramm-Schmidt method will result in the Chebyshev polynomials up to degree n (see the exercises).

## 1.14 Orthogonal Projections

Given a Hilbert space $\mathscr{H}$ and two disjoint subspaces $\mathbb{V}$ and $\mathbb{W}$, $\mathbb{V} \cap \mathbb{W} = \{\mathbf{0}\}$, a projection onto $\mathbb{V}$ is a mapping $P_v : \mathscr{H} \to \mathbb{V}$ so that $P_v \mathbf{x} = \mathbf{v}$ where $\mathbf{x} = \mathbf{v} + \mathbf{w}$, $\mathbf{v} \in \mathbb{V}$ $\mathbf{w} \in \mathbb{W}$. Alternatively, $P_w : \mathscr{H} \to \mathbb{W}$ where $P_w \mathbf{x} = \mathbf{w}$. Any projection operator $P$ must satisfy the equation $P^2 = P$, which can be used to define a projection. The Cayley-Hamilton theorem[11] can be invoked to state that the eigenvalue equation $\lambda^2 = \lambda$ must be the characteristic equation for any projection operator; thus the eigenvalues of a projection operator are 0 and 1.

If the projection is defined into the same space, i.e., $P : \mathscr{H} \to \mathscr{H}$, then the range and the null spaces $\mathscr{R}_P$ and $\mathscr{N}_P$ are disjoint linear subspaces of the space $\mathscr{H}$ (the only element common to both is the zero vector), and $\mathscr{H} = \mathscr{R}_p + \mathscr{N}_p$, i.e., the range and the null spaces are algebraic complements and are not necessarily orthogonal.

**Definition 35.** A projection $P : \mathscr{H} \to \mathscr{H}$ is said to be orthogonal if its range and null spaces are orthogonal to one another, i.e., $\mathscr{R}_P \perp \mathscr{N}_P$ [11].

**Theorem 14.** *A projection $P : \mathscr{H} \to \mathscr{H}$ is orthogonal if, and only if, it is self-adjoint.*

If $P$ is self-adjoint, then for every $x \in \mathscr{H}$ we write $x \equiv Px + (1 - P)x$ is an identity. Now clearly $Px \in \mathscr{R}_p$ and $(1 - P)x \in \mathscr{N}_p$, since $P(1 - P)x = (P - P^2) x = 0$ ($P$ being a projection operator satisfies $P = P^2$). Any member of $\mathscr{R}_p$ is orthogonal to any member of $\mathscr{N}_p$ since

$$\langle Px, (1 - P) y \rangle = \langle x, P^+ (1 - P) y \rangle = \langle x, (P - P^2) y \rangle = 0. \qquad (1.28)$$

**Theorem 15.** *A projection $P : \mathscr{H} \to \mathscr{H}$ is orthogonal if $\|Px\| \leq \|x\|$ for every $x \in \mathscr{H}$.*

To show that $\mathscr{R}_p$ is orthogonal to $\mathscr{N}_p$, assume $x \in \mathscr{N}_p^\perp$ (see definition 18). Then $y = Px - x \in \mathscr{N}_p$ since $Py = P^2x - Px = Px - Px = 0$. So $Px = x + y$

---

[11]The Cayley Hamilton theorem states that a square matrix $\mathbf{A}$ whose characteristic polynomial is defined by $\chi_{\mathbf{A}}(\lambda) \equiv |\mathbf{A} - \lambda \mathbf{1}|$ satisfies the equation $\chi_{\mathbf{A}}(\mathbf{A}) = 0$.

where $\langle x, y \rangle = 0$. Thus

$$\|x\|^2 \geq \|Px\|^2 = \|x\|^2 + \|y\|^2 \geq \|x\|^2,$$

and therefore $y = 0$, and $Px = x$, so that $x \in \mathcal{R}_p$. This shows that $\mathcal{N}_p^\perp \subset \mathcal{R}_p$. Conversely, if $x \in \mathcal{R}_p$, then $x = Px$ and we write $x = y + u$ with $y \in \mathcal{N}_p^\perp$ and $u \in \mathcal{N}_p$. Then $x = Px = Py + Pu = Py = y$, since $y \in \mathcal{N}_p^\perp$ and therefore $y \in \mathcal{R}_p$. Hence, $x \in \mathcal{N}_p^\perp$ so that $\mathcal{R}_p \subset \mathcal{N}_p^\perp$.

Orthogonal projections are used to solve the problem of approximating a given vector in $\mathcal{H}$ by a vector that lies entirely in a linear subspace of $\mathcal{H}$. For instance, consider the three-dimensional Euclidean space $\mathbb{R}^3$, a vector $\mathbf{r} = (x, y, z)^T$ where $z \neq 0$, and the $xy$-plane which is a linear subspace of $\mathbb{R}^3$. The problem here is to find a vector $\hat{\mathbf{r}}$ in the $xy$-plane that is the best approximation to the original vector $\mathbf{r}$ in the sense that $\|\mathbf{r} - \hat{\mathbf{r}}\|$ attains its minimum value. The solution is provided by the orthogonal projection of the vector $\mathbf{r}$ onto the $xy$-plane namely, $\hat{\mathbf{r}} = (x, y, 0)^T$. The error in the approximation is given by the magnitude of the error vector $(0, 0, z)^T$. The latter lies on the $z$-axis which is the subspace orthogonal to the approximating subspace, the $xy$-plane. This example illustrates the orthogonal projection theorem [11, 12].

**Theorem 16.** *Given a Hilbert space $\mathcal{H}$, a closed linear subspace $\mathbb{V}$, and a point $\mathbf{x} \in \mathcal{H}$, there exists a unique point $\mathbf{v} \in \mathbb{V}$ that is the orthogonal projection of $\mathbf{x}$ onto $\mathbb{V}$ and that minimizes the norm $\|\mathbf{x} - \mathbf{v}\|$. In addition, we can write $\mathcal{H} = \mathbb{V} \oplus \mathbb{V}^\perp$, and $\mathbb{V} = \mathbb{V}^{\perp\perp}$.*

Note that the result $\mathbb{V} \subset \mathbb{V}^{\perp\perp}$ of theorem 6 for an inner product space is now replaced by $\mathbb{V} = \mathbb{V}^{\perp\perp}$ in a Hilbert space. The difference is, of course, the completeness property of a Hilbert space.

## 1.15   Multi-Resolution Analysis Subspaces

As described in the previous section, an orthogonal projection is used in the problem of approximating a function using the closest element of a given subspace. An important concept in the theory and implementation of the discrete orthogonal wavelet transform is the concept of orthogonal projections into nested subspaces that satisfy a completeness property and a multi-resolution property [13, 14].

**Definition 36.** In $L_2(\mathbb{R})$, a set of closed subspaces $\mathcal{V}_m \subset L_2(\mathbb{R})$, $m \in \mathbb{Z}$, is said to satisfy the nesting property if

$$\mathcal{V}_m \subset \mathcal{V}_{m-1}, \ m \in \mathbb{Z}. \tag{1.29}$$

**Definition 37.** A nested set of closed subspaces $\mathscr{V}_m$ satisfying the completeness properties

$$\bigcup_{m\in\mathbb{Z}} \mathscr{V}_m = L_2(\mathbb{R}), \quad \bigcap_{m\in\mathbb{Z}} \mathscr{V}_m = \{0\}, \tag{1.30}$$

is said to have the successive approximation property. For such a set of nested subspaces we also have

$$L_2(\mathbb{R}) = \lim_{m\to-\infty} \mathscr{V}_m, \quad \{0\} = \lim_{m\to\infty} \mathscr{V}_m. \tag{1.31}$$

If $L_2(\mathbb{R})$ admits a nested set of subspaces with the successive approximation property then any square-integrable function can be approximated by an orthogonal projection onto any one of the subspaces $\mathscr{V}_m$. The process is iterative and is depicted in figure 1.1. Note that an approximation at stage $m$ produces two results: one is the coarser approximation at stage $m+1$, and the other is the detail that is lost between the approximations at stages $m$ and $m+1$. The former is an orthogonal projection into $\mathscr{V}_{m+1}$, while the latter lies in $\mathscr{V}_m^\perp$.

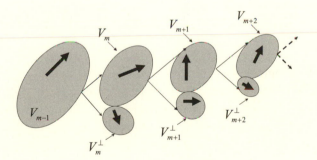

Figure 1.1: Approximation of a vector using nested subspaces and their orthogonal complements.

**Definition 38.** A nested set of successive approximation subspaces $\mathscr{V}_m$ is said to have the multi-resolution property provided that [15]

$$x(t) \in \mathscr{V}_m \Leftrightarrow x(2t) \in \mathscr{V}_{m-1}. \tag{1.32}$$

The subspaces $\mathscr{V}_m$ are then said to form a multi-resolution analysis set of subspaces and are referred to as a MRA.

We illustrate the concept of a MRA by producing a set of nested sub-spaces of $L_2(\mathbb{R}$ satisfying the properties of completeness and multi-resolution. Consider the space $\mathcal{V}_0$ consisting of all functions that are magnitude square integrable and that have constant values in all intervals of length 1 of the form $[n, n + 1]$, $n \in \mathbb{Z}$. Similarly, consider the space $\mathcal{V}_{-1}$, all functions that are magnitude square integrable and that have constant values in all intervals of length $1/2$ of the form $[n/2, n/2 + 1/2]$, in addition to the space $\mathcal{V}_1$, all functions that are square integrable and that have constant values in all intervals of length 2 of the form $[2n, 2n + 2]$, $n \in \mathbb{Z}$. Three typical functions from these three spaces are shown in figure 1.2. Each one

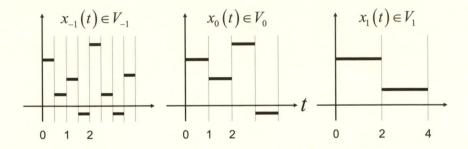

Figure 1.2: Three typical functions that are constant on intervals of length $1/2$, 1, and 2.

of these spaces is actually a subspace (the zero function is common to all of them, and a linear combination of two functions in any one space is again a function of the same type). In addition, the spaces are nested, i.e., $\mathcal{V}_{-1} \supset \mathcal{V}_0 \supset \mathcal{V}_1$. For instance, $\mathcal{V}_1 \subset \mathcal{V}_0$, since all functions that are constant on intervals $[2n, 2n + 2]$, are also constant on intervals $[n, n + 1]$, but not vice versa. The multi-resolution property, easily verified, is interpreted as doubling the approximation resolution in going from $\mathcal{V}_0$ to $\mathcal{V}_{-1}$ in the sense that if $x(t) \in \mathcal{V}_0$ then $x(2t) \in \mathcal{V}_{-1}$. Similarly the resolution is halved in going from $\mathcal{V}_0$ to $\mathcal{V}_1$ in the sense that if $x(t) \in \mathcal{V}_0$ then $x(t/2) \in \mathcal{V}_1$. Thus, we define the subspaces $\mathcal{V}_m$ of functions that have constant values in intervals of length $2^m$, of the form $[2^m n, 2^m (n + 1)]$, $m, n \in \mathbb{Z}$, and obtain a set of nested subspaces satisfying the completeness and the multi-resolution properties, also known as the Haar MRA. The Haar multi-resolution approximation to a given function will be seen to consist of orthogonal projections onto the subspaces $\mathcal{V}_m$ while the Haar wavelet transform coefficients will be the approximation errors, i.e., orthogonal transformations onto the orthogonal complement subspaces $\mathcal{V}_m^{\perp}$.

The key to the fast implementation of the discrete wavelet transform is the multi-resolution property and the existence of complete and orthonormal bases in the subspaces $\mathscr{V}_m$ and $\mathscr{V}_m^{\perp}$. We discuss complete and orthonormal bases in the next section.

## 1.16   Complete and Orthonormal Bases in $L_2(\mathbb{R})$

The concept of an orthonormal basis in a finite dimensional vector space can be extended to an infinite dimensional Hilbert space $\mathscr{H}$. Later in this chapter we study the more general concept of a frame but for now we define a complete and orthonormal basis with respect to which any element of a Hilbert space has a unique representation. The concepts of decomposition of a function (signal) in terms of a basis set (analysis) and its inverse (reconstruction or synthesis) are what transform theory in general, and wavelet theory in particular, are based on. For a complete and orthonormal basis denoted by $\phi_n$, $n \in \mathbb{Z}$, the analysis and synthesis operations are summarized in one formula

$$x = \sum_n \langle \phi_n, x \rangle \phi_n, \tag{1.33}$$

with the inner product representing the analysis operation and the entire sum representing the reconstruction or synthesis operation. The right hand side is an infinite linear combination of the form $\sum\limits_n c_n \phi_n$ and so we begin with conditions under which infinite sums of an orthonormal set converge [12].

**Theorem 17.** *For an orthonormal set* $\{\phi_n, n \in \mathbb{Z}\}$,

$$\langle \phi_n, \phi_m \rangle = \delta_{nm}, \tag{1.34}$$

*the infinite linear combination* $\sum\limits_{n=-\infty}^{\infty} c_n \phi_n$ *converges if, and only if,*

$$\underline{c} = [\ldots, c_n, \ldots]^T \in l_2(\mathbb{Z}) \quad \Leftrightarrow \quad \sum_{n=-\infty}^{\infty} |c_n|^2 < \infty.$$

The most widely used description of completeness of a basis $\phi_n$, $n \in \mathbb{Z}$, is that for every $x \in \mathscr{H}$, there is a unique expansion

$$x = \sum_{n=-\infty}^{\infty} c_n \phi_n, \quad c_n \in \mathbb{C}, \tag{1.35}$$

where the equality expressed in equation (1.35) is not a point-by-point equality, but equality in the sense of the convergence in the mean of definition 12. Thus,

$$\left\| x - \sum_{n=-N}^{N} c_n\, \phi_n \right\| \to 0, \quad \text{as } N \to +\infty. \tag{1.36}$$

The following is a more formal definition of completeness.

**Definition 39.** An orthonormal set $\{\phi_n\}$ in a Hilbert space $\mathscr{H}$ is said to be complete if it is contained in no other orthonormal set (i.e., it is a maximal orthonormal set).

The following theorems show three ways to characterize complete and orthonormal sets of functions.

**Theorem 18.** *An orthonormal set $\{\phi_n\}$ in a Hilbert space $\mathscr{H}$ is complete if, and only if, the only vector that is orthogonal to the set is the zero vector* **0**.

**Theorem 19.** *An orthonormal set $\{\phi_n\}$ in a Hilbert space $\mathscr{H}$ is complete if its closure is the entire space. The closure of the set is defined as the set of all infinite sums $\sum_{n=-\infty}^{\infty} c_n\phi_n$ whose coefficients are in $l_2\,(\mathbb{Z})$, i.e., $\sum_{n=-\infty}^{\infty} |c_n|^2 < \infty$.*

**Theorem 20.** *An orthonormal set $\{\phi_n\}$ in a Hilbert space $\mathscr{H}$ is complete if Parseval's relation holds for every member of the space,*

$$\|x\|^2 = \sum_{n=-\infty}^{\infty} |\langle \phi_n, x \rangle|^2, \quad \forall x \in \mathscr{H}. \tag{1.37}$$

The following theorem summarizes the above results [12].

**Theorem 21.** *For an orthonormal set $\{\phi_n, n \in \mathbb{Z}\}$ in a Hilbert space $\mathscr{H}$ the following statements are equivalent:*

- *The set is complete.*

- *The only vector that is orthogonal to the set is the zero vector.*

- *For every $x$ in $\mathscr{H}$ we have the expansion $x = \sum_{n=-\infty}^{\infty} \langle \phi_n, x \rangle\, \phi_n$.*

- *The closure of the set is the entire space $\mathscr{H}$.*

- *For every $x$ in $\mathscr{H}$, $\|x\|^2 = \sum\limits_{n=-\infty}^{\infty} |\langle \phi_n, x \rangle|^2$.*

- *For every $x$ and $y$ in $\mathscr{H}$, $\langle x, y \rangle = \sum\limits_{n=-\infty}^{\infty} \langle x, \phi_n \rangle \langle \phi_n, y \rangle$.*

An example of an orthogonal set that is not complete is the infinite set $\{\sin(nt), \, n \in \mathbb{Z}, \, t \in [0, 2\pi]\}$. This set, although orthogonal with respect to the usual inner product

$$\int_0^{2\pi} \sin(mt)\sin(nt)\,dt = \pi\delta_{mn}, \tag{1.38}$$

is not complete since any member of $\{\cos(kt), \, k \in \mathbb{Z}, \, t \in [0, 2\pi]\}$ is orthogonal to it.

With respect to a complete and orthonormal basis, the expansion coefficients

$$c_n = \langle \phi_n, x \rangle = \int_{-\infty}^{\infty} \phi_n^*(t) x(t)\,dt \tag{1.39}$$

are the transform coefficients of a given function $x(t)$, and fully describe that function in the sense that the function $x(t)$ can be reconstructed from the transform coefficients at almost every point $t$, i.e., all points except for a set of measure zero.[12] In other words, the transform relation is uniquely invertible in the form

$$x(t) = \sum_{n=-\infty}^{\infty} \langle \phi_n, x \rangle \phi_n(t). \tag{1.40}$$

Aronszajn's theorem 12 stated that a Hilbert space whose evaluation functional is bounded (and therefore continuous) admits a reproducing kernel. If the Hilbert space has a complete and orthonormal basis set $\{\phi_n(t), n \in \mathbb{Z}\}$ satisfying a convergence criterion then the reproducing kernel has a simple representation described in the following theorem [16].

---

[12]The Gibbs phenomenon is such an example. Given a periodic function that is piecewise continuously differentiable with a countable number of discontinuities, then its Fourier series expansion converges to the value of the function everywhere except at the points of discontinuity. If $N \gg 1$ denotes the number of terms in the Fourier series partial sum, then it will overshoot the discontinuity at one end while undershooting it at the other end by the same amount, nearly 18%.

**Theorem 22.** *For a complete and orthonormal basis* $\{\phi_n(t), n \in \mathbb{Z}\}$ *in a Hilbert space, and for a fixed* $t'$, *the sum*

$$K(t,t') = \sum_{n=-\infty}^{\infty} \phi_n^*(t)\,\phi_n(t') \tag{1.41}$$

*converges uniformly and absolutely if*

$$\sum_{n=-\infty}^{\infty} |\phi_n(t)|^2 < \infty, \quad \forall\, t \in \mathbb{R}, \tag{1.42}$$

*in which case the function* $K$ *is the unique reproducing kernel for the Hilbert space.*

We will see an example of this when we discuss the subspace of band-limited functions within $L_2(\mathbb{R})$ in section 1.22. Section 1.19 provides an example of a Hilbert space for which the condition (1.42) does not hold and hence the sum (1.41) does not exist.[13]

Completeness of an orthonormal system can now be stated in terms of the reproducing kernel [16].

**Theorem 23.** *In a Hilbert space with a reproducing kernel* $K$, *an orthonormal set* $\{\phi_n(t), n \in \mathbb{Z}\}$, *which necessarily satisfies the convergence condition* (1.42), *is complete if, and only if,*

$$K(t,t) = \sum_{n=-\infty}^{\infty} |\phi_n(t)|^2. \tag{1.43}$$

## 1.17   The Dirac Notation

The Dirac notation is a compact and useful way to describe completeness and orthonormality relations [18]. Although the evaluation functional in $L_2(\mathbb{R})$ is not bounded and so a proper reproducing kernel (i.e., square integrable) does not exist in that space, a generalized function form of the reproducing kernel can be written using the Dirac delta distribution defined in section 1.12,

$$x(t) = \int_{-\infty}^{\infty} \delta(t - t')\,x(t')\,dt'. \tag{1.44}$$

---

[13]In such Hilbert spaces (1.41) is still used as a formal representation of a generalized reproducing kernel [17].

The function $x(t)$ is normally represented by $x$ (dropping the $t$ variable) since the value of the function at $t$ is the result of using the evaluation functional (see definition 27). Dirac introduced the notation $|x\rangle$ and called it a ket vector in the corresponding Hilbert space. A generalized evaluation functional along the lines of theorem 11 is then defined so that

$$x(t) = \langle t, x \rangle, \tag{1.45}$$

in analogy with equation (1.19). Dirac chose the following notation for the inner product in equation (1.45):

$$\langle t \mid x \rangle, \tag{1.46}$$

introducing the bra vector $\langle t|$. The generalized reproducing kernel, in analogy to the definition (1.20), then becomes

$$\delta(t - t') \equiv \langle t \mid t' \rangle. \tag{1.47}$$

Equation (1.44) then becomes

$$\langle t \mid x \rangle = \int_{-\infty}^{\infty} dt' \, \langle t \mid t' \rangle \langle t' \mid x \rangle, \tag{1.48}$$

valid for all ket vectors $|x\rangle$. Thus, we have the completeness, or resolution of identity, relation

$$\int_{-\infty}^{\infty} dt' \, |t'\rangle \langle t'| = \mathbf{1}. \tag{1.49}$$

If we further introduce the Einstein integration convention that all repeated continuous variables occurring in bras and kets are integrated over (we will later introduce the same convention for repeated discrete indices, i.e., the Einstein summation convention) we have

$$|t\rangle \langle t| = \mathbf{1}. \tag{1.50}$$

Repeated continuous variables, such as the variable $t$ in the above equation, are also known as dummy variables: we can use $t'$ or any other variable in the above equation since an integration over the variable is implied.

We now make the following associations, defining the ket $|\phi_n\rangle$ and the bra $\langle \phi_n|$,

$$\phi_n(t) \leftrightarrow \langle t \mid \phi_n \rangle, \quad \phi_n^*(t) \leftrightarrow \langle \phi_n \mid t \rangle. \tag{1.51}$$

The generalized reproducing kernel property of equation (1.47) together with the completeness relation (1.49) then lead to

$$\langle \phi_n, x \rangle \equiv \int\limits_{-\infty}^{\infty} \phi_n^* \left( t \right) x \left( t \right) dt = \int\limits_{-\infty}^{\infty} \langle \phi_n \mid t \rangle \langle t \mid x \rangle \, dt \equiv \langle \phi_n \mid x \rangle . \qquad (1.52)$$

In addition to the integration convention we introduce the Einstein summation convention: An expression containing a repeated index is summed over that index. For instance, using the Dirac notation the completeness relation (1.40) becomes

$$|x\rangle = \sum_{n=-\infty}^{\infty} |\phi_n\rangle \langle \phi_n \mid x \rangle , \qquad (1.53)$$

which using the summation convention can be reduced to

$$|\phi_n\rangle \langle \phi_n| \equiv \sum_{n=-\infty}^{\infty} |\phi_n\rangle \langle \phi_n| = \mathbf{1}. \qquad (1.54)$$

Using the Dirac notation the orthonormality relation (1.34) is now written as[14]

$$\langle \phi_m \mid \phi_n \rangle = \delta_{mn}. \qquad (1.55)$$

A linear operator acting on a ket produces another ket, and together with its bra form we have the relations

$$L\,|\phi\rangle = |\psi\rangle , \quad \langle \psi| = \langle \phi| \, L^{+}. \qquad (1.56)$$

For instance, the function $(2\pi)^{-1/2} \exp \left( i\omega t \right)$ in the Dirac notation is written as $\langle t \mid \omega \rangle$ and its complex conjugate is

$$\langle \omega \mid t \rangle = (2\pi)^{-1/2} \exp \left( -i\omega t \right) .$$

Thus,

$$\langle t' \mid \omega \rangle \langle \omega \mid t \rangle = \frac{1}{2\pi} \int\limits_{-\infty}^{\infty} e^{i\omega(t'-t)} d\omega = \delta \left( t' - t \right) = \langle t' \mid t \rangle ,$$

---

[14]Using the resolution of identity (1.56), $\langle \phi_m \mid \phi_n \rangle \equiv \int\limits_{-\infty}^{\infty} \langle \phi_m \mid t \rangle \langle t \mid \phi_n \rangle \, dt = \int\limits_{-\infty}^{\infty} \phi_m^* \left( t \right) \phi_n \left( t \right) dt.$

using the completeness relation and the integration convention (integration over the repeated variable $\omega$ is assumed), again leading to the completeness relation (resolution of identity)

$$|\omega\rangle \langle\omega| = \mathbf{1}.$$

The functions $(2\pi)^{-1/2}\exp(-i\omega t)$ can be thought of as the Fourier basis functions in the expansion of any $L_2(\mathbb{R})$ function (excluding all sets of measure 0 on which functions can be badly behaved), but they are not members of $L_2(\mathbb{R})$ since they are clearly not square integrable. One way to include these functions is by considering the dual space as discussed in section 1.12. Although the existence of the Fourier transform is shown in the next section, this formal point of view has the advantage of showing the deficiencies of the continuous Fourier transform and motivating the search for other continuous basis functions (e.g., the windowed Fourier transform and the continuous wavelet transform) whose discretization will lead to frames and orthonormal bases.

## 1.18 The Fourier Transform

The continuous time Fourier transform can actually be defined for a larger class of functions than $L_2(\mathbb{R})$, namely, the class of absolutely integrable functions $L_1(\mathbb{R})$, as described in the following theorem [10, 19].

**Theorem 24.** *Let $x(t) \in L_1(\mathbb{R})$. Then $X(\omega)$ defined by*

$$X(\omega) = \int_{-\infty}^{\infty} x(t)\, e^{-i\omega t} dt \tag{1.57}$$

*is uniformly continuous and bounded for $\omega \in \mathbb{R}$, and $X(\omega) \to 0$ for $|\omega| \to \infty$. In addition, if $X(\omega) \in L_1(\mathbb{R})$, then*

$$x(t) = \frac{1}{2\pi} \int_{-\infty}^{\infty} X(\omega)\, e^{+i\omega t} d\omega, \tag{1.58}$$

*where the equality is in the sense of equality in the mean (see definition 12). Note that $X(\omega) \in L_1(\mathbb{R})$ is not implied by $x(t) \in L_1(\mathbb{R})$.*

A class of functions that are not absolutely integrable and still possess a Fourier transform also exists, e.g., the function $\sin(t)/t$. These functions are covered by the following theorem.

**Theorem 25.** *Let* $x(t) = y(t)\sin(\omega t + \phi)$. *If* $y(t + k) < y(t)$, *and if*

$$\int_{|t| > \varepsilon > 0} |x(t)/t|\, dt < \infty,$$

*then* $X(\omega)$ *exists. In addition, the inverse transform exists and is given by equation* (1.58).

For functions in $L_2(\mathbb{R})$ we have a stronger result.

**Theorem 26.** *(Plancherel) Let* $x(t) \in L_2(\mathbb{R})$. *Then* $X(\omega)$ *defined by equation* (1.57) *exists. In addition,* $X(\omega) \in L_2(\mathbb{R})$ *and equation* (1.58) *holds (in the sense of equality in the mean), i.e., we have*

$$x_\Omega(t) = (2\pi)^{-1} \int_{-\Omega}^{\Omega} X(\omega) e^{i\omega t} d\omega; \quad then \quad \lim_{\Omega \to \infty} \|x_\Omega - x\| = 0.$$

The variable $\omega$ denotes the continuous frequency. If we use a factor of $(2\pi)^{-1/2}$ on the right hand sides of equations (1.57) and (1.58), we can write both of the equations in the Dirac notation

$$X(\omega) = \langle \omega \mid t \rangle \langle t \mid x \rangle = \langle \omega \mid x \rangle, \quad x(t) = \langle t \mid \omega \rangle \langle \omega \mid x \rangle = \langle t \mid x \rangle,$$

which are equivalent to the Fourier reconstruction (inversion) formula

$$|x\rangle \equiv |\omega\rangle \langle \omega \mid x \rangle.$$

Thus, we adopt the formal view that the vectors $|\omega\rangle$, $\omega \in \mathbb{R}$ form a continuously labeled basis satisfying the completeness and orthonormality relations

$$|\omega\rangle \langle \omega| = \mathbf{1}, \quad \langle \omega \mid \omega' \rangle = \delta(\omega - \omega').$$

The orthonormality relation follows easily by insertion of the resolution of identity (1.49) or (1.50) into the inner product, i.e.,

$$\langle \omega \mid \omega' \rangle = \langle \omega \mid t \rangle \langle t \mid \omega' \rangle = (2\pi)^{-1} \int_{-\infty}^{\infty} e^{-it(\omega - \omega')} dt = \delta(\omega - \omega').$$

We also adopt the convention of using equations (1.57) and (1.58) as the definition of the forward and inverse Fourier transforms instead of their definition in terms of the normalized basis functions $\langle \omega \mid t \rangle$ and their complex conjugates, which include a factor of $(2\pi)^{-1/2}$. With this convention the equality of the energy in the time and the frequency domains known as Parseval's relation,[15] i.e.,

$$\|x\|^2 \equiv \int\limits_{-\infty}^{\infty} |x(t)|^2 \, dt = \frac{1}{2\pi} \int\limits_{-\infty}^{\infty} |X(\omega)|^2 \, d\omega, \qquad (1.59)$$

follows from theorem 27.

**Theorem 27.** *Let $x(t), y(t) \in L_2(\mathbb{R})$. Denoting their Fourier transforms by $X(\omega)$ and $Y(\omega)$ we have*

$$\int\limits_{-\infty}^{\infty} x^*(t-k) \, y(t-l) \, dt = \frac{1}{2\pi} \int\limits_{-\infty}^{\infty} X^*(\omega) \, e^{i\omega k} \, Y(\omega) \, e^{-i\omega l} \, d\omega. \qquad (1.60)$$

This can be proven using

$$x(t-k) = \frac{1}{2\pi} \int\limits_{-\infty}^{\infty} X(\omega) \, e^{i\omega(t-k)} \, d\omega.$$

The rate of decay of the Fourier transform $X(\omega)$ (as $\omega \to \infty$) is a measure of the smoothness of the function $x(t)$: the faster the decay, the smoother the function. Similarly, the rate of decay of $x(t)$ is determined by the smoothness of $X(\omega)$. The following three theorems establish these results [10, 19].

**Theorem 28.** *Given the function $x(t) \in L_2(\mathbb{R})$, assume that its Fourier transform $X(\omega)$ is $N$ times continuously differentiable and all derivatives tend to 0 as $|\omega| \to \infty$. Then $t^N x(t) \to 0$ as $t \to \infty$.*

**Theorem 29.** *Given the function $x(t) \in L_2(\mathbb{R})$, assume that its Fourier transform $X(\omega)$ decays sufficiently fast so that $\omega^N X(\omega)$ is absolutely integrable for some positive integer $N$. Then $x(t)$ is $N$ times continuously differentiable and*

$$\frac{d^n}{dt^n} x(t) = \frac{1}{2\pi} \int\limits_{-\infty}^{\infty} (i\omega)^n \, X(\omega) \, e^{i\omega t} d\omega, \quad 1 \le n \le N. \qquad (1.61)$$

---

[15]If we include the factors of $(2\pi)^{-1/2}$ on the right hand sides of equations (1.57) and (1.58), then the $2\pi$ in the last integral of Parseval's relation (1.59) must be omitted.

**Theorem 30.** *Given the function $x(t) \in L_2(\mathbb{R})$ and its Fourier transform $X(\omega)$, assume that $x(t)$ decays sufficiently fast so that $t^N x(t)$ is absolutely integrable for some positive integer $N$. Then $X(\omega)$ is $N$ times continuously differentiable and*

$$\frac{d^n}{d\omega^n} X(\omega) = \frac{1}{2\pi} \int_{-\infty}^{\infty} (-it)^n x(t) e^{-i\omega t} dt, \quad 1 \le n \le N. \tag{1.62}$$

An shortcoming of the continuously labeled vectors $|\omega\rangle$, $\omega \in \mathbb{R}$, is the absence of a discretization of the frequency variable that would produce a frame (see section 1.26), let alone an orthonormal basis. The issue can be resolved only if we construct continuously labeled functions of two variables by using time domain window functions with the vectors $|\omega\rangle$; a procedure known as the windowed Fourier transform. If, on the other hand, we limit our Hilbert space to functions of compact support then the resulting Fourier transform, when discretized appropriately, leads to an orthonormal basis and the transform is known as the Fourier series expansion.

## 1.19   The Fourier Series Expansion

The space of square-integrable functions with support in the interval $[0, T_0]$ denoted by $L_2[0, T_0]$ is an example of a Hilbert space with an orthonormal basis that results from a discretization of the exponential functions $\exp(i\omega t)$ at frequencies $\omega_n = n\omega_0$ [20]. Thus, the set of functions

$$\phi_n(t) = T_0^{-1/2} \exp(in\omega_0 t), \quad n \in \mathbb{Z}, \tag{1.63}$$

is complete (see theorem 31 below) and orthonormal[16] on the given interval provided that $\omega_0 T_0 = 2\pi$,

$$T_0^{-1} \int_0^{T_0} e^{-im\omega_0 t} e^{in\omega_0 t} dt = \delta_{mn}, \quad \omega_0 T_0 = 2\pi. \tag{1.64}$$

Using the Dirac notation and the summation convention, we have

$$\langle t \mid n \rangle \equiv \phi_n(t) \ , \quad \langle m \mid n \rangle = \delta_{mn}, \quad |n\rangle \langle n| = \mathbf{1}, \tag{1.65}$$

where the requirement $\omega_0 T_0 = 2\pi$ is understood throughout this section. This Hilbert space is an example of when the convergence condition (1.42)

---

[16] When proving orthonormality we use the result $\lim_{\varepsilon \to 0} \left[ \left( 1 - e^{-i\omega_0 T_0 \varepsilon} \right) / i\omega_0 \varepsilon \right] = T_0$.

fails to hold and hence the sum (1.41) does not converge to an ordinary function (see theorem 32 in the next section).

A function $x(t) = \langle t \mid x \rangle$ in this space has the Fourier series expansion

$$|x\rangle = c_n |n\rangle , \quad c_n \equiv \langle n \mid x \rangle . \tag{1.66}$$

Parseval's relation is now

$$\|x\|^2 = \int_0^{T_0} |x(t)|^2 \, dt = \sum_{n=-\infty}^{\infty} |c_n|^2 . \tag{1.67}$$

Thus, the transform coefficients $c_n$ have finite energy and are in the Hilbert space $l_2(\mathbb{Z})$ (definition 4). Starting with a sequence $c_n$ in the Hilbert space $l_2(\mathbb{Z})$, the Riesz-Fischer theorem ensures the existence of a function $x(t)$ defined on the interval $[0, T_0]$ and whose Fourier series coefficients are the given sequence $c_n$.

**Theorem 31.** *(Riesz-Fischer) Given a sequence $\{c_n, n \in \mathbb{Z}\}$ in the Hilbert space $l_2(\mathbb{Z})$ and the functions $\phi_n(t)$ defined in (1.63) together with the constraint $\omega_0 T_0 = 2\pi$, there exists a (Lebesgue) measurable function $x(t)$ defined on the interval $[0, T_0]$ such that*

$$\lim_{N \to \infty} \int_0^{T_0} \left| x(t) - \sum_{n=-N}^{N} c_n \phi_n(t) \right|^2 dt = 0.$$

*In addition, $c_n = \langle \phi_n, x \rangle$ and Parseval's relation, equation (1.67), holds [21].*

It is common to use

$$x(t) = T_0^{-1} \sum_{n=-\infty}^{\infty} c_n e^{+in\omega_0 t} \tag{1.68}$$

as the Fourier reconstruction formula, which shows that $x(t)$ is a periodic function of $t$ with period $T_0$.. The Fourier series coefficients are then given by

$$c_n = \int_0^{T_0} e^{-in\omega_0 t} x(t) \, dt, \tag{1.69}$$

omitting the $T_0^{-1/2}$ factor that would be present in $\langle n \mid x \rangle$, while Parseval's relation becomes

$$\|x\|^2 = \int_0^{T_0} |x(t)|^2 \, dt = T_0^{-1} \sum_{n=-\infty}^{\infty} |c_n|^2 , \tag{1.70}$$

which is a special case of the more general equation

$$\int_0^{T_0} x^* \left(t - \tau_1\right) y \left(t - \tau_2\right) dt = T_0^{-1} \sum_{n=-\infty}^{\infty} c_n^* d_n e^{+in\omega_0(\tau_1 - \tau_2)}, \qquad (1.71)$$

with $c_n$ and $d_n$ denoting the Fourier series coefficients of two functions $x(t)$ and $y(t)$.

## 1.20   The Discrete Time Fourier Transform

If we consider the Hilbert space of discrete time sequences with finite energy $l_2(\mathbb{Z})$ (see definition 4) then we have the discrete time Fourier transform coefficients [22]

$$X \left(\omega\right) = \sum_{n=-\infty}^{\infty} x \left[n\right] e^{-in\omega} \qquad (1.72)$$

where

$$x \left[n\right] = \frac{1}{2\pi} \int_{-\pi}^{\pi} X \left(\omega\right) e^{in\omega} d\omega. \qquad (1.73)$$

Clearly the Fourier transform $X \left(\omega\right)$ is periodic in the frequency variable $\omega$ with a period of $2\pi$. Consequently, the inversion formula is an integral over any interval of length $2\pi$ which, without loss of generality, we take to be the interval $[-\pi, \pi]$.[17]

An important result of (1.73) is the Poisson summation formula found by using the $2\pi$ periodic function

$$y(t) = \sum_{n=-\infty}^{\infty} x \left(t + 2n\pi\right),$$

which has a Fourier series representation of the form (1.68), namely,

$$y \left(t\right) = \frac{1}{2\pi} \sum_{n=-\infty}^{\infty} c_n e^{int}$$

---

[17]A discrete time sequence $x \left[n\right]$ is usually the result of sampling of a continuous time function $x \left(t\right)$ at times $t = n\Delta T = n/f_s$. The exponential function in the Fourier transform is then of the form $e^{-in\omega/f_s}$ where now $\omega \in [-\pi f_s, +\pi f_s,]$. It is common practice, without loss of generality, to take the sampling frequency in this case to be equal to 1 Hz, and use a normalized frequency variable $\omega$ in the range $[-\pi, +\pi]$.

with Fourier series coefficients $c_n$ given by

$$c_n = \frac{1}{2\pi} \int_0^{2\pi} e^{-int} y(t)\, dt = \frac{1}{2\pi} \sum_{k=-\infty}^{\infty} \int_0^{2\pi} e^{-int} x(t+2k\pi)\, dt =$$

$$\frac{1}{2\pi} \sum_{k=-\infty}^{\infty} \int_{2k\pi}^{2(k+1)\pi} e^{-int} x(t+2k\pi)\, dt = \int_{-\infty}^{\infty} e^{-int} x(t)\, dt = X(n).$$

**Theorem 32.** *For $x[n]$ and $X(\omega)$ defined in equations (1.72) and (1.73) the Poisson summation formula is*

$$\sum_{n=-\infty}^{\infty} x(t+2n\pi) = \sum_{n=-\infty}^{\infty} X(n) e^{int}. \tag{1.74}$$

A special case of the Poisson summation formula is

$$2\pi \sum_{n=-\infty}^{\infty} e^{in(\omega-\omega')} = \sum_{k=-\infty}^{\infty} \delta\left(\omega - \omega' + 2k\pi\right), \tag{1.75}$$

which can be used to derive Parseval's relation in the form

$$\sum_{n=-\infty}^{\infty} |s[n]|^2 = \frac{1}{2\pi} \int_{-\pi}^{\pi} |S(\omega)|^2\, d\omega, \tag{1.76}$$

or the more general form

$$\sum_{n=-\infty}^{\infty} s^*[n-k]\, q[n-l] = \frac{1}{2\pi} \int_{-\pi}^{\pi} S^*(\omega)\, e^{-i\omega k} Q(\omega)\, e^{i\omega l}\, d\omega. \tag{1.77}$$

## 1.21 The Discrete Fourier Transform

The space of discrete time sequences of finite energy that have a finite length $N$ is a finite dimensional vector space of dimension $N$. An orthonormal basis for this space is provided by the functions [22]

$$p_{kn} = N^{-1/2} e^{2i\pi kn/N}, \quad \sum_{k=0}^{N-1} p_{mk}^* p_{kn} = \delta_{mn}. \tag{1.78}$$

For $x[n]$, $n = 0, \ldots, N - 1$, in this vector space we have the discrete and finite length Fourier series representation

$$x[n] = \sum_{k=0}^{N-1} X[k]\, p_n[k] \equiv \frac{1}{\sqrt{N}} \sum_{k=0}^{N-1} X[k]\, e^{2i\pi kn/N}, \qquad (1.79)$$

where the coefficients are given by

$$X[k] = \frac{1}{\sqrt{N}} \sum_{n=0}^{N-1} x[n]\, e^{-2i\pi kn/N}. \qquad (1.80)$$

Parseval's relation is

$$\sum_{k=0}^{N-1} |X[k]|^2 = \sum_{n=0}^{N-1} |x[n]|^2. \qquad (1.81)$$

The above equations define the discrete Fourier transform (DFT), which is often implemented by (and referred to as) the fast Fourier transform (FFT). A more conventional definition for the transform pair $x[n] \leftrightarrow X[k]$ is given by [22]

$$X[k] = \sum_{n=0}^{N-1} x[n]\, e^{-2i\pi kn/N}, \quad x[n] = \frac{1}{N} \sum_{k=0}^{N-1} X[k]\, e^{2i\pi kn/N}. \qquad (1.82)$$

The difference from the original definition is in the normalization of the transform pair. Parseval's relation is now

$$\sum_{k=0}^{N-1} |X[k]|^2 = \frac{1}{N} \sum_{n=0}^{N-1} |x[n]|^2. \qquad (1.83)$$

## 1.22   Band-Limited Functions and the Sampling Theorem

A band-limited function $x_\Omega(t) \in L_2(\mathbb{R})$ is one whose Fourier transform $X_\Omega(\omega)$ has support in $[-\Omega, \Omega]$, where $\Omega$ is a positive real number. The unitarity of the Fourier transform operation ensures that all band-limited functions in $L_2(\mathbb{R})$ form a closed subspace $L_2^\Omega(\mathbb{R})$. The evaluation functional (see equations (1.17) and (1.19)) mapping $x_\Omega \to x_\Omega(t)$ is bounded (and therefore continuous) for every function in $L_2^\Omega(\mathbb{R})$ and all $t$. Aronszajn's theorem (theorem 12) then ensures that $L_2^\Omega(\mathbb{R})$ is a reproducing kernel Hilbert space. To find the kernel we use the inverse Fourier transform and the band-limited property to write

$$x_\Omega(t) = (2\pi)^{-1} \int_{-\Omega}^{\Omega} X_\Omega(\omega) e^{+i\omega t} d\omega. \tag{1.84}$$

We substitute the Fourier transform

$$X_\Omega(\omega) = \int_{-\infty}^{\infty} x_\Omega(t') e^{-i\omega t'} dt' \tag{1.85}$$

and perform the integral over $\omega$ to obtain

$$x_\Omega(t) = \int_{-\infty}^{\infty} \frac{\sin[\Omega(t-t')]}{\pi(t-t')} x_\Omega(t') \, dt'. \tag{1.86}$$

This shows the reproducing kernel property of the associated Hilbert space $L_2^\Omega(\mathbb{R})$, with a real and symmetric kernel function[18]

$$K(t,t') \equiv \sin[\Omega(t-t')]/\pi(t-t'). \tag{1.87}$$

Another way to see this result is to note that the operator projecting a function into the space $L_2^\Omega(\mathbb{R})$,

$$P_\Omega\{x(t)\} = (2\pi)^{-1} \int_{-\Omega}^{\Omega} X(\omega) e^{i\omega t} d\omega, \tag{1.88}$$

is an orthogonal projection. This operator, in the frequency domain, has the form

$$P_\Omega\{X(\omega)\} = X(\omega) \times \mathbf{1}_{[-\Omega,\Omega]} \equiv X_\Omega(\omega) \tag{1.89}$$

where the function $\mathbf{1}_{[-\Omega,\Omega]}$ is defined to be 1 on the interval $[-\Omega, \Omega]$ and 0 elsewhere, and its inverse Fourier transform is $\sin(\Omega t)/\pi t$. The inverse Fourier transform of $X_\Omega(\omega)$ is the convolution of $x_\Omega(t)$ and $\sin(\Omega t)/\pi t$, which is precisely equation (1.86) (see equations (2.1) and (2.2) of chapter

---

[18]Note that the limit $\Omega \to \infty$ and use of equation (1.23) will result in the Dirac delta generalized reproducing kernel discussed in section 1.12.

2 for theorems on convolution). Thus, the reproducing kernel (1.87) is a projection operator that maps any function in $L_2(\mathbb{R})$ onto the subspace $L_2^{\Omega}(\mathbb{R})$.

Let us define

$$p_{\tau}(t) = \sin\left[\Omega\left(t - \tau\right)\right]/\pi\left(t - \tau\right) = K\left(t, \tau\right). \tag{1.90}$$

We note that $p_{\tau}(t) = p_{\tau-t}(0)$ and so the functions $p_{\tau}(t)$ are essentially (except for an overall multiplicative constant) time-shifted versions of a "mother" function $\operatorname{sinc}(\Omega t/\pi)$, where

$$\operatorname{sinc}(t) \equiv \sin\left(\pi t\right)/\pi t. \tag{1.91}$$

Then we have

$$\int_{-\infty}^{\infty} p_{\tau}(t)\, p_{\tau'}(t)\, dt = p_{\tau}\left(\tau'\right) = p_{\tau'}\left(\tau\right). \tag{1.92}$$

The reproducing kernel equation shows that all elements of $L_2^{\Omega}(\mathbb{R})$ can be expanded in terms of the continuously parametrized functions $p_{\tau}(t)$, $\tau \in \mathbb{R}$. It is well known that a discretization of the shift parameter $\tau$,

$$\tau_n = n\pi/\Omega, \tag{1.93}$$

produces an orthogonal basis [12]. The orthogonality follows from setting $\tau = m\pi/\Omega$ and $\tau' = n\pi/\Omega$ in equation (1.92). Defining the linear bandwidth variable $B \equiv \Omega/2\pi$, we are led to the orthonormal basis functions

$$p_n(t) = \sqrt{2B}\,\operatorname{sinc}\left[2B\left(t - \frac{n}{2B}\right)\right]. \tag{1.94}$$

The Fourier transform of $p_n(t)$ is identically zero outside the closed interval $[-2\pi B, +2\pi B]$. Any band-limited function $x_{\Omega}(t)$ can be expressed in terms of this basis,

$$x_{\Omega}(t) = \sum_{n=-\infty}^{\infty} c_n p_n(t), \tag{1.95}$$

with the expansion coefficients given by

$$c_n = \langle p_n, x_{\Omega}\rangle = x_{\Omega}\left(n/2B\right). \tag{1.96}$$

Thus we arrive at the sampling theorem[19][23].

---

[19] Although the theorem is usually referred to as the Shannon sampling theorem, it was independently discovered by Nyquist, Whittaker, and Kotelnikov.

**Theorem 33.** *Let $x(t)$ be a continuous time signal whose Fourier transform $X(\omega)$ is zero outside the interval $[-2\pi B, +2\pi B]$, i.e., whose linear frequency content is limited to the band $[-B, +B]$. If $\Delta T$ is a positive constant satisfying the inequalities*

$$0 < \Delta T \le (2B)^{-1}, \tag{1.97}$$

*then $x(t)$ can be uniquely reconstructed from its sampled values $x(n\Delta T)$, $n \in \mathbb{Z}$, by the interpolation formula*

$$x(t) = 2B(\Delta T) \sum_{n=-\infty}^{\infty} x(n\Delta T) \frac{\sin[2\pi B(t - n\Delta T)]}{2\pi B(t - n\Delta T)}. \tag{1.98}$$

*Condition (1.97) is usually written in terms of the sampling frequency $f_s \equiv 1/\Delta T$ in the form $f_s \ge 2B$, and stated as the requirement that the signal sampling frequency should equal or exceed twice the highest frequency of the signal.*

The sampling theorem 33 applied to the sinc function itself, at a sampling frequency equal to twice the highest frequency, yields the following result:

$$\frac{\sin[2\pi B(t - t')]}{2\pi B(t - t')} = \sum_{n=-\infty}^{\infty} \frac{\sin\left[2\pi B\left(t - \frac{n}{2B}\right)\right]}{2\pi B\left(t - \frac{n}{2B}\right)} \frac{\sin\left[2\pi B\left(t' - \frac{n}{2B}\right)\right]}{2\pi B\left(t' - \frac{n}{2B}\right)}, \tag{1.99}$$

which shows the reproducing kernel as an infinite sum over the products of the orthogonal basis functions. This is an example of theorem 22 and equation (1.41).

## 1.23 The Basis Operator in $L_2(\mathbb{R})$

The correspondence between a function and its transform coefficients with respect to an orthonormal basis expressed by equations (1.39) and (1.40) can be interpreted in terms of linear operators [24, 25, 15]. Consider the linear operator

$$T_\phi : L_2(\mathbb{R}) \to l_2(\mathbb{Z}), \tag{1.100}$$

associated with a basis (not necessarily orthonormal) $\{\phi_n(t), n \in \mathbb{Z}\}$, defined by

$$T_\phi\{x(t)\} = [\ldots, \langle \phi_n \mid x \rangle, \ldots]^T. \tag{1.101}$$

It is common to refer to this expansion as the analysis formula, or the forward transform, and to the linear operator as the analysis operator of the

associated basis. The corresponding adjoint operator, the synthesis operator, is a linear map

$$T^{+}{}_{\phi} : l_2\left(\mathbb{Z}\right) \to L_2\left(\mathbb{R}\right), \tag{1.102}$$

defined by the synthesis formula, or the inverse transform,

$$T^{+}{}_{\phi}\left\{\underline{c}\right\} = \sum_{n=-\infty}^{\infty} c_n \phi_n\left(t\right), \quad \underline{c} \equiv [\ldots, c_n, \ldots]^T. \tag{1.103}$$

The basis operator is then defined by

$$T^{+}{}_{\phi}T_{\phi} : L_2\left(\mathbb{R}\right) \to L_2\left(\mathbb{R}\right) \tag{1.104}$$

and can be expressed in the Dirac notation together with the summation
convention as

$$T^{+}{}_{\phi}T_{\phi} = |\phi_n\rangle\langle\phi_n|, \tag{1.105}$$

or can be defined by its action on any function $x(t)$ (note the summation
convention in the Dirac notation),

$$T^{+}{}_{\phi}T_{\phi}\left\{x\left(t\right)\right\} = \sum_{n=-\infty}^{\infty} \langle\phi_n, x\rangle\,\phi_n\left(t\right) = |\phi_n\rangle\langle\phi_n \mid x\rangle. \tag{1.106}$$

The basis operator $T^{+}{}_{\phi}T_{\phi}$ is clearly self-adjoint and positive definite, and
so we have

$$\left\|T^{+}{}_{\phi}T_{\phi}\right\|_2 = \rho_{T^{+}{}_{\phi}T_{\phi}} = \lambda_{\max}, \quad \left\|\left(T^{+}{}_{\phi}T_{\phi}\right)^{-1}\right\|_2 = \lambda_{\min}^{-1}, \tag{1.107}$$

where $\lambda_{\max}$ and $\lambda_{\min}$ are the largest and smallest eigenvalues of the basis
operator $T_{\phi}^{+}T_{\phi}$. The ratio $\lambda_{\max}/\lambda_{\min}$ is the condition number of the same
operator and its size is crucial in calculating the inverse operator in a nu-
merically stable manner.

   If the basis functions are not orthonormal, as is the case in an overcomplete set in which the elements are not linearly independent, the numerical inversion of the basis operator $T^{+}{}_{\phi}T_{\phi}$ depends on the size of its
condition number. If, on the other hand, the analysis operator $T_{\phi}$ is unitary
then $T^{+}{}_{\phi}T_{\phi} = \mathbf{1}$, all the eigenvalues of the basis operator are equal to 1,
and the inverse is guaranteed to exist. A unitary analysis operator indicates
an orthonormal basis and for such an operator we have

$$\|x\|^2 = \langle x, x\rangle = \langle x, T^{+}{}_{\phi}T_{\phi}x\rangle = \langle T_{\phi}x, T_{\phi}x\rangle = \sum_{n=-\infty}^{\infty} |\langle\phi_n, x\rangle|^2, \tag{1.108}$$

which is Parseval's relation (1.37).[20]

---

[20]In the Dirac notation $\|x\|^2 = \langle x| T_{\phi}^{+}T_{\phi} |x\rangle = \langle x \mid \phi_n\rangle\langle\phi_n \mid x\rangle \equiv \sum_{n=-\infty}^{\infty} |\langle\phi_n \mid x\rangle|^2.$

# 1.24 Biorthogonal Bases and Representations in $L_2(\mathbb{R})$

More general than an orthogonal basis is the concept of biorthogonal bases that occur in the definition of a Riesz basis. As we shall see later, biorthogonal bases allow more freedom in choosing wavelet functions of compact support. The idea of biorthogonality is best explained for finite dimensional vector spaces. Consider the $xy$-plane and a pair of vectors $\mathbf{v}_1 \equiv [0,1]^T$ and $\mathbf{v}_2 \equiv [1, \tan\theta]^T$ and a second pair $\mathbf{w}_1 \equiv [-\tan\theta, 1]^T$ and $\mathbf{w}_2 \equiv [1,0]^T$ for some angle $0 < \theta < \pi/2$. Then $\langle \mathbf{w}_i, \mathbf{v}_j \rangle \equiv \mathbf{w}_i^T \mathbf{v}_j = \delta_{ij}$ and we call the two pairs biorthogonal. The two pairs of vectors are shown in figure 1.3 for $\theta = 60°$. Now any vector $\mathbf{x} = [x,y]^T$ can be expanded in terms of either

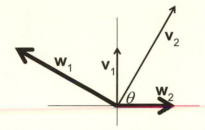

Figure 1.3: Two biorthogonal pairs of vectors.

the first or the second pair. The coefficients of expansion are then given by the inner product of the given vector with the other pair. For instance, if $\mathbf{x} = \alpha_1 \mathbf{v}_1 + \alpha_2 \mathbf{v}_2$, then $\alpha_j = \langle \mathbf{w}_j, \mathbf{x} \rangle$. Alternatively, if $\mathbf{x} = \beta_1 \mathbf{w}_1 + \beta_2 \mathbf{w}_2$, then $\beta_j = \langle \mathbf{v}_j, \mathbf{x} \rangle$ for $j = 1, 2$. Neither the pair $\{\mathbf{v}_1, \mathbf{v}_2\}$, nor the pair $\{\mathbf{w}_1, \mathbf{w}_2\}$, however, is an orthogonal set.

**Definition 40.** Given a complete and orthonormal basis $\{\phi_n(t),\ n \in \mathbb{Z}\}$, a new set of basis functions, a Riesz basis, is defined through a nonunitary but invertible continuous linear operator $L$ (also known as a topological isomorphism) by the relations [26]

$$\xi_n(t) = L\phi_n(t) \leftrightarrow |\xi_n\rangle = L|\phi_n\rangle. \tag{1.109}$$

Since $L$ is not unitary, the new basis functions are not orthonormal.[21] We have

$$\langle \phi_m, \phi_n \rangle = \delta_{mn} = \left\langle L^{-1} L\phi_m, \phi_n \right\rangle = \left\langle L\phi_m, \left(L^{-1}\right)^+ \phi_n \right\rangle. \tag{1.110}$$

---

[21]If $L$ were unitary, i.e., $L^+L = 1$, then $\langle \phi_m, \phi_n \rangle = \langle L^+L\phi_m, \phi_n \rangle = \langle \xi_m, \xi_n \rangle = \delta_{mn}$.

Since $L$ is a topological isomorphism, its inverse is also a continuous linear map and it can be used to define a second Riesz basis set $\{\chi_n(t),\ n \in \mathbb{Z}\}$

$$\chi_n(t) \equiv \left(L^{-1}\right)^+ \phi_n(t) \leftrightarrow |\chi_n\rangle = \left(L^{-1}\right)^+ |\phi_n\rangle. \qquad (1.111)$$

It is easy to see that

$$\langle \xi_m \mid \chi_n \rangle = \delta_{mn}, \qquad (1.112)$$

which shows the biorthogonal property of the two Riesz bases $\{\xi_n(t)\}$ and $\{\chi_n(t)\}$. The corresponding analysis operators are denoted by $T_\xi$ and $T_\chi$, where

$$T_\xi x(t) = [\ldots, \langle \xi_n, x \rangle, \ldots]^T, \quad T_\chi x(t) = [\ldots, \langle \chi_n, x \rangle, \ldots]^T. \qquad (1.113)$$

Any element of $L_2(\mathbb{R})$ can now be represented in two equivalent ways in terms of the two biorthogonal sets. Using the Dirac notation and the summation convention, we have the following expansion in terms of the original orthonormal set:

$$|x\rangle = |\phi_n\rangle \langle \phi_n \mid x \rangle,$$

which when acted on by the operator $L$ and with use of equation (1.109) becomes

$$L|x\rangle = L|\phi_n\rangle \langle \phi_n \mid x \rangle = |\xi_n\rangle \langle \phi_n \mid x \rangle.$$

Now using the adjoint of equation (1.111) we write

$$\langle \phi_n| = \langle \chi_n| L,$$

and so for any $x(t)$ we have

$$L|x\rangle = |\xi_n\rangle \langle \chi_n| L |x\rangle,$$

from which we arrive at the completeness relation

$$|\xi_n\rangle \langle \chi_n| = \mathbf{1}, \qquad (1.114)$$

with the associated expansion formula

$$|x\rangle = \langle \xi_n \mid x \rangle |\chi_n\rangle. \qquad (1.115)$$

On the other hand, starting with the identity

$$L^+ |x\rangle = |\phi_n\rangle \langle \phi_n| L^+ |x\rangle,$$

we arrive at the completeness relation

$$|\chi_n\rangle \langle \xi_n| = \mathbf{1},$$ (1.116)

with the associated expansion formula

$$|x\rangle = \langle \chi_n \mid x \rangle |\xi_n\rangle .$$ (1.117)

Thus, biorthogonal Riesz bases are more general than orthogonal bases. We will use biorthogonal bases in the construction of biorthogonal wavelets of compact support whose implementation does not require filter sequences of even length; both even and odd lengths will be permitted in this context. Even more general is the concept of a frame which is introduced in the next two sections. It is important to note, however, that both orthogonal and Riesz bases have the property that removal of a single element from either set will destroy their completeness; such bases are known as exact. Riesz bases are also exact frames [27]. Before we introduce the concept of a frame in an infinite dimensional Hilbert space we look, in the next section, at the problem of vector representation in a finite dimensional vector space, e.g., $\mathbb{C}^n$.

## 1.25   Frames in a Finite Dimensional Vector Space

A vector $\mathbf{x} \in \mathbb{C}^n$ is represented by the column $[x_1, \ldots, x_n]^T$ where $x_k \in \mathbb{C}$, $k = 1, \ldots, n$. The inner product is defined by

$$\langle \mathbf{z}, \mathbf{x} \rangle \equiv \mathbf{z}^+ \mathbf{x} = \sum_{k=1}^{n} \mathbf{z}_k^* \mathbf{x}_k, \quad \mathbf{z}, \mathbf{x} \in \mathbb{C}^n.$$

Consider a linear mapping (operator) $\mathbb{C}^n \to \mathbb{C}^m$, $n \geq m$, represented by a complex $m \times n$ matrix $\mathbf{A}$, i.e., $\mathbf{A}\mathbf{x} = \mathbf{y}$, $\mathbf{x} \in \mathbb{C}^n$, $\mathbf{y} \in \mathbb{C}^m$. Denoting the columns of $\mathbf{A}$ by $\mathbf{c}_k$, $k = 1, \ldots, n$, it is easy to see that

$$\mathbf{A}\mathbf{x} = \sum_{k=1}^{n} x_k \mathbf{c}_k,$$ (1.118)

that is, the range space of the matrix $\mathbf{A}$, denoted by $\mathscr{R}(\mathbf{A})$, is the space spanned by the columns of the same matrix. If we assume that $\mathscr{R}(\mathbf{A})$ is all of $\mathbb{C}^m$, $m \leq n$, then there are $m$ independent columns of $\mathbf{A}$. The column rank of the matrix $\mathbf{A}$ is the number of independent columns of the matrix, and defines the dimension of the range space $\mathscr{R}(\mathbf{A})$.

The adjoint map (operator) denoted by $\mathbf{A}^+$ is defined $\mathbb{C}^m \to \mathbb{C}^n$ and is the $n \times m$ matrix $(\mathbf{A}^*)^T$ (the Hermitian conjugate of $\mathbf{A}$). The dimension of the range space $\mathscr{R}(\mathbf{A}^+)$ is the column rank of the matrix $\mathbf{A}^+$, which is equal to the row rank of the matrix $\mathbf{A}$,[22] which is also equal to the column rank of the same matrix. The rank $r_{\mathbf{A}}$ of the matrix $\mathbf{A}$ is the number of linearly independent columns (or rows) of that matrix. The following theorem specifies the dimensions of the various range and null spaces [28].

**Theorem 34.** *If a complex $m \times n$ matrix $\mathbf{A}$ has rank $r_{\mathbf{A}}$ then*

- *$\mathscr{R}(\mathbf{A})$ and $\mathscr{R}(\mathbf{A}^+)$ have dimension $r_{\mathbf{A}}$,*

- *$\mathscr{N}(\mathbf{A})$ has dimension $(n - r_{\mathbf{A}})$ while $\mathscr{N}(\mathbf{A}^+)$ has dimension $(m - r_{\mathbf{A}})$.*

When $m = n = r_{\mathbf{A}}$, the matrix $\mathbf{A}$ is square and has full rank. It is, therefore, nonsingular (has nonzero determinant) and invertible. Consequently, an equation of the form

$$\mathbf{A}\mathbf{x} = \mathbf{y}, \quad \mathbf{x} \in \mathbb{C}^n, \quad \mathbf{y} \in \mathbb{C}^m, \tag{1.119}$$

has the unique solution $\mathbf{x} = \mathbf{A}^{-1}\mathbf{y}$ when $m = n = r_{\mathbf{A}}$. More generally, a solution exists only if $\mathbf{y} \in \mathscr{R}(\mathbf{A})$, i.e., if $\mathbf{y}$ is a linear combination of the columns of $\mathbf{A}$. By theorem 9 we have $\mathscr{R}(\mathbf{A}) = (\mathscr{N}(\mathbf{A}^+))^\perp$ and so if $\mathbf{y} \in \mathscr{R}(\mathbf{A})$ then $\mathbf{y}$ must be orthogonal to $\mathscr{N}(\mathbf{A}^+)$. In other words, the matrix equation (1.119) has a solution if, and only if, $\langle \mathbf{z}, \mathbf{y} \rangle = 0$ for every $\mathbf{z} \in \mathbb{C}^m$ that satisfies $\mathbf{A}^+\mathbf{z} = \mathbf{0}$.

If a solution to (1.119) exists then it is unique if, and only if, $\mathscr{N}(\mathbf{A}) = \{\mathbf{0}\}$. Thus, when $m \le n$ and $r_{\mathbf{A}} < \min(m, n)$ the dimension of the null space $\mathscr{N}(\mathbf{A})$ is greater than zero and so if a solution to (1.119) exists, it is not unique. When multiple solutions exist, a unique solution whose norm is a minimum can be found using the following theorem [29, 30].

**Theorem 35.** *Let $\mathbf{A}$ be a complex $m \times n$ matrix, $m \le n$. Then the unique solution to the equation $\mathbf{A}\mathbf{x} = \mathbf{y}$, $\mathbf{x} \in \mathbb{C}^n$ and $\mathbf{y} \in \mathbb{C}^m$, whose norm $\|x\|$ is the smallest among all the solutions is $\hat{\mathbf{x}} = \mathbf{A}^+\mathbf{u}$ where $\mathbf{u}$ is any solution to the equation $\mathbf{A}\mathbf{A}^+\mathbf{u} = \mathbf{y}$. In particular, when $r_{\mathbf{A}} = m \le n$, the minimum norm solution is*

$$\hat{\mathbf{x}} = \mathbf{A}^+ \left(\mathbf{A}\mathbf{A}^+\right)^{-1} \mathbf{y} \equiv \mathbf{A}_R^\dagger \mathbf{y}, \tag{1.120}$$

---

[22]To be precise, the row rank of $\mathbf{A}^*$, which is the same as the row rank of $\mathbf{A}$.

*which also defines the right pseudoinverse*[23] $\mathbf{A}_R^\dagger$ *satisfying*

$$\mathbf{A}\mathbf{A}_R^\dagger = \mathbf{1}. \tag{1.121}$$

*Furthermore, if $\mathbf{B}$ is a complex $n \times m$ matrix, $n \geq m$, whose rank is $m$, then the equation*

$$\mathbf{B}\mathbf{y} = \mathbf{x}, \quad \mathbf{y} \in \mathbb{C}^m, \quad \mathbf{x} \in \mathbb{C}^n, \tag{1.122}$$

*has no solution. However, there is a unique vector $\hat{\mathbf{y}}$ that minimizes the norm $\|\mathbf{B}\mathbf{y} - \mathbf{x}\|$ (for the $l_2$ norm this is known as the least squares solution), namely,*

$$\hat{\mathbf{y}} = \left(\mathbf{B}^+\mathbf{B}\right)^{-1}\mathbf{B}^+\mathbf{x} \equiv \mathbf{B}_L^\dagger\mathbf{x}, \tag{1.123}$$

*which also defines the left pseudoinverse $\mathbf{B}_L^\dagger$ satisfying*

$$\mathbf{B}_L^\dagger\mathbf{B} = \mathbf{1}. \tag{1.124}$$

Now consider the identity

$$\mathbf{y} = \mathbf{A}\mathbf{A}^+\left(\mathbf{A}\mathbf{A}^+\right)^{-1}\mathbf{y} = \mathbf{A}\mathbf{A}_R^\dagger\mathbf{y}, \quad \mathbf{y} \in \mathbb{C}^m, \tag{1.125}$$

or equivalently

$$y_k = \sum_{j=1}^{n}\sum_{l=1}^{m} A_{kj}\left[\mathbf{A}_R^\dagger\right]_{jl} y_l, \quad k = 1,\ldots,m. \tag{1.126}$$

We write

$$\sum_{l=1}^{m}\left[\mathbf{A}_R^\dagger\right]_{jl} y_l = \sum_{l=1}^{m}\left[\left(\mathbf{c}_j^\dagger\right)^*\right]_l y_l \equiv \left\langle \mathbf{c}_j^\dagger, \mathbf{y} \right\rangle,$$

where

$$\left[\left(\mathbf{c}_j^\dagger\right)\right]_l = \left[\left(\mathbf{A}_R^\dagger\right)^*\right]_{jl} = \left[\left(\mathbf{A}_R^\dagger\right)^+\right]_{lj}.$$

Thus, $\mathbf{c}_j^\dagger$ is the $j$th column of the $n \times m$ matrix $\left(\mathbf{A}_R^\dagger\right)^+$,

$$\left(\mathbf{A}_R^\dagger\right)^+ \equiv \left[\mathbf{c}_1^\dagger,\ldots,\mathbf{c}_n^\dagger\right],$$

---

[23] A Moore-Penrose pseudoinverse or generalized inverse of an $m \times n$ matrix $\mathbf{A}$ is an $n \times m$ matrix $\mathbf{A}^\dagger$ satisfying the following conditions: $\mathbf{A}^\dagger\mathbf{A}$ and $\mathbf{A}\mathbf{A}^\dagger$ are Hermitian, $\mathbf{A}\mathbf{A}^\dagger\mathbf{A} = \mathbf{A}$, and $\mathbf{A}^\dagger\mathbf{A}\mathbf{A}^\dagger = \mathbf{A}^\dagger$ [30]. The left and right generalized inverses defined in this theorem are used to find solutions to the linear equation $\mathbf{A}\mathbf{x} = \mathbf{y}$ in two cases: (a) an over-determined set ($m > n$), in which case a least squares solution is found, and (b) an under-determined set ($m < n$), in which case a minimum-norm solution is found.

and equation (1.126) can be written as

$$y_k = \sum_{j=1}^{n} \mathbf{A}_{kj} \left\langle \mathbf{c}_j^\dagger, \mathbf{y} \right\rangle, \quad k = 1, \ldots, m,$$

which in view of the relation (1.118) becomes

$$\mathbf{y} = \sum_{j=1}^{n} \left\langle \mathbf{c}_j^\dagger, y \right\rangle \mathbf{c}_j, \tag{1.127}$$

where $\mathbf{c}_j$ are the columns of the $m \times n$ matrix $\mathbf{A}$

$$\mathbf{A} \equiv [\mathbf{c}_1, \ldots, \mathbf{c}_n].$$

Equation (1.127) expresses a reconstruction formula for a vector $\mathbf{y}$ in terms of a sum over the product of a set of vectors $\mathbf{c}_k$ (columns of the matrix $\mathbf{A}$) multiplied by a set of scalar transform coefficients, obtained in the analysis stage by taking inner products of $\mathbf{y}$ with the columns of the Hermitian conjugate of the right pseudoinverse, and represented by $\left\langle \mathbf{c}_k^\dagger, \mathbf{y} \right\rangle$.

A dual reconstruction formula to (1.127) can be obtained starting with the identity

$$\mathbf{y} = \left(\mathbf{A}\mathbf{A}^+\right)^{-1} \left(\mathbf{A}\mathbf{A}^+\right)\mathbf{y} = (\mathbf{A}_R^\dagger)^+ \mathbf{A}^+ \mathbf{y}, \quad \mathbf{y} \in \mathbb{C}^m, \tag{1.128}$$

instead of (1.125). Now we have

$$y_k = \sum_{j=1}^{n} \sum_{l=1}^{m} \left[(\mathbf{A}_R^\dagger)^+\right]_{kj} (\mathbf{A}^+)_{jl} \, y_l, \quad k = 1, \ldots, m, \tag{1.129}$$

while this time

$$\sum_{l=1}^{m} \left[\mathbf{A}^+\right]_{jl} y_l = \langle \mathbf{c}_j, \mathbf{y} \rangle,$$

where $\mathbf{c}_j$ are the columns of the matrix $\mathbf{A}$, and we find

$$\mathbf{y} = \sum_{j=1}^{n} \langle \mathbf{c}_j, y \rangle \, \mathbf{c}_j^\dagger, \tag{1.130}$$

where $\mathbf{c}_j^\dagger$ are the columns of the $m \times n$ matrix $(\mathbf{A}_R^\dagger)^+$.

The reconstruction formula, however, is only useful when the right pseudoinverse matrix $\mathbf{A}_R^\dagger$ can be computed in a stable fashion, which is possible only when the matrix $\mathbf{A}\mathbf{A}^+$ has a reasonable condition number, that is, the magnitude of the ratio of the largest eigenvalue to the smallest eigenvalue is not very large.[24] But the eigenvalues of $\mathbf{A}\mathbf{A}^+$ are the singular values of the matrix $\mathbf{A}$.[25] Denoting the extreme singular values by $\sigma_{\min}$ and $\sigma_{\max}$ we have the following inequalities:

$$\sigma_{\min}\|\mathbf{y}\|^2 \leq \langle \mathbf{A}\mathbf{A}^+\mathbf{y}, \mathbf{y}\rangle \leq \sigma_{\max}\|\mathbf{y}\|^2, \quad \forall \mathbf{y} \in \mathbb{C}^m. \tag{1.131}$$

Using the reconstruction formula (1.130) we have

$$\langle \mathbf{A}\mathbf{A}^+\mathbf{y}, \mathbf{y}\rangle = \langle \mathbf{y}, \mathbf{A}\mathbf{A}^+\mathbf{y}\rangle = \sum_{j=1}^{n} \langle \mathbf{c}_j, y\rangle \left\langle y, \mathbf{A}\mathbf{A}^+\mathbf{c}_j^\dagger \right\rangle.$$

Using the definition of the right pseudoinverse we have [29]

$$\left(\mathbf{A}\mathbf{A}^+\mathbf{c}_j^\dagger\right)_k = (\mathbf{A}\mathbf{A}^+)_{ki}\left[\left(\mathbf{A}_R^\dagger\right)^+\right]_{ij} = (\mathbf{A}\mathbf{A}^+)_{ki}\left[(\mathbf{A}\mathbf{A}^+)^{-1}\right]_{il}\mathbf{A}_{lj} = \mathbf{A}_{kj},$$

which is the $j$th column of $\mathbf{A}$ and so

$$\left(\mathbf{A}\mathbf{A}^+\mathbf{c}_j^\dagger\right) = \mathbf{c}_j.$$

The inequalities (1.131) then become

$$\sigma_{\min}\|\mathbf{y}\|^2 \leq \langle \mathbf{A}\mathbf{A}^+\mathbf{y}, \mathbf{y}\rangle = \sum_{j=1}^{n} |\langle \mathbf{c}_j, y\rangle|^2 \leq \sigma_{\max}\|\mathbf{y}\|^2, \quad \forall \mathbf{y} \in \mathbb{C}^m, \tag{1.132}$$

---

[24] Note that $\mathbf{A}\mathbf{A}^+$ is Hermitian and positive semi definite and so has real and positive (or possibly zero) eigenvalues [30]. The same holds for $\mathbf{A}^+\mathbf{A}$.

[25] Any real or complex $m \times n$ matrix $\mathbf{A}$ can be factored as $\mathbf{A} = \mathbf{U}\mathbf{\Sigma}\mathbf{V}^+$ where $\mathbf{\Sigma}$ is a $m \times n$ real diagonal matrix of singular values $\sigma_i \geq 0$, $i = 1,\ldots,k$ and $k = \text{Min}(m, n)$ [31]. When $m \neq n$, $\mathbf{\Sigma}$ is constructed from the singular values by placing them along the main diagonal and then placing zeros in other locations to preserve the $m \times n$ size of the matrix, which in such cases will inevitably result in some zero columns (when $m < n$), or some zero rows (when $m > n$). The singular values are the square roots of the intersection of the two sets of eigenvalues of the Hermitian and positive semidefinite matrices $\mathbf{A}\mathbf{A}^+$ and $\mathbf{A}^+\mathbf{A}$. Assuming, without loss of generality, that $m < n$ and defining the square $m \times m$ diagonal matrix $\mathbf{\Lambda}_m \equiv \text{Diag}\left[\sigma_1^2, \ldots, \sigma_m^2\right]$ and the square $n \times n$ diagonal matrix $\mathbf{\Lambda}_n \equiv \text{Diag}\left[\sigma_1^2, \ldots, \sigma_n^2\right]$, we have $\mathbf{A}\mathbf{A}^+\mathbf{U} = \mathbf{U}\mathbf{\Lambda}$ and $\mathbf{A}^+\mathbf{A}\mathbf{V} = \mathbf{V}\mathbf{\Lambda}$. Thus, the eigenvectors of $\mathbf{A}\mathbf{A}^+$ are the columns of $\mathbf{U}$, $\mathbf{U} = [\mathbf{u}_1, \ldots, \mathbf{u}_n]$, while the eigenvectors of $\mathbf{A}^+\mathbf{A}$ are the columns of $\mathbf{V}$, $\mathbf{V} = [\mathbf{v}_1, \ldots, \mathbf{v}_n]$. In addition, $\mathbf{U}\mathbf{U}^+ = 1$, $\mathbf{V}\mathbf{V}^+ = 1$. If the rank $r_\mathbf{A} < m < n$ then $\sigma_i > 0$, $1 \leq i \leq r_\mathbf{A}$, and $\sigma_i = 0$, $r_\mathbf{A} < i \leq m$, and we have the representation $\mathbf{A} = \sum_{i=1}^{r_\mathbf{A}} \sigma_i \mathbf{u}_i \mathbf{v}_i^+$ which can now be used to compute any function of a matrix, namely, $f(\mathbf{A}) = \sum_{i=1}^{r_\mathbf{A}} f(\sigma_i)\mathbf{u}_i\mathbf{v}_i^+$.

also known as frame inequalities. Any set of $n$ vectors $\{\mathbf{c}_k\}$ satisfying the above inequalities is a frame for $\mathbb{C}^m$ and the quantities $\sigma_{\min}$ and $\sigma_{\max}$ are the associated frame bounds. When $\sigma_{\min} = \sigma_{\max}$, i.e., the condition number of $\mathbf{A}$ is 1, the frame is tight and it mimics an orthogonal basis even though the frame vectors might be linearly dependent. If $\sigma_{\min} = \sigma_{\max} = 1$ the frame becomes an orthogonal basis. Given nontight frame vectors $\mathbf{c}_k$, a tight frame can be constructed by choosing

$$\mathbf{c}'_k \equiv \left(\mathbf{A}\mathbf{A}^+\right)^{-1/2}\mathbf{c}_k,$$

where the square root operation is defined using the singular value decomposition (SVD) of the associated matrix. The matrix $\mathbf{A}\mathbf{A}^+$ is the finite dimensional analog of the basis operator $T^+{}_\phi T_\phi$ defined in a Hilbert space, (1.104).

## 1.26   Frames in $L_2(\mathbb{R})$

In analogy to the previous section we generalize the concept of a Riesz basis in a Hilbert space by relaxing the exactness property of a Riesz basis (a Riesz basis is an exact frame) which allows for overcomplete (i.e., linearly dependent) sets that will lead to the concept of frames [26]. We begin with the definition of a basis and an unconditional basis in $L_2(\mathbb{R})$.

**Definition 41.** A set of functions $\{\phi_n(t)\}$, $n \in \mathbb{Z}$, whose closure is the entire space $L_2(\mathbb{R})$, is a basis (or a Schauder basis) if every $x \in L_2(\mathbb{R})$ has the representation

$$x = \sum_{n=-\infty}^{\infty} c_n \phi_n,$$

where the coefficients $c_n$ are unique and depend on $x$. Furthermore, if, for this basis, two positive constants $A$ and $B$ exist such that

$$A\|\underline{c}\|^2 \leq \left\|\sum_{n=-\infty}^{\infty} c_n \phi_n\right\|^2 \leq B\|\underline{c}\|^2$$

for every $\underline{c} \in l_2(\mathbb{Z})$, then the basis is an unconditional basis.

The concept of a frame can be introduced by relaxing the uniqueness of the expansion coefficients in an unconditional basis.

**Definition 42.** A set of functions $\{\phi_n(t), n \in \mathbb{Z}\}$ is a frame in $L_2(\mathbb{R})$ if positive constants $A$ and $B$, the frame bounds, exist such that

$$A\|x\|^2 \leq \sum_{n=-\infty}^{\infty} |\langle \phi_n, x \rangle|^2 \leq B\|x\|^2 \qquad (1.133)$$

for every $x \in L_2(\mathbb{R})$. If $A = B$ the frame is tight.

**Theorem 36.** *With reference to definition 42 and equation* (1.133), *if the frame is exact then it is a Riesz basis. If $A = B = 1$ then the frame is an orthonormal basis [27, 15].*

Thus, an othonormal basis is exact and tight with unity frame bounds. Tight frames that are not orthonormal can be constructed by the union of two orthonormal bases

$$\{\xi_n\} = \{\phi_n\} \cup \{\psi_n\}, \quad \langle \xi_n \mid \xi_m \rangle = \langle \phi_n \mid \phi_m \rangle = \delta_{nm}$$

that are not orthogonal to each other. The resulting frame is not orthogonal and for any $x(t)$ we have

$$\sum_{n=-\infty}^{\infty} |\langle \xi_n, x \rangle|^2 = \sum_{n=-\infty}^{\infty} |\langle \phi_n, x \rangle|^2 + \sum_{n=-\infty}^{\infty} |\langle \psi_n, x \rangle|^2 = 2\|x\|^2,$$

which clearly shows that the corresponding frame bounds are equal to 2. If we choose the second basis to be an orthogonal but unnormalized set, e.g., $\|\phi_n\| = 2, \forall n$, then the resulting union is a frame that is neither tight, nor exact.

In analogy to equations (1.100) and (1.101), we define the frame analysis operator $T_\phi$ and its adjoint $T^+{}_\phi$, the frame synthesis operator. Thus, we arrive at the definition of the frame operator.

**Definition 43.** Given the frame functions $\{\phi_n(t), n \in \mathbb{Z}\}$ we define the frame operator $T^+{}_\phi T_\phi$ by the equation

$$T^+{}_\phi T_\phi \{x(t)\} = \sum_{n=-\infty}^{\infty} \langle \phi_n, x \rangle \phi_n(t), \quad \forall\, x(t) \in L_2(\mathbb{R}). \qquad (1.134)$$

Using the definition of the frame operator and its adjoint, the frame definition equation (1.133) can be rewritten as

$$A\|x\|^2 \leq \langle x, T^+{}_\phi T_\phi x \rangle \equiv \sum_{n=-\infty}^{\infty} |\langle \phi_n, x \rangle|^2 \leq B\|x\|^2. \qquad (1.135)$$

The analysis operator $T_\phi$ is not unitary since the frame elements are not, in general, orthonormal. The frame definition of equation (1.135) implies that the frame operator $T^+{}_\phi T_\phi$ is bounded and positive definite, and hence invertible. The operator norms of $T^+{}_\phi T_\phi$ and its inverse are bounded by the frame bounds. Using equation (1.10) we have

$$\left\|T^+{}_\phi T_\phi\right\| \le B, \quad \left\|\left(T^+{}_\phi T_\phi\right)^{-1}\right\| \le A^{-1}. \tag{1.136}$$

The ratio $B/A$ is known as the frame redundancy ratio and can be used as an approximation to the condition number of the operator $T^+{}_\phi T_\phi$. A finite frame ratio that is not too large and not too small is required for stable synthesis or recovery, i.e., reconstruction of a signal from its transform coefficients, as will be seen below in the frame representation equation. An orthonormal basis $\hat{\phi}_n(t)$ can be constructed for a frame with a reasonable frame ratio, using the square root of the inverse of the frame operator $T^+{}_\phi T_\phi$,

$$\hat{\phi}_n(t) \equiv \left(T^+{}_\phi T_\phi\right)^{-1/2} \phi_n(t). \tag{1.137}$$

**Definition 44.** Given a frame $\{\phi_n,\ n \in \mathbb{Z}\}$ with an invertible frame operator, we define the dual frame elements by

$$\xi_n(t) = \left(T^+{}_\phi T_\phi\right)^{-1} \phi_n(t), \quad n \in \mathbb{Z}. \tag{1.138}$$

The inverse of this relation shows the linear dependence of the original frame elements (vectors), i.e.,

$$\phi_n(t) = T^+{}_\phi\, T_\phi\, \xi_n(t) = \sum_{m=-\infty}^{\infty} \langle \phi_m, \xi_n \rangle\, \phi_m(t). \tag{1.139}$$

The dual frame is a new frame with frame bounds $B^{-1}$ and $A^{-1}$, i.e., it satisfies the defining frame equation (1.133) with the new bounds [15]

$$B^{-1}\left\|x\right\|^2 \le \sum_{n=-\infty}^{\infty} \left|\langle \xi_n, x \rangle\right|^2 \le A^{-1}\left\|x\right\|^2. \tag{1.140}$$

The frame representation for functions in $L_2(\mathbb{R})$ is now expressed in the dual frame reconstruction equations

$$x(t) = \sum_{n=-\infty}^{\infty} \langle \phi_n, x \rangle\, \xi_n(t) = \sum_{n=-\infty}^{\infty} \langle \xi_n, x \rangle\, \phi_n(t). \tag{1.141}$$

The last equation recast in terms of frame operators is simply

$$x(t) = T^+{}_\xi T_\phi \{x(t)\} = T^+{}_\phi T_\xi \{x(t)\}. \tag{1.142}$$

This equation is valid for all $x(t) \in L_2(\mathbb{R})$, and therefore

$$T^+{}_\phi T_\xi = T^+{}_\xi T_\phi = 1. \tag{1.143}$$

The dual reconstruction equations (1.141) are the solutions to two dual minimization problems as stated in the following theorem [15].

**Theorem 37.** *The coefficients $c_n \in l_2(\mathbb{Z})$ in the expansion*

$$x(t) = \sum_{n=-\infty}^{\infty} c_n \phi_n(t)$$

*that minimize the square of the norm of the difference between the left and the right hand sides are not unique. A minimum norm solution, i.e., a solution for which the quantity $\sum_{n=-\infty}^{\infty} |c_n|^2$ is a minimum among the set of all possible solutions, is given by $c_n = \langle \xi_n, x \rangle$. Similarly, the coefficients $c'_n \in l_2(\mathbb{Z})$ that minimize the square of the norm of the difference between the two sides of the equation*

$$x(t) = \sum_{n=-\infty}^{\infty} c'_n \xi_n(t)$$

*and have minimum norm are given by $c'_n = \langle \phi_n, x \rangle$.*

The reconstruction of a function from its coefficients via equations (1.141) must be unique, i.e., the equality $\langle \phi_n, x_1 \rangle = \langle \phi_n, x_2 \rangle$ must imply that the two underlying functions are the same, $x_1(t) = x_2(t)$, where the equality is understood to mean $\|x_1 - x_2\| = 0$. Equivalently, we must have

$$\langle \phi_n, e \rangle = 0 \Leftrightarrow e(t) = 0. \tag{1.144}$$

The frame inequalities (1.133) and (1.140) ensure that the above uniqueness condition is satisfied [15].

**Theorem 38.** *A function $x(t)$ in $L_2(\mathbb{R})$ can be uniquely reconstructed from the coefficients $\langle \phi_n, x \rangle$ or $\langle \xi_n, x \rangle$ if, and only if, the functions $\phi_n(t)$ and $\xi_n(t)$ form dual frames.*

The following theorem shows the close connection between a frame and a reproducing kernel in a reproducing kernel Hilbert space [15, 32].

**Theorem 39.** *A Hilbert space with a frame and frame bounds that allow the construction of dual frames* $\{\phi_n, \xi_n\}$ *admits a unique reproducing kernel that has the representation*

$$K\left(t, t'\right) = \sum_{n=-\infty}^{\infty} \xi_n^*\left(t\right) \phi_n\left(t'\right) \qquad (1.145)$$

*and that is self-adjoint, i.e.,* $K\left(t, t'\right) = K^*\left(t', t\right)$.

Equation (1.41) is a special case of this result applied to an orthonormal basis.

## 1.27   Dual Frame Construction Algorithm

The dual frame reconstruction formulas expressed in equation (1.141) can only be used when the dual frame elements are known. If only the frame elements are available then an iterative scheme can be used to approximate the dual frame elements. The approximation is useful only when the frame ratio $B/A$ is not too much larger than 1.

Using the definition of the dual frame in equation (1.138), the reconstruction formula (1.141) can be written as

$$x\left(t\right) = \sum_{n=-\infty}^{\infty} \langle\phi_n, x\rangle \left(T^+{}_\phi T_\phi\right)^{-1} \phi_n\left(t\right),$$

which is equivalent to

$$\left(1 - \frac{2}{A+B} T^+{}_\phi T_\phi\right) x\left(t\right) = x\left(t\right) - \frac{2}{A+B} \sum_{n=-\infty}^{\infty} \langle\phi_n, x\rangle \phi_n\left(t\right). \qquad (1.146)$$

It can be shown that [15]

$$\left\| 1 - \frac{2}{A+B} T^+{}_\phi T_\phi \right\| \le \frac{B-A}{B+A} < 1, \qquad (1.147)$$

and so according to theorem 8 and equation (1.13) the operator

$$\left(1 - \left(1 - \frac{2}{A+B} T^+{}_\phi T_\phi\right)\right)^{-1} = \frac{A+B}{2} \left(T^+{}_\phi T_\phi\right)^{-1}$$

has a convergent Neumann expansion. Thus,

$$\left(T^{+}{}_{\phi}T_{\phi}\right)^{-1} = \frac{2}{A+B} \sum_{k=0}^{\infty} \left(1 - \frac{2}{A+B}T^{+}{}_{\phi}T_{\phi}\right)^{k}. \qquad (1.148)$$

An approximation to the dual frame vectors is obtained by using the first $N$ terms of the Neumann series in equation (1.138). Thus,

$$\xi_{nN}(t) \equiv \frac{2}{A+B} \sum_{k=0}^{N-1} \left(1 - \frac{2}{A+B}T^{+}{}_{\phi}T_{\phi}\right)^{k} \phi_{n}(t). \qquad (1.149)$$

The approximated dual frame when used in equation (1.142) gives the approximate reconstructed signal. The norm of the error in the reconstruction can be shown to decay exponentially,

$$\left\| x - \sum_{n=-\infty}^{\infty} \langle \phi_{n}, x \rangle \xi_{nN} \right\| \leq \left(\frac{B-A}{B+A}\right)^{N} \|x\|. \qquad (1.150)$$

Frames provide a powerful method of signal representation because of their redundancy. Among important advantages of a redundant basis are better approximations of the continuous wavelet transform coefficients, robustness to quantization effects, and fewer restrictions in the choice of analyzing basis functions which makes it easier to design functions matching specific signal characteristics. In addition, a signal can be successfully reconstructed even when some coefficients are missing. However, frames can be difficult to use for signal reconstruction since the reconstruction formula requires the computation of the dual frame elements, which, as has been noted above, is a cumbersome and approximate process. Orthonormal basis functions (tight frames with bounds $A = B = 1$), on the other hand, are far easier to deal with. For wavelets of compact support we will study both orthonormal and biorthogonal basis functions, since, as we shall see later, the transform coefficients can be calculated using finite impulse response (FIR) filters, and without the need to compute any of the basis functions.

Localization of signals in both time and frequency, or in time and scale, is a major motivations for studying the windowed Fourier transform and the continuous wavelet transform: both are linear maps between $L_{2}(\mathbb{R})$ and $L_{2}(\mathbb{R}^{2})$ and both transforms use two continuous parameters, time shift and frequency for the windowed Fourier transform, and time shift and scale for the continuous wavelet transform. The property that is common to both is that the transform coefficients for a function $x(t)$ are obtained by taking the inner product between $x(t)$ and a family of analyzing functions that are constructed from the same mother window, in the case of the windowed Fourier transform, and the same mother wavelet, in the case of the continuous wavelet transform.

The process of building frames and orthonormal bases in either case is based on discretization of the continuous variables on which the analyzing functions depend. This results in analyzing functions that carry two integer indices: one for time translation and one for frequency, in the case of the windowed Fourier transform, and one for time translation and one for scale, in the continuous wavelet transform case. In either case we may represent the basis functions by $\xi_{mn}(t)$ and the corresponding inner products by $c_{mn}$ with the following analysis and synthesis (reconstruction) equations

$$c_{mn} = \langle \xi_{mn}, x \rangle, \quad x(t) = \sum_{n=-\infty}^{\infty} c_{mn}\xi_{mn}(t), \tag{1.151}$$

and, in the Dirac notation and with summation over repeated indices, the completeness relation

$$|\xi_{mn}\rangle \langle \xi_{mn}| = \mathbf{1}. \tag{1.152}$$

If the associated frames are actually orthonormal bases then we will have the additional orthonormality relations

$$\langle \xi_{mn} \mid \xi_{lk} \rangle = \delta_{ml}\delta_{nk}. \tag{1.153}$$

## 1.28  Exercises

**1**  Consider the set of functions $\{1, t, t^2, \ldots, t^{N-1}\}$ defined on the closed interval $[-1, 1]$. The members of this set are $N$ linearly independent functions. Using the inner product $\langle f, g \rangle \equiv \int_{-1}^{1} f(t)g(t)\, dt$ and the Gram-Schmidt procedure, find a set of $N$ orthonormal polynomials (the latter are known as the Legendre polynomials).

**2**  Repeat the last problem using the same set of functions, but this time use the inner product $\langle f, g \rangle \equiv \int_{-1}^{1} f(t)\left(1 - t^2\right)^{-1/2} g(t)\, dt$. The resulting orthonormal functions are known as the Chebyshev polynomials $T_0(t) \equiv \pi^{-1/2}$, and $T_n(t) \equiv (\pi/2)^{-1/2} \cos\left[n\cos^{-1}(t)\right]$, $n = 1, 2, \ldots$.

**3** Prove equation (1.9).

**4** Prove the completeness relation (1.116).

**5** Consider the function defined as $e^{-|n|}$ on intervals $[n, n+1]$, $n \in \mathbb{Z}$. Show that this function is square integrable and is in the subspace $\mathcal{V}_0$.

**6** Consider a continuous time function $f(t) = \sum_{n=-\infty}^{\infty} h[n] \phi(t-n)$. Calculate the Fourier transform of the function, expressing your result in terms of the Fourier transforms $H(\omega)$ and $\Phi(\omega)$.

**7** Prove the result

$$\int_{-\infty}^{\infty} \frac{\sin(a\pi t)}{a\pi t} e^{-i\omega t} dt = \begin{cases} 1/a, & -a\pi < \omega < a\pi, \\ 0, & \text{otherwise}, \end{cases}$$

by calculating the inverse Fourier transform of the right hand side. Use this result to find the Fourier transform of $p_k(t)$ as defined in equations (1.94) and (1.91) [31].

**8** Given a signal $s(t)$ whose band is limited to $[-B, +B]$, we use the completeness of the functions $p_k(t)$ (see the last problem and equations (1.94) and (1.91)) to write $s(t) = \sum_{k=-\infty}^{\infty} c_k p_k(t)$. Using the orthonormality of the functions $\{p_k(t)\}$, we find the coefficients $c_k = \int_{-\infty}^{\infty} s(t) p_k(t) dt$. Use theorem 27 and the Fourier transform result of the previous problem to calculate this coefficient. This is a proof of the sampling theorem (theorem 33).

**9** Using the reproducing kernel function of equation (1.87), show that the kernel is a projection operator from $L_2(\mathbb{R})$ into $L_2^{\Omega}(\mathbb{R})$.

**10** Verify that the two generalized inverses defined in equations (1.120) and (1.121) satisfy the Moore-Penrose conditions for a pseudoinverse of an $m \times n$ matrix (footnote 23).

# CHAPTER 2

# Linear Time-Invariant Systems

## 2.1 Introduction

The mathematical characterization of linear time-invariant systems rests on the concept of convolution, in both continuous time and discrete time. The concepts of linear system theory play important roles in wavelet transforms and their implementation. We review these concepts in the following sections.

## 2.2 Convolution in Continuous Time

An important result in the continuous time Fourier transform is the convolution theorem [19].

**Definition 45.** The convolution product of two functions $x(t), y(t) \in L_2(\mathbb{R})$ is defined by

$$x * y(t) \equiv \int\limits_{-\infty}^{\infty} x(s) y(t-s) ds. \tag{2.1}$$

**Theorem 40.** *The convolution product of two functions $x(t), y(t) \in L_2(\mathbb{R})$ is commutative, i.e., $x * y(t) = y * x(t)$, and it is uniformly bounded for all $t \in \mathbb{R}$.*[1]

Taking the Fourier transform of the right hand side of equation (2.1), we find

$$\int\limits_{-\infty}^{\infty} dt\, e^{-i\omega t} \int\limits_{-\infty}^{\infty} x(s) y(t-s)\, ds = X(\omega) Y(\omega), \tag{2.2}$$

---

[1]Even though $x(t), y(t) \in L_2(\mathbb{R})$, the convolution product of the two may not be in $L_2(\mathbb{R})$.

which is the statement of the convolution theorem.

**Theorem 41.** *The Fourier transform of the convolution of two functions is the product of their Fourier transforms.*

A related and useful result is the inverse to the convolution theorem.

**Theorem 42.** *The Fourier transform of the product of two functions $x(t), y(t) \in L_2(\mathbb{R})$ is the convolution (in the frequency domain) of their respective Fourier transforms*

$$\int_{-\infty}^{\infty} x(t)y(t) e^{-i\omega t} dt = \int_{-\infty}^{\infty} X(\sigma)Y(\omega - \sigma) d\sigma. \qquad (2.3)$$

## 2.3   Convolution in Discrete Time

The convolution of two discrete time sequences is defined as [22]

$$x * y[n] \equiv \sum_{m=-\infty}^{\infty} x[m]y[n-m] \qquad (2.4)$$

and the convolution theorem now states that the discrete time Fourier transform of the convolution of two discrete time sequences is the product of the corresponding Fourier transforms. Thus,

$$\sum_{n=-\infty}^{\infty} e^{-in\omega} \sum_{m=-\infty}^{\infty} x[m] y[n-m] = X(\omega) Y(\omega), \qquad (2.5)$$

which can be proven by exchanging the two sums and changing the summation index. A similar equation relates the discrete time Fourier transform of the product of two discrete time sequences to the convolution of their Fourier transforms,

$$\sum_{n=-\infty}^{\infty} e^{-in\omega} f[n] g[n] = \int_{-\pi}^{\pi} F(\sigma) G(\omega - \sigma) d\sigma, \qquad (2.6)$$

which can be proven by using equation (1.72).

When using a software package FFT routine to compute and plot spectra of discrete data samples, it is important to know if equations (1.82) are in fact used (as opposed to other normalizations such as equations

(1.79) and (1.80)). Assuming the conventional normalization of (1.82), the spectrum of a sequence $x[n]$, $n = 0, 1, \ldots, N-1$, is then given by $|X[k]|^2$, $k = 0, 1, \ldots, N-1$. Assuming $N$ to be an even integer (normally it is a power of 2), to plot the spectrum as a function of frequency, a frequency axis is constructed using $f[k] = kf_s/N$, $k = 0, \ldots, N-1$, where $f_s$ is the sampling frequency. To properly map to negative and positive frequencies the frequency axis should be constructed with $k = -N/2, \ldots, 0, \ldots, N/2 - 1$. The positive frequencies then correspond to elements $X[0], \ldots, X[N/2-1]$, whereas the negative frequencies now correspond to the elements $X[N/2], \ldots, X[N-1]$. If the original sequence is real then the spectrum will be a symmetric function of the frequency, i.e., $|X[k]|^2 = |X[-k]|^2$, and so one usually plots only the positive frequency portion of the spectrum.

## 2.4 Convolution of Finite Length Sequences

The convolution theorem in the case of two finite sequences requires a little further explanation to help with implementation. The convolution of two finite duration discrete time signals $s[n]$, $n = 0, 1, \ldots, N-1$, and $h[m]$, $m = 0, 1, \ldots, M-1$, assuming $N \geq M$, is defined by[2][22]

$$s * h[n] = \sum_{m=0}^{M-1} s[n-m]h[m], \quad m \leq n, \; n = 0, \ldots, N+M-2, \qquad (2.7)$$

which is a sequence of length $N + M - 1$. It is often convenient to use the inverse Fourier transform of the product of two Fourier transforms (see equation (2.5)) to compute the convolution sequence for two finite length time series. When one or both time series has a large number of samples, computation of Fourier transforms using the FFT algorithm provides a much faster way to obtain convolutions between the given time series than use of equation (2.7).[3]

---

[2]In our discussion of orthogonal wavelets we will use only finite length filters whose support is on integers in an interval $[0, M-1]$, for some positive integer $M$. In general, the support of a finite length filter will be on integers in an interval $[N_1, N_2]$, $N_1, N_2 \in \mathbb{Z}$, as we shall see when we construct biorthogonal wavelets of compact support in section 7.6.

[3]When the number of samples is huge, e.g., billions or more, or if a real time system is producing the signal $s[n]$, and assuming the filter $h[m]$ is fixed and available, then the input samples have to be divided up into sections, Fourier transformed and multiplied by the Fourier transform of the filter, inverse Fourier transformed, and then used in

This convolution result can be obtained using the discrete Fourier transform by first appending zeros to both sequences to produce two sequences of length $N + M - 1$ [22],

$$\hat{s}[0,\ldots,N+M-2] \equiv [s_0,\ldots,s_{N-1},0,\ldots,0]$$
$$\hat{h}[0,\ldots,N+M-2] \equiv [h_0,\ldots,h_{M-1},0,\ldots,0] \ ,$$

and next computing the $(N + M - 1)$-point discrete Fourier transforms using the first of equations (1.82), and an FFT algorithm, multiplying the Fourier transforms, and finally taking the inverse Fourier transform using the second of equations (1.82), again using an FFT algorithm. Thus, for $0 \leq l \leq N + M - 2$ we have

$$s * h[l] = \frac{1}{N + M - 1} \sum_{k=0}^{N+M-2} \hat{S}[k]\hat{H}[k]\, e^{2i\pi kl/(N+M-1)}. \qquad (2.8)$$

Discrete convolution can be written as matrix multiplication. Consider equation (2.4) written for a finite length filter (also known as a finite impulse response or FIR filter) $h_m$, $m = 0,\ldots,M-1$, and an input signal $x_n$ resulting in the output sequence $y_n$ for $n \in \mathbb{Z}$,

$$y_n = \sum_{m=0}^{M-1} h_m x_{n-m}, \quad n \in \mathbb{Z},$$

which can be expressed as the (infinite dimensional) matrix and vector multiplication below

$$\begin{bmatrix} \vdots \\ y_{-1} \\ y_0 \\ y_1 \\ \vdots \end{bmatrix} = \begin{bmatrix} \ddots & \ddots & \ddots & \ddots & \ddots & 0 & \ddots & \ddots & \ddots \\ \ddots & 0 & h_{M-1} & \cdots & h_0 & 0 & 0 & 0 & \ddots \\ \ddots & 0 & 0 & h_{M-1} & \cdots & h_0 & 0 & 0 & \ddots \\ \ddots & 0 & 0 & 0 & h_{M-1} & \cdots & h_0 & 0 & \ddots \\ \ddots & \ddots & \ddots & 0 & 0 & \ddots & \ddots & \ddots & \ddots \end{bmatrix} \begin{bmatrix} \vdots \\ x_{-1} \\ x_0 \\ x_1 \\ \vdots \end{bmatrix}.$$

For signals of finite length, say, $n = 0,\ldots,N-1$, we appropriately modify the size of the matrices and vectors to conform with equation (2.7). For instance,

combination with two standard methods of Overlap and Add or Overlap and Save, to construct the convolution time series.

consider a filter given by two coefficients $h_0$ and $h_1$, and $x_n$, $n = 0, 1, 2$. Then the output sequence $y_k$, $k = 0, \ldots, 3$, has four elements. Thus,

$$
\begin{bmatrix} y_0 \\ y_1 \\ y_2 \\ y_3 \end{bmatrix} = \begin{bmatrix} h_0 & 0 & 0 \\ h_1 & h_0 & 0 \\ 0 & h_1 & h_0 \\ 0 & 0 & h_1 \end{bmatrix} \begin{bmatrix} x_0 \\ x_1 \\ x_2 \end{bmatrix}.
$$

For this example the transformation is a $4 \times 3$ matrix. In chapter 6 when implementing the discrete wavelet transform (DWT) we require the input and output sequences to be of the same length (and always a power of 2) and so the matrix will is square.

Although the above matrix equation represents a convolution, a correlation equation of the form

$$
y_n = \sum_{m=0}^{M-1} h_m x_{n+m}, \quad n \in \mathbb{Z}, \tag{2.9}
$$

can be written in a similar matrix form, since correlation with a discrete filter $h_m$, $m = 0, \ldots, M - 1$, is identical to a convolution with the time-reversed discrete filter $h_{M-m}$, $m = 0, \ldots, M - 1$.

## 2.5 Linear Time-Invariant Systems and the Z Transform

The mathematical characterization of a linear time-invariant (LTI) system is an impulse response, system function, or filter $h(t)$ in continuous time, or $h[m]$ in discrete time: when the system input to a linear time-invariant system is $x(t)$ or $x[n]$, the output is given by the convolution of the input function and the system function, $y(t) = x * h(t)$, or, in discrete time, $y[n] = x * h[n]$ [33]. The system function $h$ is also referred to as a filter. If in discrete time the filter has a finite number of nonzero elements, namely, $h[m]$, $m = 0, \ldots, M - 1$, then it is a finite impulse response (FIR) filter. Otherwise, it is an infinite impulse response (IIR) filter. Figure 2.1 shows the definition of a linear time-invariant system in both continuous and discrete time.[4] The eigenfunctions of a linear time-invariant system for continuous

---

[4]The requirement for linearity is that $y(t) = \int\limits_{-\infty}^{\infty} h(t, \tau) x(\tau) d\tau$, while time invariance is $h(t, \tau) \equiv h(t - \tau)$; for discrete time sequences they become $y_n = \sum\limits_{m=-\infty}^{\infty} h_{nm} x_m$ and $h_{nm} \equiv h[n - m]$.

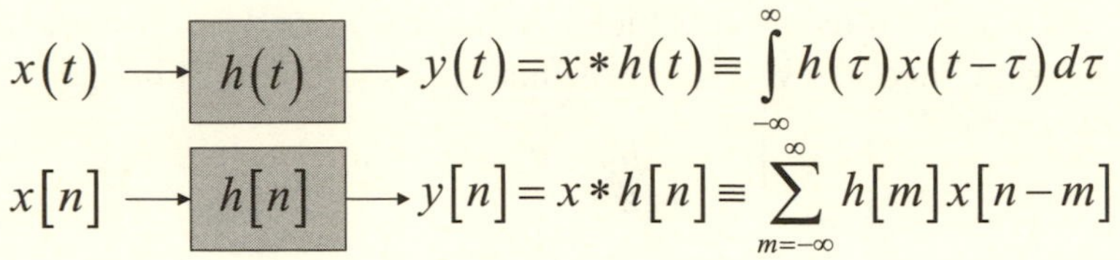

$$x(t) \longrightarrow \boxed{h(t)} \longrightarrow y(t) = x * h(t) \equiv \int_{-\infty}^{\infty} h(\tau) x(t-\tau) d\tau$$

$$x[n] \longrightarrow \boxed{h[n]} \longrightarrow y[n] = x * h[n] \equiv \sum_{m=-\infty}^{\infty} h[m] x[n-m]$$

Figure 2.1: Linear time-invariant (LTI) system convolution: continuous and discrete time.

time are the functions $e^{-st}$ where $s \in \mathbb{C}$, since the output for such an input is given by the convolution

$$\int_{-\infty}^{\infty} h(t-\tau) e^{-s\tau} d\tau = \int_{-\infty}^{\infty} h(\tau) e^{-s(t-\tau)} d\tau \equiv H(s) e^{-st}, \qquad (2.10)$$

and $H(s)$ is the Laplace transform of the system function. Evaluation of the Laplace transform along the imaginary axis in the complex $s = \sigma + i\omega$ plane yields the Fourier transform.

For a linear time-invariant system in discrete time, the corresponding eigenfunctions are $z^{-n}, n \in \mathbb{Z}$. Consequently, a useful quantity for analysis of a discrete time signal $x[n]$ is the **Z** transform defined as the following Laurent series [33]:

$$\mathbf{Z}\{x[n]\} \equiv X(z) \equiv \sum_{n=-\infty}^{\infty} x[n] z^{-n}. \qquad (2.11)$$

It is easy to see that the **Z** transform of the convolution of two infinite duration sequences is the product of the respective **Z** transforms,

$$\sum_{n=-\infty}^{\infty} z^{-n} \sum_{k=-\infty}^{\infty} x[n-k] h[k] = X(z) H(z). \qquad (2.12)$$

Some other **Z** transform properties are listed in table 2.1. If a sequence $x[n]$ has a finite number of nonzero elements, e.g., $x[n] = 0, n \leq -M, n \geq N$, for positive integers $M, N$, then its time-reversed version $x[-n] = 0, n \leq -N$, $n \geq M$, as illustrated in figure 2.2. The **Z** transform of $x[-n]$ is exactly as shown in table 2.1. In addition, the **Z** transforms of both $x[n]$ and $x[-n]$ become Laurent polynomials, since there are only a finite number of nonzero terms in the series. If, on the other hand, we use a minimum index of 0

| Function | **Z** transform |
|----------|-----------------|
| $x[n-k]$ | $z^{-k}X(z)$ |
| $x[-n]$ | $X(1/z)$ |
| $x^*[n]$ | $X^*(z^*)$ |
| $x^*[-n]$ | $X^*(1/z^*)$ |
| $a^{-n}x[n],\ a>0$ | $X(az)$ |

Table 2.1: Properties of Z transform.

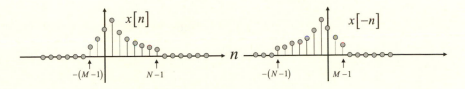

Figure 2.2: Finite length discrete sequence $x[n]$, $n = -(M-1),\ldots,N-1$, and its time-reversed form $x[-n]$.

for all finite length sequences, including the time-reversed version, then we must include a time shift operation when calculating the **Z** transform. For instance, a finite length sequence $x[n]$, $n = 0,\ldots,N-1$ whose transform is denoted by $X(z)$, has a time-reversed version now given by $x[N-1-n]$, $n = 0,\ldots,N-1$, and whose **Z** transform is

$$\sum_{n=0}^{N-1} x[N-1-n]\,z^{-n} = z^{-(N-1)}\sum_{n=0}^{N-1} x[n]\,z^{n} \equiv z^{-(N-1)}X(1/z). \quad (2.13)$$

Evaluating the **Z** transform on the unit circle $z = e^{i\omega}$ in the complex $z$-plane amounts to computing the discrete time Fourier transform, which we denote equivalently as either $X(e^{i\omega})$ or $X(\omega)$. Table 2.2 lists the correspondences between the two transforms.

| $X(\omega)$ | $X(z)$ |
|-------------|--------|
| $X(\omega+\pi)$ | $X(-z)$ |
| $X^*(\omega)$ | $X(1/z)$ |
| $X^*(\omega+\pi)$ | $X(-1/z)$ |

Table 2.2: Relations between Fourier and **Z** transforms.

We can depict a correlation of an input function $x$ with a filter function

$h$ using the result that correlation with $h$ is identical to convolution with the time-reversed version of $h$. Figure 2.3 shows this result for both continuous and discrete time functions [33].

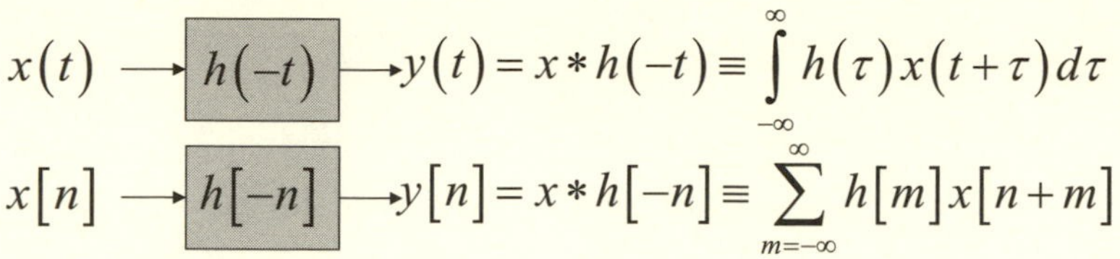

$$x(t) \longrightarrow \boxed{h(-t)} \longrightarrow y(t) = x * h(-t) \equiv \int_{-\infty}^{\infty} h(\tau) x(t + \tau) d\tau$$

$$x[n] \longrightarrow \boxed{h[-n]} \longrightarrow y[n] = x * h[-n] \equiv \sum_{m=-\infty}^{\infty} h[m] x[n + m]$$

Figure 2.3: Linear time-invariant (LTI) system correlation: continuous and discrete time.

## 2.6   Spectral Factorization for Finite Length Sequences

Given a sequence $h[n]$, $n \in \mathbb{Z}$, we use the term spectrum to refer to the quantity $S(\omega) \equiv |H(\omega)|^2$, i.e., the square of the magnitude of its Fourier transform. This is, of course, the same as $H(z) H(1/z)$ evaluated on the unit circle. Clearly $H(\omega)$ is a periodic function of $\omega$ and so is the spectrum. Here we will limit our discussion to discrete time sequences of finite length (FIR filters) and their spectra. According to table 2.2 and given a finite length sequence $h[n]$, $n = 0, \ldots, N - 1$, we have

$$S(\omega) \equiv |H(\omega)|^2 = \sum_{n=0}^{N-1} \sum_{m=0}^{N-1} h_n h_m^* e^{-i(n-m)\omega} = \sum_{k=-(N-1)}^{N-1} s_k e^{-ik\omega}, \qquad (2.14)$$

where

$$s_k = \begin{cases} \sum_{n=k}^{N-1} h_n h_{n-k}^*, & \text{if } k \geq 0, \\ \sum_{n=0}^{N-1-k} h_n h_{n-k}^*, & \text{if } k < 0. \end{cases} \qquad (2.15)$$

Thus, $S(\omega)$ is a Laurent polynomial in $z = \exp(i\omega)$. It is easy to see that

$$s_{-k} = s_k^*, \qquad (2.16)$$

which is required for $S(\omega)$ to be real and non-negative. Note that if $h_n$ are real then so are $s_k$ which together with equation (2.16) implies that

$s_{-k} = s_k$. In this case the Laurent polynomial on the right hand side of equation (2.14) reduces to the trigonometric form

$$s_0 + 2 \sum_{k=1}^{N-1} s_k \cos(k\omega).$$

All real and non-negative Laurent polynomials in $z = \exp(i\omega)$, such as $S(\omega)$, can be factored as described by the spectral factorization theorem [34].

**Theorem 43.** *(Fejer-Riesz) Any real and non-negative trigonometric polynomial function*

$$S(\omega) = \sum_{k=-(N-1)}^{N-1} s_k e^{-ik\omega} \tag{2.17}$$

*can be written in the form*

$$S(\omega) \equiv |H(\omega)|^2, \tag{2.18}$$

*where the polynomial*

$$H(z) \equiv \sum_{n=0}^{N-1} h_n z^{-n}$$

*is unique except for an arbitrary phase, and has all of its roots inside or on the unit circle, i.e., the region $|z| \leq 1$.*

For, consider the function $(S(1/z^*))^*$,

$$(S(1/z^*))^* = \sum_{k=-(N-1)}^{N-1} s_k^* z^k = \sum_{k=-(N-1)}^{N-1} s_{-k} z^k = \sum_{k=-(N-1)}^{N-1} s_k z^{-k} = S(z),$$

which, using equation (2.16), shows that if $z_j$ is a solution to $S(z) = 0$, then so is $\left(z_j^*\right)^{-1}$. In addition, we can write

$$S(z) = z^{-(N-1)} s_{-(N-1)} \left( z^{2(N-1)} + \cdots + s_{(N-1)} \big/ s_{-(N-1)} \right),$$

which clearly shows that the product of all the roots is $s_{(N-1)} \big/ s_{-(N-1)}$. The latter has unit magnitude because of equation (2.16). Hence, the roots must occur in pairs of the form $\left( z_j, 1 \big/ z_j^* \right)$, $j = 0, \ldots, N-1$, where we assume that the roots have been suitably ordered to have $|z_j| \leq 1$, $\forall j$. Thus,

$$S(z) = s_{-(N-1)} \left( \prod_{l=0}^{N-1} z_l^* \right)^{-1} \prod_{j=0}^{N-1} \left[ (1 - z_j/z)(zz_j^* - 1) \right]. \tag{2.19}$$

The theorem follows if we form $H(z)$ by choosing only the roots that are on or inside the unit circle, i.e.,

$$H\left(z\right) = \sqrt{s_{-(N-1)}} \left(\prod_{l=0}^{N-1} z_l^*\right)^{-1/2} \prod_{j=0}^{N-1} (1 - z_j/z), \qquad (2.20)$$

for which we find

$$S\left(z\right) = H\left(z\right)\left(H\left(1/z^*\right)\right)^* \;\Rightarrow\; S\left(\omega\right) \equiv \left|H\left(\omega\right)\right|^2.$$

We use spectral factorization in chapter 7 to construct orthonormal wavelet families of compact support.

## 2.7   Perfect Reconstruction Quadrature Mirror Filters

Orthogonal wavelets of compact support are very closely related, indeed equivalent, to perfect reconstruction quadrature mirror FIR filter banks (PR-QMFs) that have been used in multi-rate signal processing for more than three decades. Filter banks are used to decompose a signal into low-pass and high-pass components at half the original sampling rate with exact signal reconstruction from the components. The fields of subband coding and multi-rate signal processing began when it was shown that quadrature mirror filter banks could be used for exact signal reconstruction [35]. The reduction to half-rate and the doubling of the sample rate of a signal are also known as down-sampling, or decimation, and up-sampling, or interpolation, by a factor of 2 in our discussion. For practical reasons, finite impulse response filters are sought that must satisfy certain conditions to ensure the unitarity of the analysis/synthesis to have exact reconstruction: the latter is also known as perfect reconstruction (PR).

The simplest filter bank structure is that of a pair of FIR filters of equal and even length: $h_0[n]$ and $h_1[n]$, $0 \le n \le N-1$, with $N$ an even integer. The quadrature mirror filter property refers to the following relation between the two filters:

$$h_1\left[n\right] = (-1)^n\, h_0\left[N - 1 - n\right]. \qquad (2.21)$$

This PR-QMF structure is shown in figure 2.4 in which discrete time data $x[n]$ is filtered (convolved) with FIR sequences $h_0[-n]$ and $h_1[-n]$ (or equivalently, $x[n]$ is correlated with $h_0[n]$ and $h_1[n]$) and then sampled at half the

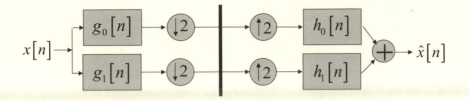

Figure 2.4: PR-QMF analysis and synthesis filter banks.

original rate (down-sampled by a factor of 2, depicted by the down arrow). The outputs are denoted by $x_0[n]$ and $x_1[n]$.

To reconstruct the signal we proceed to sample the outputs at twice their rate (up-sampled by a factor of 2, depicted by the up arrow) producing the sequences $x_0'[n]$ and $x_1'[n]$ which are then filtered (convolved) with FIR sequences $h_0'[n]$ and $h_1'[n]$, and finally combined (added) to produce the output $\hat{x}[n]$. The QMF property relates $h_0$ and $h_1$ through equation (2.21). The PR (perfect reconstruction) property is the requirement that $\hat{x}$ is identical to the original input $x$. For a finite impulse response implementation using filter sequences with a finite number of nonzero coefficients whose minimum time index is assumed to be 0 (see the discussion relating to figure 2.2), perfect reconstruction is achieved to within a known delay between $x$ and $\hat{x}$. We use **Z** transform analysis to obtain the PR-QMF conditions, which will be equivalent to signal decomposition and reconstruction using discrete orthogonal wavelets of compact support. The problem now is to find the relationships among all four filters to ensure a PR-QMF bank of filters [36].

We begin by analyzing the down-sampling operation (down arrow), also known as decimation. Consider the discrete time signal $x[n]$ and let $y[n] = x[2n]$, $n \in \mathbb{Z}$. Taking **Z** transforms of both sides we have

$$Y(z) = \sum_{n=-\infty}^{\infty} y[n]\,z^{-n} = \sum_{n=-\infty}^{\infty} x[2n]\,z^{-n}$$

$$= \frac{1}{2} \sum_{n=-\infty}^{\infty} (1 + (-1)^n)\,x[n]\,z^{-n/2}$$

$$= \frac{1}{2}\left(X\left(\sqrt{z}\right) + X\left(-\sqrt{z}\right)\right). \tag{2.22}$$

Next we consider the **Z** transform of an up-sampled (by a factor of 2 — up arrow) version of the discrete time signal, defined by

$$y[2n] = x[n], \quad y[2n+1] = 0, \quad n \in \mathbb{Z}. \tag{2.23}$$

Taking **Z** transforms we have

$$Y\left(z\right) = \sum_{n=-\infty}^{\infty} y\left[2n\right] z^{-2n} = \sum_{n=-\infty}^{\infty} x\left[n\right] z^{-2n} = X\left(z^2\right). \qquad (2.24)$$

Using the convolution property of the **Z** transform, equation (2.12), and equations (2.22) and (2.24), we have

$$\hat{X}\left(z\right) = \frac{1}{2}\left\{G_0\left(z\right) H_0\left(z\right) + G_1\left(z\right) H_1\left(z\right)\right\} X\left(z\right)$$

$$+ \frac{1}{2}\left\{G_0\left(-z\right) H_0\left(z\right) + G_1\left(-z\right) H_1\left(z\right)\right\} X\left(-z\right). \qquad (2.25)$$

The perfect reconstruction requirement, i.e., $\hat{X}(z) = X(z)$, is equivalent to

$$G_0\left(z\right) H_0\left(z\right) + G_1\left(z\right) H_1\left(z\right) = 2, \qquad (2.26a)$$
$$G_0\left(-z\right) H_0\left(z\right) + G_1\left(-z\right) H_1\left(z\right) = 0. \qquad (2.26b)$$

Before continuing on with further analysis we will write this requirement in the time domain. The output on the right hand side of figure 2.4 is

$$\hat{x}\left[n\right] = \sum_{j=0}^{1}\sum_{k} h_j\left[n - 2k\right] \sum_{m} g_j\left[m\right] x\left[2n + m\right]$$

$$= \sum_{j=0}^{1}\sum_{l}\sum_{k} h_j\left[n - 2k\right] g_j\left[2k - l\right] x\left[l\right]. \qquad (2.27)$$

Therefore perfect reconstruction is equivalent to the following requirement on all four filters

$$\sum_{k}\left\{h_0\left[n - 2k\right] g_0\left[2k - l\right] + h_1\left[n - 2k\right] g_1\left[2k - l\right]\right\} = \delta_{nl}. \qquad (2.28)$$

We now proceed to find the quadrature mirror filter conditions. We will consider both infinite impulse response and finite impulse response filters for completeness. For IIR filters we make the following choices for the pair $G_0$ and $G_1$, to satisfy (2.26b)

$$G_0\left(z\right) = H_0\left(1/z\right) \quad \Leftrightarrow \quad g_0\left[n\right] = h_0\left[-n\right], \qquad (2.29a)$$
$$G_1\left(z\right) = H_1\left(1/z\right) \quad \Leftrightarrow \quad g_1\left[n\right] = h_1\left[-n\right]. \qquad (2.29b)$$

In addition, we enforce the QMF property by relating the pair $h_0$ and $h_1$,

$$H_1\left(z\right) = -z^{-1} H_0\left(-1/z\right) \quad \Leftrightarrow \quad h_1\left[n\right] = \left(-1\right)^n h_0\left[1 - n\right], \qquad (2.30)$$

which when substituted into (2.26a) will result in a single equation for the single unknown filter function $H_0(z)$,

$$H_0(z) H_0(1/z) + H_0(-z) H_0(-1/z) = 2. \tag{2.31}$$

When $h_0[n]$ is a finite length sequence (an FIR filter) of length $N$, $0 \leq n \leq N - 1$, equations (2.29) and (2.30) refer to negative indices. This can be avoided by a simple time shift that, as shown in the following equations, maintains the same relationships between the pairs of sequences involved:

$$G_0(z) = z^{-(N-1)} H_0(1/z) \quad \Leftrightarrow \quad g_0[n] = h_0[N - 1 - n],$$
$$G_1(z) = z^{-(N-1)} H_1(1/z) \quad \Leftrightarrow \quad g_1[n] = h_1[N - 1 - n],$$

together with the QMF condition

$$H_1(z) = (-z)^{-(N-1)} H_0(-1/z) \quad \Leftrightarrow \quad h_1[n] = (-1)^n h_0[N - 1 - n]. \tag{2.32}$$

This finite length analog of the relation between $h_0$ and $h_1$ is known as the "alternating flip" [36] (the coefficients $h_1[n]$ are obtained by reversing the coefficients $h_0[n]$ and changing the signs of every other element starting at index 1). Equation (2.26b), which was identically satisfied in the IIR filter case, is now reduced to

$$\left[ 1 + (-1)^{N-1} \right] H_0(z) H_0(-1/z) = 0,$$

which can hold only when $N$ is an even integer. Thus, the coefficient of $X(-z)$ on the right hand side of (2.25) vanishes when $N$ is even, while the coefficient of $X(z)$ becomes

$$2^{-1} z^{-(N-1)} \left( H_0(z) H_0(1/z) + H_0(-z) H_0(-1/z) \right),$$

which upon choosing $H_0(z)$ to satisfy equation (2.31) will result in the following perfect reconstruction relation

$$\hat{X}(z) = z^{-(N-1)} X(z) \quad \Leftrightarrow \quad \hat{x}[n] = x[n - (N - 1)], \tag{2.33}$$

which implies a time delay but no signal distortion.

Thus, the PR-QMF bank for finite impulse response filters must incorporate even length filters and the PR condition is as shown above. The implementation only requires the finite (even length) set $h_0[n]$ with all other filters in the bank given in terms of $h_0[n]$ as shown:

$$h_1[n] \equiv h_1[n] = (-1)^n h_0[N - 1 - n], \tag{2.34a}$$
$$g_0[n] \equiv h_0[N - 1 - n], \tag{2.34b}$$
$$g_1[n] \equiv h_1[N - 1 - n], \quad n = 0, \ldots, N - 1. \tag{2.34c}$$

The PR-QMF equations (2.31) and (2.30) evaluated on the unit circle take the forms

$$H_1(\omega) = -e^{-i\omega} H_0^*(\omega + \pi), \tag{2.35a}$$

$$|H_0(\omega)|^2 + |H_0(\omega + \pi)|^2 = 2. \tag{2.35b}$$

As we shall see later, orthogonal wavelets of compact support are characterized by two finite even length filters (also known as low-pass and high-pass filter coefficients in the context of wavelets) $h_0[n]$ and $h_1[n]$, that satisfy the same PR-QMF conditions expressed in (2.35). The implementation of the discrete wavelet transform is then identical to the filter bank structure of figure 2.4.

We end our discussion of PR-QMF filter banks with two solutions of the PR-QMF equation (2.35b). The first example is provided by the Haar filter, $N = 2$, for which $H_0(z) = a + bz^{-1}$. Equation (2.35b) is then equivalent to $a^2 + b^2 = 1$. A further equation is found by demanding that $H_1(z)$ be a high-pass function, i.e., it should vanish at $\omega = 0$ or $z = 1$. This is, of course, equivalent to the requirement that the filter bank splits the signal into two components: one whose spectrum is the low half-band, and the second whose spectrum is the high half-band. Thus, $H_1(1) = a - b = 0$, and the solution is given by $a = b = 1/\sqrt{2}$. The three other filters for the filter bank are found from these values and using equations (2.34).

For the second example consider $N = 4$, which is equivalent to the Daubechies compact support wavelet of length 4. Now we have $H_0(z) = a + bz^{-1} + cz^{-2} + dz^{-3}$, which when substituted into Equation (2.35b) and equating coefficients of equal powers of $z^{-1}$ will lead to two equations, viz., $a^2 + b^2 + c^2 + d^2 = 1$ and $ac + bd = 0$. Two more equations are needed to solve for the unknown coefficients. The high-pass nature of $H_1(z)$ is again used to write $H_1(z = 1) = 0$, and to provide a third equation $a - b + c - d = 0$. In the case of wavelets of compact support, the fourth equation is found by demanding some regularity condition (a zero moment condition to be discussed later), e.g., that the first derivative of $H_1(z)$ should vanish at $z = 1$ ($\omega = 0$). Using the QMF condition (2.34a),

$$\frac{d}{dz} H_1(z) = \sum_{n=0}^{3} (-1)^{n+1} n h_0[3 - n] z^{-n-1},$$

which upon setting $z = 1$ becomes $-0\,d + 1\,c - 2\,b + 3\,a = 0$. Thus, we find the unique solution

$$h_0 = \left\{ 1 + \sqrt{3}, 3 + \sqrt{3}, 3 - \sqrt{3}, 1 - \sqrt{3} \right\} \Big/ 4\sqrt{2}.$$

This corresponds to the Daubechies compact support wavelet of order 4
[15]. A systematic method to solve for all higher-order Daubechies wavelets
of compact support is shown in chapter 7.

## 2.8 Exercises

**1** Prove the results in table 2.1.

**2** Prove the results in table 2.2.

**3** Let $A(z)$ and $B(z)$ denote the **Z** transforms of two sequences $a_n$, $N_1 \leq n \leq N_2$, and $b_n$, $M_1 \leq n \leq M_2$, $N_1, N_2, M_1, M_2 \in \mathbb{Z}$. Prove that the product $C(z) \equiv A(z)B(z)$ can be written as

$$\sum_{k=N_1+M_1}^{N_2+M_2} c_k z^{-k}$$

where the coefficients are given by the discrete convolution

$$c_k = \sum_m a_{k-m} b_m = \sum_m a_m b_{k-m}$$

with appropriate limits for the summation index $m$. Now consider the functions

$$C(z) = \sum_{k=-6}^{11} c_k z^{-k}, \quad B(z) = \sqrt{2} \left( \frac{1+z^{-1}}{2} \right)^6,$$

with the coefficients $c_k$, $-6 \leq k \leq 11$, given by

$$c_k = [-0.00379352, \ 0.00778260, \ 0.0220695, \ -0.0657719,$$
$$-0.0528240, \ 0.405177, \ 0.773029, \ 0.428484,$$
$$-0.0441353, \ -0.0823019, \ 0.0138065, \ 0.0158805,$$
$$-0.000708563, \ -0.00257452, \ -0.000265718, \ 0.000466218,$$
$$-7.09834 \times 10^{-5}, \ -3.45999 \times 10^{-5}].$$

Determine the coefficients $a_k$ of the function $A(z)$ where

$$A(z) = \sum_{k=-6}^{5} a_k z^{-k} = C(z)/B(z)$$

by writing out the above convolution equation in matrix form with a lower
triangular matrix composed of the coefficients of $B(z)$.

**4**   Consider a function $H(z)$ of the form

$$H(z) = z^L (1+z)^K \prod_{m=1}^{M} (z - z_m)(z - 1/z_m), \qquad (2.36)$$

where $L, K, M \in \mathbb{Z}$ and $z_k \in \mathbb{C}$. Show that

$$H(1/z) = z^{-2L-K-2M} H(z), \qquad (2.37)$$

where $H(z)$ is a polynomial in $z$ of the form

$$z^L \left( h_0 + \cdots + h_{N-1} z^{N-1} \right), \quad N = K + 2M + 1. \qquad (2.38)$$

The coefficients $h_n$ are said to be symmetric, and $H(z)$ is said to represent a symmetric filter, if $h_n = h_{N-n-1}$, $0 \leq n \leq N - 1$. Show that the symmetry condition is equivalent to

$$H(1/z) = z^{-(2L+N-1)} H(z). \qquad (2.39)$$

Hence show that functions of the form (2.36) are symmetric.

**5**   Consider the function $S(\omega) = 1 + \cos(\omega)$. Use theorem 43 to find the corresponding spectral factors of $S(\omega)$.

# CHAPTER 3

# Time, Frequency, and Scale Localizing Transforms

## 3.1 Introduction

The analysis of nonstationary signals requires methods that represent signal features simultaneously in time and frequency. The Fourier transform decomposes signals into frequency components but does not provide a time history of when the frequencies actually occur. When the frequency content of a signal is time varying, the Fourier transform as defined by equation (1.57) is incapable of capturing any local time variations and so would not be suitable for the analysis of nonstationary signals [37].

Consider, for instance, the bowhead whale sound depicted in figure 3.1: a 1024-point time series sampled at 1 Hz. The spectrum of this signal (magnitude squared of its Fourier transform) shows several peaks but no indication of the underlying complex set of nonlinear chirps that are actually present in the data. The simplest way to obtain a time-dependent spectrum, also

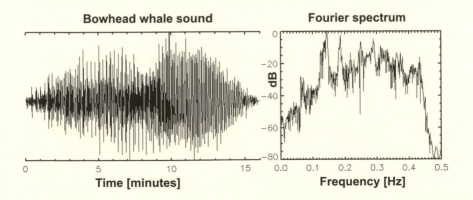

Figure 3.1: Bowhead whale sound and its Fourier spectrum.

known as a spectrogram, is to divide the data into (possibly overlapping) shorter sections, compute the Fourier transform of each section, and assign that spectrum (magnitude squared of the Fourier transform) to the center time of that section. If we assume that each section has duration $T$ (seconds), then the time resolution is $T$ (seconds) and the frequency resolution for that section is $1/T$ (Hz). Three spectrograms of the above data using sections of length 32, 64, and 128 points, overlapped 50%, are depicted in figure 3.2. Note that increasing the section duration produces sharper frequency lines at the expense of time resolution. The spectrogram is an implementation of the general concept of a windowed Fourier transform (WFT), also known as the short time Fourier transform (STFT) [37]. The short time Fourier

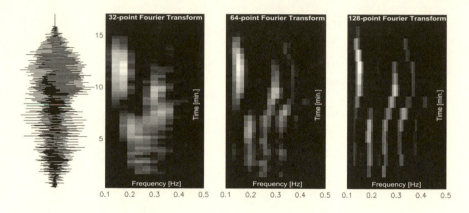

Figure 3.2: Three spectrograms of the bowhead whale sound of figure 3.1.

transform is an example of a class of methods to construct time-frequency representations of signals based on signal expansion in terms of collections of elementary analyzing functions.[1] Since the existence of an inverse is essential in almost all applications, e.g., denoising, compression, function approximation, the analyzing functions must form a frame (to ensure that the inverse can be numerically computed), or an orthonormal basis. A time-frequency energy density function for a signal is found by taking inner products of the signal with each of the analyzing functions used in the expansion. As described later in this section the STFT offers a fixed resolution method of time-frequency analysis.

---

[1] A very different approach to this problem is to construct, from the data, functions of both time and frequency that satisfy a number of, often conflicting, mathematical requirements. One important method is the Wigner distribution function [37]. These approaches, however, are nonlinear and are not discussed in this text.

There are many circumstances, however, under which a variable resolution is preferred: short bursts of high frequencies need good time resolution, whereas longer duration low frequencies can be analyzed using longer times (worse time resolutions). The simplest way to achieve this variable resolution is to change the scale of the analyzing functions: stretching or compressing them. This defines a property of the signal known as scale. Whereas the windowed Fourier functions (the STFT) analyze frequency components of a signal around each time location, the continuous wavelet functions, defining the continuous wavelet transform, analyze a signal at different resolutions, or scales, as a function of time. The same observations regarding the frame, or orthonormal basis, property of the STFT apply here, too. A time-scale energy density function is found by the same inner product procedure described above. Scale and frequency attributes of a signal are inversely related: higher scales contain lower frequencies and lower scales contain higher frequencies. This follows from the Fourier transform result that if $x(t) \leftrightarrow X(\omega)$ is a Fourier transform pair then so is $x(t/\eta) \leftrightarrow |\eta| X(\eta\omega)$.

We consider analyzing functions that are parametrized by two continuous variables: one variable to indicate time localization and the other to indicate localization of frequency or scale. In order for all functions to be representable by the superposition of a collection of analyzing functions, the latter must satisfy the completeness relation. Practical implementation requires a discretization of the two parameters of the analyzing functions so that the discretized functions form either a frame (a redundant basis), or an orthogonal basis. We have already seen an example of such a procedure in section 1.22, the reproducing kernel Hilbert space of band-limited functions $x_\Omega(t)$, in which a discretization of the kernel led to an orthogonal basis. The range space of the windowed Fourier transform, and the continuous wavelet transform, will turn out to be reproducing kernel Hilbert subspaces of $L_2(\mathbb{R}^2)$ and so it is natural to discretize the associated continuous parameters of two-dimensional analyzing functions that define the reproducing kernel, in search of frames and orthogonal bases.

In order to analyze the local time-frequency properties of a signal we must use localized functions of time. Construction of such a set of functions is best done systematically using a single localized function, for instance, a window function $w(t)$ that is equal to 1 for $0 \leq t \leq T_0$ and 0 otherwise. Gabor [38] described the windowed Fourier transform in which a fixed duration window over the time function extracts all the frequency content in that time interval. Gabor functions are time-shifted and frequency-modulated versions of the Gaussian window function $w(t) = \exp\left(-t^2/2\right)$. It turns out that the discretized Gabor functions form a frame only when the discretization parameters, $\tau = m\tau_0$ and $\omega = n\omega_0$, satisfy the inequality $\omega_0\tau_0 < 2\pi$. In particular, when $\omega_0\tau_0 = 2\pi$, they do not form a frame and so cannot be used for signal reconstruction. Thus, the windowed Fourier transform will be described by the family of functions

$$\xi_{\tau\omega}(t) \equiv w^*(t - \tau)\exp(+i\omega t) \qquad (3.1)$$

for continuous values of $(\tau, \omega)$, and also on a discrete grid of values, $\tau_m = m\tau_0$ and $\omega_n = n\omega_0$.

Although it is impossible to find general conditions under which a combination of $w(t)$, and discretization constants $\tau_0$ and $\omega_0$, would lead to a frame and its dual, specific results for compact windows exist [15]. Clearly the functions shown in equation (3.1) are constructed from a "mother" window function $w(t)$ by continuous time shifts and frequency modulations. Although the window function itself does not have to satisfy any properties except for being square integrable, it is always assumed to be symmetric. If such functions are to be used in the time and frequency analysis of a signal, the window function must also have an effective time duration that is roughly equal to the desired time resolution. This last requirement fixes both the time and the frequency resolution of the analyzing signal completely and so this is a fixed resolution method of time and frequency analysis.

There are many circumstances, however, under which a variable resolution, as opposed to the fixed resolution offered by the windowed Fourier transform, is preferred. This requirement follows from the observation that short bursts of high frequency need good time resolution, whereas low frequencies can be analyzed using longer times (worse time resolutions). A family of analyzing functions that have variable resolution in time and frequency can be constructed by continuous scaling and time shifts of a single mother function $\psi(t)$ [39, 1], also known as the mother wavelet, of the form

$$\psi_{\tau\eta}(t) = |\eta|^{-1/2}\,\psi\left(\frac{t - \tau}{\eta}\right). \qquad (3.2)$$

The mother wavelet must satisfy an admissibility condition if the resulting continuous wavelet transform (CWT) is to be invertible. A discretized version of this transform, calculated on a dyadic grid, can lead to an orthogonal basis for $L_2(\mathbb{R})$ provided that the mother wavelet satisfies additional constraints [15].

Another perspective on the windowed Fourier transform and the continuous wavelet transform is the problem of range and velocity determination for a localized target using radar. In narrow band radar signal analysis a pulse

$\xi(t)$ of very narrow band frequency content is transmitted and the return signal $s(t)$ is matched against time-shifted and frequency-modulated replicas of the transmitted pulse, i.e., functions of the form shown in equation (3.1). The resulting correlation, a function of lag $\tau$ and Doppler frequency $\omega$, is known as the narrow band ambiguity function and is identical in form to the windowed Fourier transform of the signal $s(t)$. If the ambiguity function had a single outstanding maximum magnitude, the corresponding time shift and (Doppler) frequency would translate to a unique range and velocity for the localized target. In general, the ambiguity function has a nearly flat region in the $(\tau, \omega)$ space (hence the name "ambiguity") and so a major problem of radar signal analysis is the design of transmitted functions that reduce that region of ambiguity to as small an area as possible given the system parameters that are present. In wide band applications, however, the Doppler frequency of the return varies in the signal band, resulting in a stretching or compression of the transmitted signal and so the matching (correlation) of the return is performed against a time-shifted and scaled (stretched or compressed) replica of the transmitted signal, i.e., functions of the form shown in equation (3.2). The corresponding correlation is known as the wide band ambiguity function in the lag $(\tau)$ and scale $(\eta)$ domain, and is identical in form to the continuous wavelet transform [40].

## 3.2 The Windowed Fourier Transform

**Definition 46.** Given a time function $s(t)$ and a real or complex window function $w(t)$, both in $L_2(\mathbb{R})$, the windowed Fourier transform is defined by

$$S_{\tau\omega} \equiv \int_{-\infty}^{\infty} s(t)\, w(t - \tau)\, e^{-i\omega t} dt, \tag{3.3}$$

where the time shift $\tau$ denotes the position of the window center, $\omega$ is the frequency, and $\tau, \omega \in \mathbb{R}$. The magnitude of the window, $|w(t)|$, is assumed to be symmetric about $t = 0$.

Gabor [38] used a Gaussian window function $w(t) \propto e^{-t^2/2}$, but any window function, satisfying the condition in definition 46 is permitted.[2]

---

[2]Equation (2.3) shows that the Fourier transform of a windowed function is the convolution of the Fourier transform of the original signal with the Fourier transform of the window. Consequently, the choice of the shape of a window is based on a compromise between the width of the main lobe of its spectrum versus the height of the first side lobe: side lobes are reduced (reduced leakage) at the expense of a broader main lobe (worse resolution).

The quantity $|S_{\tau\omega}|^2$ is known as the spectrogram and is routinely calculated in signal processing applications where time and frequency localization is of importance (see figure 3.2).

We can view equation (3.3) as an inner product of the original signal with continuously indexed basis functions which are generated by translation (parameter $\tau$) of a window function followed by modulation (multiplication by $e^{-i\omega t}$). Thus, defining

$$\xi_{\tau\omega}(t) \equiv w^*(t-\tau)\,e^{+i\omega t}, \tag{3.4}$$

we have

$$S_{\tau\omega} = \langle \xi_{\tau\omega}(t), \ s(t) \rangle \equiv \langle \xi_{\tau\omega} \mid s \rangle. \tag{3.5}$$

Three examples of the basis functions defined in equation (3.4) for a triangular window are shown in figure 3.3.

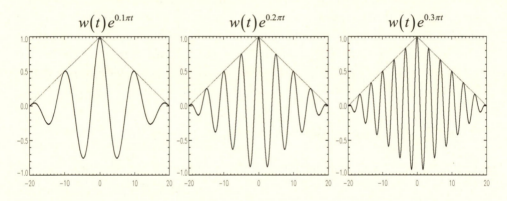

Figure 3.3: Basis functions $\xi_{\tau\omega}(t)$ for a triangular window, $\tau = 0$ and $\omega = 0.1\pi$, $0.2\pi$, and $0.3\pi$.

**Definition 47.** For a function $w(t)$, the translation by $\tau$ is a linear operator $\mathscr{T}_\tau$ defined by

$$\mathscr{T}_\tau w(t) \equiv w(t-\tau), \tag{3.6}$$

while the modulation by $\omega$ is a linear operator $\mathscr{M}_\omega$ defined by

$$\mathscr{M}_\omega w(t) \equiv e^{i\omega t} w(t). \tag{3.7}$$

Thus the functions $\xi_{\tau\omega}$ can be written as

$$\xi_{\tau\omega}(t) \equiv \mathscr{M}_\omega \mathscr{T}_\tau w^*(t). \tag{3.8}$$

## 3.3   The Windowed Fourier Transform Inverse

The original signal can be recovered from the windowed Fourier transform by applying the inverse Fourier transform (equation (1.58)) to the function $s(t)w(t - \tau)$,

$$s(t)\,w(t - \tau) = (2\pi)^{-1} \int\limits_{-\infty}^{\infty} S_{\tau\omega} e^{+i\omega t} d\omega, \qquad (3.9)$$

which upon multiplication with $w^*(t - \tau)$ and integration over $\tau$ will result in

$$2\pi \left\| w \right\|^2 s(t) = \int\limits_{-\infty}^{\infty} \int\limits_{-\infty}^{\infty} S_{\tau\omega} w^*(t - \tau)\, e^{i\omega t} d\omega d\tau. \qquad (3.10)$$

Using the Dirac notation and the integration convention we find

$$|s\rangle = \left\| \sqrt{2\pi} w \right\|^{-2} |\xi_{\tau\omega}\rangle \langle \xi_{\tau\omega} \mid s\rangle. \qquad (3.11)$$

In analogy to dual frames, let us define the dual functions

$$\chi_{\tau\omega}(t) \equiv \left\| \sqrt{2\pi} w \right\|^{-2} \xi_{\tau\omega}(t). \qquad (3.12)$$

Thus, we have the completeness relation (using the Dirac notation and the convention to integrate over repeated continuous variables $\tau$ and $\omega$)

$$|\chi_{\tau\omega}\rangle \langle \xi_{\tau\omega}| = \mathbf{1}. \qquad (3.13)$$

Note that $\xi_{\tau\omega}$ and $\chi_{\tau\omega}$ do not form an orthogonal set, in the sense that the inner product $\langle \chi_{\tau\omega} \mid \xi_{\tau'\omega'} \rangle \neq 0$ when $\tau \neq \tau'$ or $\omega \neq \omega'$. The inner product is the reproducing kernel function discussed in the next section. Also, if the window function is normalized so that $\|w\| = (2\pi)^{-1/2}$, then the dual functions $\chi_{\tau\omega}$ are identical to the original analyzing functions $\xi_{\tau\omega}$.

## 3.4   The Range Space of the Windowed Fourier Transform

Let us assume the normalization $\|w\| = (2\pi)^{-1/2}$ for the window function so that the dual functions of $\xi_{\tau\omega}(t)$ are identical to themselves, i.e., $\chi_{\tau\omega} = \xi_{\tau\omega}$.

The space $L_2\left(\mathbb{R}^2\right)$ of all (magnitude) square integrable functions of two continuous variables $\tau, \omega \in \mathbb{R}$ has the following inner product:

$$\langle X, Y \rangle \equiv \int\limits_{-\infty}^{\infty} \int\limits_{-\infty}^{\infty} X_{\tau\omega}^* Y_{\tau\omega} d\omega d\tau. \tag{3.14}$$

**Theorem 44.** *For a normalized window function if the functions* $x(t), y(t) \in L_2(\mathbb{R})$ *have windowed Fourier transforms* $X_{\tau\omega}$ *and* $Y_{\tau\omega}$, *respectively, then Parseval's relation is*

$$\langle x, y \rangle = \langle X, Y \rangle. \tag{3.15}$$

This can be shown by using the completeness relation (3.13)

$$\langle X, Y \rangle = \langle x \mid \xi_{\tau\omega} \rangle \langle \xi_{\tau\omega} \mid y \rangle = \langle x \mid y \rangle.$$

An immediate result of the above theorem is Parseval's relation for a function $s(t)$ and its windowed Fourier transform $S_{\tau\omega}$,

$$\int\limits_{-\infty}^{\infty} \int\limits_{-\infty}^{\infty} |S_{\tau\omega}|^2 d\tau d\omega = \|s\|^2. \tag{3.16}$$

The range space of the windowed Fourier transform is not the entire space $L_2(\mathbb{R}^2)$, but a proper subspace that can be characterized by the reproducing kernel Hilbert space property. Defining a kernel function[3]

$$K_{\tau\omega;\tau'\omega'} \equiv \langle \xi_{\tau\omega} \mid \xi_{\tau'\omega'} \rangle, \tag{3.17}$$

we have the following result.

**Theorem 45.** *For a normalized window function, the associated kernel of equation (3.17) is Hermitian, i.e.,* $K_{\tau\omega;\tau'\omega'} = K_{\tau'\omega';\tau\omega}^*$, *and for any function* $X_{\tau\omega}$ *that is the windowed Fourier transform of a function* $x(t) \in L_2(\mathbb{R})$, *i.e.,* $X$ *is in the range space of the transform, we must have*

$$X_{\tau\omega} = \int\limits_{-\infty}^{\infty} \int\limits_{-\infty}^{\infty} K_{\tau\omega;\tau'\omega'} X_{\tau'\omega'} d\tau' d\omega'. \tag{3.18}$$

---

[3]If the window is not normalized then we should define $K_{\tau\omega;\tau'\omega'} \equiv \langle \xi_{\tau\omega} \mid \chi_{\tau'\omega'} \rangle$.

The completeness relation (3.13) again can be used to show

$$\langle \xi_{\tau\omega} \mid x \rangle \equiv \langle \xi_{\tau\omega} \mid \xi_{\tau'\omega'} \rangle \langle \xi_{\tau'\omega'} \mid x \rangle. \tag{3.19}$$

The left hand side of the above equation is precisely $X_{\tau,\omega}$, while the right hand side (remembering the integration convention over repeated indices) is the right hand side of (3.18).

The reproducing kernel, therefore, can be used as a projection operator from $L_2(\mathbb{R}^2)$ onto the range space of the windowed Fourier transform operator, as shown in figure 3.4.

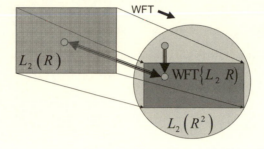

Figure 3.4: The range space of the windowed Fourier transform.

**Theorem 46.** *Given a function $X_{\tau\omega} \in L_2(\mathbb{R}^2)$ that is not in the range space of the windowed Fourier transform operator, the function $\hat{X}$ defined by*

$$\hat{X}_{\tau\omega} \equiv \langle \xi_{\tau\omega} \mid \xi_{\tau'\omega'} \rangle X_{\tau'\omega'} \equiv \int\limits_{-\infty}^{\infty} \int\limits_{-\infty}^{\infty} K_{\tau\omega;\tau'\omega'} X_{\tau'\omega'} d\tau' d\omega' \tag{3.20}$$

*is in the range space of the windowed Fourier transform.*

That is, $\hat{X}$ satisfies equation (3.17) which characterizes the range space, as can be seen by the following

$$\hat{X}_{\tau\omega} = \langle \xi_{\tau\omega} \mid \xi_{\tau'\omega'} \rangle X_{\tau'\omega'} = \langle \xi_{\tau\omega} \mid \xi_{\tau_1\omega_1} \rangle \langle \xi_{\tau_1\omega_1} \mid \xi_{\tau'\omega'} \rangle X_{\tau'\omega'}$$
$$= \langle \xi_{\tau\omega} \mid \xi_{\tau_1\omega_1} \rangle \hat{X}_{\tau_1\omega_1}. \tag{3.21}$$

## 3.5 The Discretized Windowed Fourier Transform

Here we follow the example of the reproducing kernel Hilbert space of band-limited functions, in which a frame (actually an orthogonal basis) resulted from a discretization of the shift parameter $\tau$ in the complete set of functions $p_\tau(t)$.

In order to study the conditions under which a discretization of the functions $\xi_{\tau\omega}(t)$ in the form $\tau_m = m\tau_0$ and $\omega_n = n\omega_0$, $n, m \in \mathbb{Z}$, would lead to a frame or an orthogonal basis for the range space of the windowed Fourier transform, we consider the windowed function $w(t - \tau)s(t)$, and take the window to be an unnormalized function of compact support in the time domain,[4] whose support is the interval $[0, T_0]$. Using the Fourier series expansion of equation (1.68) for the function $w(t - \tau)s(t)$ we have

$$w(t - \tau)s(t) = T_0^{-1} \sum_{n=-\infty}^{\infty} c_n(\tau) e^{in\omega_0 t} \qquad (3.22)$$

where $\omega_0 = 2\pi/T_0$ and the series coefficients $c_n$ depend on the window center parameter $\tau$ and are found from equation (1.69) to be

$$c_n(\tau) = \int_{\tau-T_0/2}^{\tau+T_0/2} e^{-in\omega_0 t} w(t - \tau)s(t)\, dt. \qquad (3.23)$$

The finite region of integration in the above integral can be extended to the entire real line since we have assumed a window of compact support. Thus,

$$c_n(\tau) = \int_{-\infty}^{\infty} e^{-in\omega_0 t} w(t - \tau)s(t)\, dt = \langle \xi_{\tau, n\omega_0} \mid s \rangle, \qquad (3.24)$$

where we have used the definition of the windowed Fourier transform coefficients in equation (3.3). Equation (3.24) shows that the Fourier series coefficients of the function $w(t - \tau)s(t)$, whose support is the closed interval $[0, T_0]$, are obtained by discretizing the windowed Fourier transform coefficients $\langle \xi_{\tau\omega} \mid s \rangle$ at discrete frequency values $n\omega_0 = 2\pi n/T_0$. To reconstruct the signal we use the same method we used to derive equation (3.10), i.e., we multiply (3.22) on both sides with $w^*(t - \tau)$ and integrate over $\tau$ to obtain

$$s(t) = T_0^{-1} \|w\|^{-2} \sum_{n=-\infty}^{\infty} e^{in\omega_0 t} \int_{-\infty}^{\infty} \langle \xi_{\tau, n\omega_0} \mid s \rangle w^*(t - \tau)\, d\tau. \qquad (3.25)$$

---

[4]A function cannot be simultaneously time and frequency limited. For instance, if the function has compact support in time then its Fourier transform is not limited to any compact region of the frequency axis, and vice versa.

However, a further discretization in the time domain is still necessary if we are to find a frame representation. To this end we discretize the variable $\tau$ in the form $\tau_m = m\tau_0$, multiply equation (3.22) by $w^*(t - \tau_m)$, and sum over all integers $m \in \mathbb{Z}$ to obtain

$$s(t) \sum_{m=-\infty}^{\infty} |w(t - m\tau_0)|^2 = T_0^{-1} \sum_{n=-\infty}^{\infty} c_n(m\tau_0) w^*(t - m\tau_0) e^{in\omega_0 t}$$

$$= T_0^{-1} \sum_{m=-\infty}^{\infty} \sum_{n=-\infty}^{\infty} e^{in\omega_0 t} w^*(t - m\tau_0) \langle \xi_{m\tau_0, n\omega_0} \mid s \rangle. \tag{3.26}$$

Evidently, the function $s(t)$ can be reconstructed from its transform coefficients if we can divide both sides of the above equation by the quantity

$$\mathscr{W}(t, \tau_0) \equiv \sum_{m=-\infty}^{\infty} |w(t - m\tau_0)|^2. \tag{3.27}$$

The expression above is a periodic function of $t$ with period $\tau_0$ and so it need only be calculated in the interval $[0, \tau_0]$. In addition, for each value of $t$ only a finite number of terms contribute to this sum of squares due to the fact that the window function has support only in the closed interval $[0, T_0]$; if $t \in [0, \tau_0]$ the values of $m$ range from 0 to $T_0/\tau_0$. Thus, we redefine the periodic quantity $\mathscr{W}$ as follows

$$\mathscr{W}(t, \tau_0) \equiv \sum_{m=0}^{T_0/\tau_0} |w(t - m\tau_0)|^2, \quad t \in [0, \tau_0], \quad \mathscr{W}(t, \tau 0) = \mathscr{W}(t + \tau_0, \tau_0). \tag{3.28}$$

If $\mathscr{W}$ is finite for all $t$ we recover the signal from the equation

$$s(t) = T_0^{-1} \sum_{m=-\infty}^{\infty} \sum_{n=-\infty}^{\infty} S_{\tau\omega} \mathscr{M}_{n\omega_0} \mathscr{W}^{-1}(t, \tau_0) \mathscr{T}_{m\tau_0} w^*(t), \tag{3.29}$$

where we have used definition 47 to write

$$e^{in\omega_0 t} w^*(t - m\tau_0) \equiv \mathscr{M}_{n\omega_0} \mathscr{T}_{m\tau_0} w^*(t).$$

Now, on account of the periodicity of $\mathscr{W}$ as a function of $t$, we have

$$\mathscr{W}^{-1}(t, \tau_0) \mathscr{T}_{m\tau_0} w^*(t) = \mathscr{W}^{-1}(t - m\tau_0, \tau_0) \mathscr{T}_{m\tau_0} w^*(t)$$

$$= \mathscr{T}_{m\tau_0} \mathscr{W}^{-1}(t, \tau_0) w^*(t), \tag{3.30}$$

and so we may define a dual window function $\tilde{w}(t)$,

$$\tilde{w}(t) \equiv T_0^{-1} \mathscr{W}^{-1}(t, \tau_0) w^*(t), \tag{3.31}$$

whose shifted and modulated versions are used in the reconstruction formula
and will form the dual frame. This is actually a more general result: for a
general window function, not of compact support, if the dual synthesis functions exist then they are generated by time translation and frequency shifts
of a single mother dual window function since the operator $T_w^+ T_w$, whose
inverse applied to the analysis functions produces the synthesis functions,
commutes with time translation and frequency shift operators, separately
[15]. Figure 3.5 shows a trapezoidal window $w$ defined in the interval $[0, 1]$,
i.e. $T_0 = 1$, $\tau_0 = 0.25$, and $\omega_0 \tau_0 = 0.5\pi$, together with its dual $\tilde{w}$.

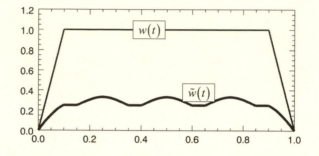

Figure 3.5: Trapezoidal window and its dual.

A necessary condition for $\mathscr{W}(t, \tau_0)$ to be strictly positive is $0 < \tau_0 \leq T_0$.
Since $\omega_0 = 2\pi/T_0$ this means that a necessary condition for the inverse of
the discretized windowed Fourier transform to exist is that $0 < \omega_0 \tau_0 \leq 2\pi$.
Nevertheless, to ensure that the inversion can be performed without any
numerical errors we demand that there exist two positive constants $A'$ and
$B'$ that are independent of $t$ and for which

$$0 < A' \leq \mathscr{W}(t, \tau_0) \leq B' < \infty, \quad \forall t. \tag{3.32}$$

Again the periodicity of $\mathscr{W}(t, \tau_0)$ can be used to find a closed expression for
its integral over $t$ over the range $[0, \tau_0]$ as follows.

$$\int_0^{\tau_0} \mathscr{W}(t, \tau_0) dt = \sum_{m=-\infty}^{\infty} \int_{-m\tau_0}^{-(m-1)\tau_0} |w(t)|^2 dt = \int_0^{T_0} |w(t)|^2 dt \equiv \|w\|^2, \tag{3.33}$$

where we have used the fact that the window function has support only in
$[0, T_0]$. Integration of the inequalities (3.32) between 0 and $\tau_0$ then gives

$$0 < A'\tau_0 \leq \|w\|^2 \leq B'\tau_0 < \infty. \tag{3.34}$$

This set of inequalities (compare these with the frame inequalities (1.135)) can be written in terms of two new positive constants $A \equiv 2\pi A'/\omega_0$ and $B \equiv 2\pi B'/\omega_0$, and hence we arrive at the following theorem [15].

**Theorem 47.** *Given a compact window with support in $[0, T_0]$, a function $s(t)$ can be recovered from its discretized windowed Fourier transform coefficients $S_{m\tau_0, n\omega_0}$ if $\tau_0 \omega_0 \leq 2\pi$ and two positive constants $A$ and $B$ exist for which the following inequalities hold*

$$0 < A \leq \frac{2\pi}{\omega_0 \tau_0} \|w\|^2 \leq B < \infty. \tag{3.35}$$

*Furthermore, defining the functions $\xi_{mn}$ and their duals $\chi_{mn}$ by*

$$\xi_{mn}(t) = \mathscr{M}_{n\omega_0} \mathscr{T}_{m\tau_0} w^*(t), \quad \chi_{mn}(t) = \mathscr{M}_{n\omega_0} \mathscr{T}_{m\tau_0} \tilde{w}(t), \tag{3.36}$$

*where $m, n \in \mathbb{Z}$, $\omega_0 T_0 = 2\pi$, $\tau_0 \leq T_0$, and the dual window $\tilde{w}$ is defined in equation (3.31), they form a frame and dual frame in $L_2(\mathbb{R})$. The completeness relation (with summation over repeated indices) is*

$$|\chi_{mn}\rangle \langle \xi_{mn}| = \mathbf{1}, \tag{3.37}$$

*and the reconstruction formula is*

$$|s\rangle = \langle \xi_{mn} \mid s \rangle |\chi_{mn}\rangle. \tag{3.38}$$

The frame inequalities (3.35) are a necessary condition for the existence of a frame. They do not place any restrictions on the window function except that it should be square integrable. Arbitrary choices of the window function $w(t)$, $\omega_0$ and $\tau_0$ with $\omega_0 \tau_0 < 2\pi$ do not, in general, produce a frame (unless, of course, the window is a compact one). Equation (3.35) shows that for a tight frame, when using a normalized window with $\|w\|^2 = 1$, the frame bounds are given by

$$A = B = 2\pi/\omega_0 \tau_0. \tag{3.39}$$

In particular, a tight frame reduces to an orthonormal basis when $A = B = 1$, i.e., when $\omega_0 \tau_0 = 2\pi$, while $\omega_0 \tau_0 > 2\pi$ destroys the frame property. Although in the critical case, defined by $\omega_0 \tau_0 = 2\pi$, windowed Fourier transform frames exist, the following theorem suggests that their (mother) window function must be badly behaved in the time or the frequency domain [15].

**Theorem 48.** *(Balian-Low) If the family of functions $\xi_{mn}(t)$, constructed through equation (3.36) from a window function $w(t)$ whose Fourier transform is denoted by $W(\omega)$, form a frame in $L_2(\mathbb{R})$, and the transform parameters satisfy the critical condition $\omega_0\tau_0 = 2\pi$ then*

$$\text{either} \quad \textbf{(a)} \int_{-\infty}^{\infty} t^2 \left|w(t)\right|^2 dt = \infty, \quad \text{or} \quad \textbf{(b)} \int_{-\infty}^{\infty} \omega^2 \left|W(\omega)\right|^2 d\omega = \infty. \quad (3.40)$$

Two examples for windowed Fourier transform functions that form frames in $L_2(\mathbb{R})$ for the critical case $\omega_0\tau_0 = 2\pi$ are the rectangular window function $w(t) = 1$, $0 \leq t \leq T_0$, consistent with (b), and the sinc window function $w(t) = \sin(\pi t)/(\pi t)$, consistent with (a).

Gabor's original choice of unit norm window $w(t) = \pi^{-1/4}e^{-t^2/2}$ turns out not to produce a frame when $\omega_0\tau_0 = 2\pi$! Although the Gabor transform coefficients satisfy the completeness condition $\langle g_{mn}, e \rangle = 0 \Leftrightarrow e(t) = 0$, the lower bound $A$ of inequalities (3.35) does not exist. Consequently (see equation (1.138)) the function $\mathcal{W}$ of equation (3.28) cannot be inverted and so the dual window of equation (3.31) does not exist. Dual frames for Gabor functions exist when $\omega_0\tau_0 < 2\pi$, i.e., the lower bound $A$ is numerically computable and is greater than 0. In fact, for $\omega_0\tau_0 = \pi/2$ or $\omega_0\tau_0 = \pi$ both $A$ and $B$ can be calculated exactly. For instance, for $T_0 = 2$, $\omega_0 = \pi$ and $\tau_0 = 0.5$, the bounds are $A = 1.221$ and $B = 7.091$. These correspond to a frame with a redundancy ratio of 5.896, while $T_0 = 6$, $\omega_0 = \pi/3$ and $\tau_0 = 1.5$ produce a nearly tight frame with $A = 3.899$ and $B = 4.101$, and redundancy ratio of 1.052 [15].

## 3.6  Time-Frequency Resolution of the Windowed Fourier Transform

The main problem with the windowed Fourier transform (and its discretized version) is that the fixed duration window function is accompanied by a fixed frequency resolution (a record length of $T_0$ seconds can resolve two neighboring frequencies that are at least $T_0^{-1}$ Hz apart) [37]. To see this let us define the time width $\sigma_t$ and the frequency width $\sigma_f$ of a window function $w(t)$ whose Fourier transform, in terms of linear frequency $f = \omega/2\pi$, is

$W(f)$,

$$\sigma_t^2 \equiv \frac{\int\limits_{-\infty}^{\infty} (t - \bar{t})^2 \, |w^2(t)| \, dt}{\int\limits_{-\infty}^{\infty} |w^2(t)| \, dt}, \quad \sigma_f^2 \equiv \frac{\int\limits_{-\infty}^{\infty} (f - \bar{f})^2 \, |W^2(f)| \, df}{\int\limits_{-\infty}^{\infty} |W^2(f)| \, df}, \quad (3.41)$$

where $\bar{t}$, the mean time of the square of the window function, and $\bar{f}$, the mean frequency of the window spectrum, are defined by

$$\bar{t} \equiv \frac{\int\limits_{-\infty}^{\infty} t \, |w^2(t)| \, dt}{\int\limits_{-\infty}^{\infty} |w^2(t)| \, dt}, \quad \bar{f} \equiv \frac{\int\limits_{-\infty}^{\infty} f \, |W^2(f)| \, df}{\int\limits_{-\infty}^{\infty} |W^2(f)| \, df}. \quad (3.42)$$

It is easy to show that $\sigma_t \sigma_f \geq (4\pi)^{-1}$, with equality achieved for a Gaussian window function. A change of the parameter $\tau$ merely translates the window in time while its spread is kept fixed. Similarly, as the modulation parameter $f$ increases, the transform is translated in frequency, retaining a constant width. Thus, the resolution cells in the time and frequency plane have fixed dimensions $\sigma_t$ and $\sigma_f$ as seen in figure 3.6 which displays the constant-resolution cells in which a sliding time window is centered at integral multiples of $\tau$ and the transforms are evaluated at bin frequencies centered at integral multiples of $f$.

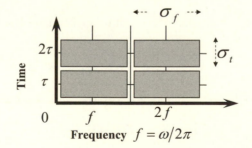

Figure 3.6: Resolution cells for the windowed Fourier transform.

The windowed Fourier transform allows only a fixed time and frequency resolution. In the next section we describe the continuous wavelet transform as a variable-resolution technique to construct a time and frequency representation of a signal.

## 3.7   The Continuous Wavelet Transform

In analogy to the wide band ambiguity function we define the continuous wavelet transform (CWT) of a signal $x(t)$ as follows[5] [39]:

$$X_{\tau\eta} \equiv |\eta|^{-1/2} \int_{-\infty}^{\infty} \psi^* \left( \frac{t-\tau}{\eta} \right) x(t)\, dt, \quad \eta, \tau \in \mathbb{R},\ \eta \neq 0. \tag{3.43}$$

The continuous wavelet transform coefficients are then the inner product of the signal (time function) with a set of analyzing functions that are formed by dilations (change of scale) followed by translations of a prototype mother wavelet $\psi(t)$. Defining the analyzing functions by

$$\psi_{\tau\eta}(t) \equiv |\eta|^{-1/2} \psi \left( \frac{t-\tau}{\eta} \right) \tag{3.44}$$

we have

$$X_{\tau\eta} = \langle \psi_{\tau\eta} \mid x \rangle \equiv \int_{-\infty}^{\infty} \psi_{\tau\eta}^*(t)\, x(t)\, dt. \tag{3.45}$$

The normalization in equation (3.44) ensures that the dilated wavelet functions have the same energy at each scale, i.e.,

$$\|\psi\|^2 = \|\psi_{\tau\eta}\|^2. \tag{3.46}$$

Three examples of the basis functions defined in equation (3.44) for $\tau = 0$ and a generic $\psi(t)$ are shown in figure 3.7.

**Definition 48.** For a function $\psi(t)$, dilation by a factor $\eta \neq 0$ is a linear operator $\mathscr{D}_\eta$ defined by

$$\mathscr{D}_\eta \psi(t) \equiv |\eta|^{-1/2} \psi(t/\eta). \tag{3.47}$$

Thus the basis functions (3.44) can be written as

$$\psi_{\tau\eta}(t) = \mathscr{T}_\tau \mathscr{D}_\eta \psi(t), \tag{3.48}$$

where the translation operator $\mathscr{T}_\tau$ was introduced in definition 47 and equation (3.6). The dilation and translation operators do not commute, i.e., $\mathscr{D}_\eta \mathscr{T}_\tau \neq \mathscr{T}_\tau \mathscr{D}_\eta$. A useful result on the product of the translation and the dilation operators is the following theorem.

---

[5]The scaling of $t$ is usually defined by $t/\eta$ instead of $\eta t$. The integral measure for scale is, therefore, slightly more complicated using the usual definition that is adopted here: $|\eta|^{-2}\, d\eta$ instead of $d\eta$.

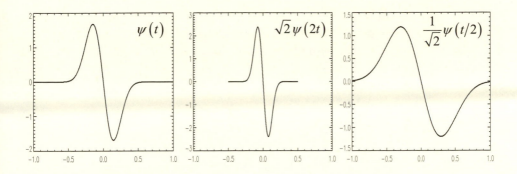

Figure 3.7: Unit-norm basis functions $\psi_{\tau\eta}(t)$ for a generic $\psi(t)$, $\tau = 0$ and $\eta = 1$, $0.5$, and $2$.

**Theorem 49.** *The dilation and translation operators satisfy the relation*

$$\mathscr{D}_\eta \mathscr{T}_\tau = \mathscr{T}_{\eta\tau} \mathscr{D}_\eta. \tag{3.49}$$

For

$$\mathscr{T}_{\eta\tau} \mathscr{D}_\eta \psi(t) = \mathscr{T}_{\eta\tau} |\eta|^{-1/2} \psi(t/\eta) = |\eta|^{-1/2} \psi((t - \eta\tau)/\eta) = |\eta|^{-1/2} \psi(t/\eta - \tau),$$
$$\mathscr{D}_\eta \mathscr{T}_\tau \psi(t) = \mathscr{D}_\eta \psi(t - \tau) = |\eta|^{-1/2} \psi(t/\eta - \tau).$$

If the mother function is localized in time, its dilation produces short-duration high-frequency, and long-duration low-frequency functions. These functions are clearly better suited for representing short bursts of high frequency or long-duration slowly varying signals.

**Theorem 50.** *If $\Psi(\omega)$ is the Fourier transform of the mother wavelet function $\psi(t)$, then the Fourier transform of $\psi(t/\eta)$ is $\eta\Psi(\eta\omega)$, where $\eta$ is the scale parameter.*

Thus, a contraction in time, $|\eta| < 1$, results in an expansion in frequency and vice versa. This procedure of dilating and translating is analogous to the construction of constant-$Q$ filter banks [36]. To overcome the fixed resolution limitation of the windowed Fourier transform one allows the two resolutions to vary in the time and frequency plane. Constant-$Q$ filter banks have the property that the ratios of their root mean square bandwidth to the center frequency of all the filters in the bank are the same. The continuous wavelet transform follows this same idea while simplifying the procedure of filter construction by choosing the filters to be dilated versions of the same prototype function. If the mother wavelet has a frequency spread $\sigma_f$ and a center frequency $f_0$, then each dilated wavelet has a frequency spread $\sigma_f/\eta$ and a center frequency $f/\eta$ satisfying the constant-$Q$ filter condition, i.e., the resolution cells in the time frequency plane have a fixed ratio $\sigma_f/f$ as shown in figure 3.8.

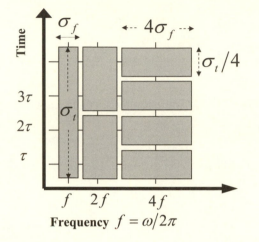

Figure 3.8: Resolution cells for the continuous wavelet transform (CWT).

Mother wavelet functions of interest are band-pass filters that are oscillatory and decay rapidly in the time domain to have effectively compact support and thus provide good time resolution. So for values of $\eta$ whose magnitudes are smaller than 1, the basis functions become stretched versions of the mother wavelet, i.e., low-frequency functions, whereas for those values of $\eta$ whose magnitudes are larger than 1, the basis functions are contracted versions of the mother wavelet, i.e., short-duration, high-frequency functions. The parameter $\tau$ defines a translation of the wavelet and provides for time localization. Some of the properties of the CWT are listed in table 3.1.

| Function | CWT |
|---|---|
| $x(t)$ | $X_{\tau\eta}$ |
| $x(t) + y(t)$ | $X_{\tau\eta} + Y_{\tau\eta}$ |
| $x(t - T_0)$ | $X_{\tau - T_0, \eta}$ |
| $\|\eta'\|^{-1/2} x(t/\eta')$ | $X_{\tau/\eta', \eta/\eta'}$ |

Table 3.1: Properties of the continuous wavelet transform (CWT).

## 3.8    The Continuous Wavelet Transform Inverse

The continuous wavelet transform has an inverse provided that the mother wavelet $\psi(t)$ satisfies a certain admissibility condition [39, 15]. To derive the inverse we begin by taking the Fourier transform of both sides of equation (3.43) with respect to $\tau$ to obtain

$$\hat{X}(\omega, \eta) \equiv \int\limits_{-\infty}^{\infty} e^{-i\omega\tau} X_{\tau\eta} d\tau =$$
$$|\eta|^{-1/2} \int\limits_{-\infty}^{\infty} \int\limits_{-\infty}^{\infty} e^{-i\omega\tau} \psi^* \left(\frac{t-\tau}{\eta}\right) x(t) \, dt d\tau. \tag{3.51}$$

In the $\tau$ integral on the right hand side, we change variables $\tau = t + \eta\sigma$ and perform the integral over $\sigma$ and then the integral over $t$ to obtain

$$\hat{X}(\omega, \eta) = |\eta|^{+1/2} X(\omega) \Psi^*(\eta\omega), \tag{3.52}$$

where $X(\omega)$ and $\Psi(\omega)$ denote the Fourier transforms of $x(t)$ and $\psi(t)$, respectively. Next we multiply both sides of equation (3.52) by $|\eta|^{-3/2} \Psi(\eta\omega)$ and integrate both sides with respect to the scaling parameter $\eta$ to obtain

$$\int\limits_{-\infty}^{\infty} \hat{X}(\omega, \eta) |\eta|^{-3/2} \Psi(\eta\omega) \, d\eta = X(\omega) \int\limits_{-\infty}^{\infty} |\eta|^{-1} |\Psi(\eta\omega)|^2 d\eta. \tag{3.53}$$

The integral on the right hand side is actually independent of $\omega$, as can be shown by a simple change of the integration variable $\omega\eta \to \sigma$, $d\eta \to \omega^{-1} d\sigma$. Thus,

$$\int\limits_{-\infty}^{\infty} |\eta|^{-1} |\Psi(\eta\omega)|^2 d\eta = \int\limits_{-\infty}^{\infty} |\sigma|^{-1} |\Psi(\sigma)|^2 \, d\sigma \equiv C_\psi. \tag{3.54}$$

Finally, equations (3.53) and (3.54), together with the additional stipulation known as the mother wavelet admissibility condition (changing integration variables from $\sigma$ to $\omega$ in the second integral of equation (3.54))

$$C_\psi \equiv \int\limits_{-\infty}^{\infty} |\omega|^{-1} |\Psi(\omega)|^2 \, d\omega < \infty, \tag{3.55}$$

result in

$$X(\omega) = C_\psi^{-1} \int\limits_{-\infty}^{\infty} \hat{X}_\psi(\omega, \eta) |\eta|^{-3/2} \Psi(\eta\omega) \, d\eta. \tag{3.56}$$

Taking inverse Fourier transform of both sides of (3.56) and using the Fourier transform pair (deduced from the convolution equation (2.2)),

$$\hat{X}\left(\omega,\eta\right)\Psi\left(\eta\omega\right) \leftrightarrow \int\limits_{-\infty}^{\infty} X_{\tau\eta}\left|\eta\right|^{-1}\psi\left(\frac{t-\tau}{\eta}\right)d\tau. \tag{3.57}$$

and the definition in (3.44) we find the final expression for the inverse CWT as stated in the following theorem.

**Theorem 51.** *Given a mother wavelet function $\psi(t)$ with finite energy that satisfies the admissibility condition expressed in equation (3.55), a function $x(t) \in L_2(\mathbb{R})$ can be reconstructed from its continuous wavelet transform coefficients $X_{\tau\eta}$ by the equation*

$$x\left(t\right) = C_{\psi}^{-1}\int\limits_{-\infty}^{\infty}\int\limits_{-\infty}^{\infty} X_{\tau\eta}\psi_{\tau\eta}\left(t\right)\left|\eta\right|^{-2}d\eta d\tau . \tag{3.58}$$

The admissibility condition (3.55) is always satisfied by a square-integrable $\psi(t)$ provided that

$$\int\limits_{-\infty}^{\infty}\psi\left(t\right)dt = \Psi\left(\omega = 0\right) = 0. \tag{3.59}$$

An admissible mother wavelet must, therefore, integrate to zero, or have no DC Fourier component [39].

Equation (3.58) of theorem 51 describes the reconstruction of a function from its CWT coefficients and the dual wavelet functions defined by

$$\chi_{\tau\eta} \equiv C_{\psi}^{-1}\left|\eta\right|^{-2}\psi_{\tau\eta}, \tag{3.60}$$

in the compact form (integration over repeated indices $\left|\eta\right|,\left|\tau\right| < \infty$ is implied)

$$\left|x\right\rangle = \left\langle\psi_{\tau\omega} \mid x\right\rangle \left|\chi_{\tau\omega}\right\rangle . \tag{3.61}$$

Evidently we have the completeness relation

$$\left|\chi_{\tau\eta}\right\rangle \left\langle\psi_{\tau\eta}\right| = \mathbf{1}. \tag{3.62}$$

Taking the inner product of the inverse equation (3.58) with another function $y(t)$, represented by the bra $\left\langle y\right|$, leads to the general Parseval relation.

**Theorem 52.** *Let $x(t), y(t) \in L_2(\mathbb{R})$ have continuous wavelet transform coefficients $X_{\tau\eta}$ and $Y_{\tau\eta}$, respectively. Then*

$$\langle y \mid x \rangle = \langle Y \mid X \rangle \equiv C_\psi^{-1} \int\limits_{-\infty}^{\infty} \int\limits_{-\infty}^{\infty} Y_{\tau\eta}^* X_{\tau\eta} |\eta|^{-2} \, d\eta d\tau. \qquad (3.63)$$

If the mother wavelet $\psi(t)$ is a real function, then its Fourier transform $\Psi(\omega)$ satisfies the equation $\Psi(\omega) = \Psi^*(-\omega)$, and the admissibility condition (3.55) reduces to an integral over positive frequencies only,

$$C_\psi \equiv 2 \int\limits_0^\infty \omega^{-1} |\Psi(\omega)|^2 d\omega < \infty, \qquad (3.64)$$

while the inverse CWT, equation (3.58) reduces to an integral over positive scales,

$$x(t) = 2C_\psi^{-1} \int\limits_{\tau=-\infty}^{\infty} \int\limits_{\eta=0}^{\infty} X_{\tau\eta} \psi_{\tau\eta}(t) \eta^{-2} d\eta d\tau. \qquad (3.65)$$

In the Dirac notation, the integration ranges over repeated indices are given by $\{-\infty < \tau < \infty\}$ and $\{0 < \eta < \infty\}$.

If we restrict the scales to a positive range $0 < \eta < \infty$ and assume a complex mother wavelet $\psi(t)$, then we must require a stronger version of the admissibility condition (cf., equation (3.55)), viz.,

$$C_\psi \equiv 2 \int\limits_0^\infty \omega^{-1} |\Psi(\omega)|^2 \, d\omega = 2 \int\limits_0^\infty \omega^{-1} |\Psi(-\omega)|^2 \, d\omega < \infty. \qquad (3.66)$$

## 3.9 The Range Space of the Continuous Wavelet Transform

Given a mother wavelet function $\psi(t)$, the set of all CWT coefficients $X_{\tau\eta}$ of $x \in L_2(\mathbb{R})$, for $\tau, \eta \in \mathbb{R}$ and $\eta \neq 0$, is a proper subspace of $L_2(\mathbb{R}^2)$. Just as in the case of the windowed Fourier transform, this subspace is a reproducing kernel Hilbert space. Defining the kernel function

$$K_{\tau\eta;\tau'\eta'} \equiv \langle \psi_{\tau\eta} \mid \psi_{\tau'\eta'} \rangle, \qquad (3.67)$$

and using the resolution of identity (3.62) we have

$$X_{\tau\eta} = \langle \psi_{\tau\eta} \mid x \rangle = \langle \psi_{\tau\eta} \mid \psi_{\tau'\eta'} \rangle \langle \psi_{\tau'\eta'} \mid x \rangle \equiv$$
$$C_\psi^{-1} \int\limits_{-\infty}^{\infty} \int\limits_{-\infty}^{\infty} K_{\tau\eta;\tau'\eta'} X_{\tau'\eta'} \left| \eta' \right|^{-2} d\eta' d\tau'. \tag{3.68}$$

Again, as in the case of the windowed Fourier transform, the kernel function is a projection onto the range space of the continuous wavelet transform $L_2(\mathbb{R}^2) \to \mathscr{R}\{CWT\}$, as depicted in figure 3.9. In analogy to equations

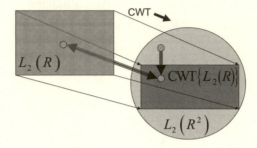

Figure 3.9: The range space of the continuous wavelet transform (CWT).

(3.20) and (3.21), if $X_{\tau\eta}$ is not in the range space of the continuous wavelet transform, then the quantity

$$K_{\tau\eta;\tau'\eta'} X_{\tau'\eta'} \equiv \int\limits_{-\infty}^{\infty} \int\limits_{-\infty}^{\infty} K_{\tau\eta;\tau'\eta'} X_{\tau'\eta'} \left| \eta' \right|^{-2} d\eta' d\tau'$$

is its projection onto the range space of the continuous wavelet transform. This projection now has a unique inverse as shown in equation (3.58) [15].

## 3.10   The Morlet, the Mexican Hat, and the Haar Wavelets

The Morlet mother wavelet for the construction of the continuous wavelet transform is a complex function defined by [39, 41]

$$\psi(t) = a \left( e^{i\omega_0 t} - e^{-\omega_0^2 \sigma^2 / 2} \right) e^{-t^2 / 2\sigma^2}, \tag{3.69}$$

where

$$a \equiv \left( \pi \sigma^2 \right)^{-1/4} \left( 1 - 2e^{-3\omega_0^2 \sigma^2 / 4} + e^{-\omega_0^2 \sigma^2} \right)^{-1/2},$$

or, equivalently, by its Fourier transform

$$\Psi(\omega) = \sqrt{2\pi}a\sigma\left(e^{-\sigma^2(\omega-\omega_0)^2/2} - e^{-\omega_0^2\sigma^2/2}e^{-\omega^2\sigma^2/2}\right). \tag{3.70}$$

This wavelet function has unit norm and it satisfies the admissibility condition (3.55) because it has no DC Fourier component ($\Psi(0) = 0$). The parameter $\sigma$ is used to control the number of oscillations within the Gaussian envelope while $\omega_0$ determines the center frequency of the mother wavelet. Figure 3.10 shows the real and imaginary parts of a Morlet mother wavelet computed using the following three parameters: the center frequency $\omega_0 = 0.1\pi$, the total number of cycles within the Gaussian envelope $N_{cyc} = 3$, and the number of samples per cycle $N_{spc} = 51$. The center frequency is then $f_0 = 2\pi\omega_0 = 0.05$ Hz and the corresponding period is $P_0 = 1/f_0 = 20$ seconds. The total time span is $T = N_{cyc}P_0$ and the sampling frequency is $f_s = f_0 N_{spc} = 2.5$ Hz. The value 18.75 was used for the width parameter $\sigma$ and it can be adjusted to reduce the height of the peaks on either side of the central peak. Figure 3.10 was produced using parameter values $N_{cyc} = 3$, $\omega_0 = 0.1\pi$ ($f_0 = 0.05$ Hz), and $N_{spc} = 51$. The spectrum of the corresponding complex wavelet with its peak (center frequency) at 0.05 Hz is also shown.

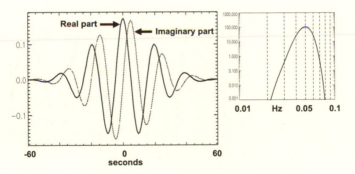

Figure 3.10: Real and imaginary parts of the Morlet mother wavelet ($f_s = 1.25$ Hz and $\omega_0 = 0.1\pi$, $\sigma = 0.8$ seconds) and its spectrum.

The Morlet wavelet is often used for time and frequency analysis of signals since its oscillatory nature allows for good frequency resolution while its time decay allows for good time resolution. The Morlet CWT coefficients are given by

$$X_{\tau\eta} = a|\eta|^{-1/2}\int_{-\infty}^{\infty}\left(e^{-i\omega_0(t-\tau)/\eta} - e^{-\omega_0^2\sigma^2/2}\right)e^{-(t-\tau)^2/2\sigma^2\eta^2}x(t)dt. \tag{3.71}$$

If the signal is a complex sinusoid $x(t) = \exp{(i\omega t)}$ we use equation (3.71) with a transformation of variable $(t - \tau)/\eta \to t'$ to find

$$
\begin{aligned}
X_{\tau\omega} &= |\eta|^{1/2}\,e^{i\omega\tau}\,\Psi^*\,(\eta\omega)\\
&= a\sigma\sqrt{2\pi}\,|\eta|^{1/2}\,e^{i\omega\tau}\left(e^{-0.5\eta^2\sigma^2(\omega-\omega_0/\eta)^2} - e^{-0.5\omega_0^2\sigma^2}e^{-0.5\omega^2\eta^2\sigma^2}\right)\\
&\approx a\sigma\sqrt{2\pi}\,|\eta|^{1/2}\,e^{i\omega\tau}e^{-0.5\omega^2\sigma^2(\eta-\omega_0/\omega)^2},
\end{aligned}
\tag{3.72}
$$

where we have neglected the second term in parentheses in the second line. Thus, $|X_{\tau\eta}|$ peaks at a scale inversely proportional to the signal's frequency $\eta = \omega_0/\omega$ with an approximate width (at half the maximum) of $2.3/\sigma\omega$. The ratio of peak scale to width is then proportional to $\omega_0\sigma$ which is a constant. The constant-$Q$ filter bank property of the Morlet wavelets follows from the fact that each scaled function $\psi\,(t/\eta)$ has a Fourier transform that is a Gaussian function centered at $\omega_0/\eta$ and a width parameter $1/\eta\sigma$ and so the ratio of the center frequency to the width parameter is fixed for each function. This is illustrated in figure 3.11 for the spectra of a Morlet wavelet family with parameters $f_0 = 0.4$ Hz and $\eta = 1, 2, 4, 8, 16$.

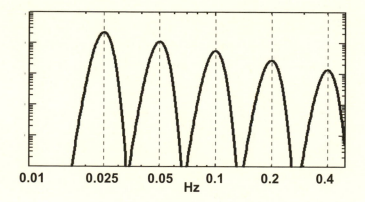

Figure 3.11:   The Morlet constant-$Q$ filter bank:   $f_0 = 0.4$ Hz, $\eta = 1, 2, 4, 8, 16$.

The Mexican hat wavelet is a real function defined by [15]

$$
\psi\,(t) = 3^{-1/2}2\pi^{-1/4}\sigma^{-1/2}\left(1 - t^2/\sigma^2\right)e^{-t^2/2\sigma^2}.
\tag{3.73}
$$

It has unit norm, its integral over the entire real line vanishes, and it is shown in figure 3.12 for the parameter value $\sigma = 1$. The Fourier transform of the Mexican hat wavelet is given by

$$
\Psi\,(\omega) = 2^{3/2}3^{-1/2}\pi^{1/4}\sigma^{5/2}\omega^2 e^{-\sigma^2\omega^2/2},
\tag{3.74}
$$

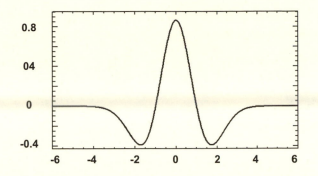

Figure 3.12: The Mexican hat mother wavelet.

which has a peak at $\omega = \pm\sqrt{2}/\sigma$. The dilated Mexican hat wavelets are, therefore, band-pass filters centered at $\omega = \pm\sqrt{2}/\sigma\eta$ which, together with the fact that the widths of the Fourier transforms of dilated wavelets are proportional to $\eta^{-1}$, imply the constant-$Q$ filter bank property of the transform. Four members of the Mexican hat wavelet filter bank are shown in figure 3.13. These filters are centered at 1, 0.5, 0.25 and 0.125 and 0.0625 Hz, respectively. They were computed using a sampling frequency of 100 Hz, and $\sigma = 0.225$ that ensures the undilated mother wavelet is centered at 1 Hz. The Mexican hat continuous wavelet transform coefficients for $\sigma = 1$

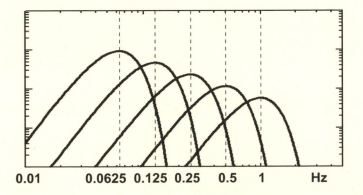

Figure 3.13: The Mexican hat mother wavelet filter bank members: $\eta = 1, 2, 2^2, 2^3$.

for an arbitrary function are given by

$$X_{\tau\eta} = 3^{-1/2} 2\pi^{-1/4} |\eta|^{-1/2} \int_{-\infty}^{\infty} \left(1 - \left(\frac{t-\tau}{\eta}\right)^2\right) e^{-(t-\tau)^2/2\eta^2} x(t)\, dt. \quad (3.75)$$

If we use the Mexican hat wavelet for the analysis of a complex sinusoid $\exp(i\omega t)$, we find

$$X_{\tau\eta} = 3^{-1/2} 2^{3/2} \pi^{1/4} |\eta|^{5/2} \omega^2 e^{i\tau\omega - \eta^2\omega^2/2}, \quad (3.76)$$

whose peak occurs at a scale inversely proportional to the frequency of the signal, i.e., $\eta = \sqrt{2}/\omega$.

The Haar wavelet [42] is the simplest mother function for the calculation of the continuous wavelet transform. It was introduced more than seventy years before the advent of wavelets as an example of a function whose expansion or compression followed by translations on an irregular grid would produce an orthonormal basis for $L_2(\mathbb{R})$. The Haar wavelet is defined by

$$\psi(t) = \begin{cases} 1, & 0 \leq t \leq 1/2, \\ -1, & 1/2 \leq t \leq 1. \end{cases} \quad (3.77)$$

It has very good time resolution, but its discontinuities in the time domain allow for no frequency resolving power. Its Fourier transform is

$$\Psi(\omega) = ie^{-i\omega/2}\sin^2(\omega/4)/(\omega/4), \quad (3.78)$$

and its spectrum, shown in figure 3.14, has its peak at $\omega \approx 4.66$, corresponding to linear frequency value $f = \omega/2\pi = 0.74$, and has a significant side lobe. The Haar continuous wavelet transform coefficients are

$$X_{\tau\eta} = |\eta|^{-1/2} \left\{ \int_{\tau}^{\tau+\eta/2} x(t)\, dt - \int_{\tau+\eta/2}^{\tau+\eta} x(t)\, dt \right\}, \quad (3.79)$$

which for a complex exponential $\exp(i\omega t)$ reduce to

$$X_{\tau\eta} = -4ie^{i\omega(\tau+\eta/2)}\sin^2(\eta\omega/4)/\omega. \quad (3.80)$$

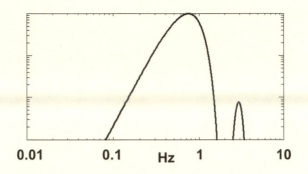

Figure 3.14: Spectrum of the Haar wavelet.

## 3.11 Discretizing the Continuous Wavelet Transform

The discretization of the CWT is a not as straightforward as the windowed Fourier transform where a regular grid was simply given by the discretization of the time shift and the frequency variables [15]. The scale parameter in the CWT is multiplicative and we must ensure that the scale invariance of the CWT coefficients, as described by the last entry of table 3.1, holds for the discretized coefficients. This can be achieved only by adjusting the time shift sample rate by the scale sample rate which means that the resulting grid will not be regular. An irregular time grid will necessarily mean that the resulting discretized coefficients will not be time invariant. Scale invariance, however, can be preserved after discretization only if we use the irregular grid defined by

$$\{\eta_m = \eta_0^m, \quad \tau_n = n\eta_0^m\tau_0; \ \eta_0 > 1, \tau_0 > 0; \ m, n \in \mathbb{Z}\}. \tag{3.81}$$

The above irregular grid includes only positive scales which is sufficient for signal reconstruction if the mother wavelet, in addition to satisfying the admissibility condition, has both positive and negative frequencies. If the mother wavelet has only positive frequencies we must also include negative scales $-\eta_0^m$ in order to be able to reconstruct the signal.

To see the time shift properties of the discretized wavelet transform on the above grid consider a time shifted function $y(t) \equiv x(t - K\tau_0)$ for which

$$
Y_{\tau_n \eta_m} = 2^{-m/2} \int\limits_{-\infty}^{\infty} \psi^* \left( \eta_0^{-m} t - \left( n - K\eta_0^{-m} \right) \tau_0 \right) x(t) \, dt
$$

$$
= X_{\tau_n - K\eta_0^{-m}\tau_0, \eta_m},
$$

(3.82)

which is not a time translation of the coefficients $X_{\tau_n \eta_m}$ unless $K$ is an integer multiple of $\eta_0^{2m}$. To achieve the time translation invariance property we use the regular grid

$$
\{ \eta_m = \eta_0^m, \quad \tau_n = n\tau_0; \; \eta_0 > 1, \tau_0 > 0; \; m, n \in \mathbb{Z} \}.
$$

(3.83)

For then

$$
Y_{\tau_n \eta_m} = 2^{-m/2} \int\limits_{-\infty}^{\infty} \psi^* \left( \eta_0^{-m} t - (n - K) \tau_0 \right) x(t) \, dt
$$

$$
= X_{(n-K)\tau_0, \eta_m},
$$

(3.84)

i.e., the discretized wavelet coefficients for a time shifted function $x(t - K\tau_0)$ are simply the shifted discretized wavelet coefficients of the original function $x(t)$. When dealing with orthogonal discrete wavelet transform (see section 5.4) and using the grid defined by $\eta_0 = 2$ and $\tau_0 = 1$, the associated transform is known as the undecimated (UDWT), the shift-invariant (SIDWT), or the redundant (RDWT) discrete wavelet transform.

The reconstruction of a function $x(t) \in L_2(\mathbb{R})$ from the sample values of the continuous wavelet transform coefficients $X_{\tau_n \eta_m}$ on the irregular grid, however, depends on whether the discretized functions

$$
\psi_{mn}(t) \equiv \psi_{\tau_n \eta_m}(t) = \eta_0^{-m/2} \psi \left( \eta_0^{-m} t - n\tau_0 \right), \quad m, n \in \mathbb{Z},
$$

(3.85)

form a frame. If an admissible mother wavelet is discretized according to (3.81) and with arbitrary values of $\eta_0$ and $\tau_0$, the resulting functions are not necessarily a frame in $L_2(\mathbb{R})$. There is, however, a converse shown in the following theorem [15].

**Theorem 53.** *If the discretized wavelet functions $\psi_{mn}(t)$ of equation (3.85), for both positive and negative values of the scale parameter $\eta_0$, form a frame then the corresponding mother wavelet $\psi(t)$ satisfies the admissibility condition (3.55) through the inequalities*

$$
0 < \frac{\tau_0 \ln \eta_0}{\pi} A \leq C_\psi \equiv \int\limits_{-\infty}^{\infty} \frac{|\Psi(\omega)|^2}{|\omega|} \, d\omega \leq \frac{\tau_0 \ln \eta_0}{\pi} B.
$$

(3.86)

*In particular, for a tight frame $A = B = \pi C_\psi / \tau_0 \ln \eta_0$.*

Daubechies has shown that so long as $\psi(t)$ satisfies good decay properties in both time and frequency then there are large ranges of values of $\eta_0$ and $\tau_0$ for which the corresponding discretized wavelets are a frame. In addition, as $\eta_0 \to 1$ and $\tau_0 \to 0$, then $|(A - B)/(A + B)| \to 0$. Thus, discretizing the continuous wavelet transform on a very dense grid, according to (3.81), will produce a nearly tight frame [15].

Given a wavelet frame $\psi_{mn}(t)$, any function $x(t) \in L_2(\mathbb{R})$ can be reconstructed from its frame coefficients $\langle \psi_{mn} \mid x \rangle$ using the dual frame functions $\tilde{\psi}_{mn}(t)$ as discussed in section 1.26 and theorem 38. Although the construction of the dual frame functions is, in general, impossible in practice (a little less so for very tight frames!), special mother wavelets that lead to biorthogonal bases exist (even for discretization values $\eta_0 = 2$ and $\tau_0 = 1$). In the latter case both the frame and the dual frame functions are found by scaling and translations of the mother wavelet and its dual. Thus, we have the transform and the reconstruction formulas

$$|x\rangle = \langle \psi_{mn} \mid x \rangle \left| \tilde{\psi}_{mn} \right\rangle , \quad \tilde{\psi}_{mn}(t) \equiv \eta_0^{-m/2} \psi \left( \eta_0^{-m} t - n\tau_0 \right).$$

We are often interested in using the irregular grid (3.81) with the particular values $\eta_0 = 2$ and $\tau_0 = 1$. The corresponding dyadic grid of discretized scales is very coarse and an effective dense grid of scales can be achieved by choosing many voices per octave, i.e., fractionally dilated versions of the same admissible mother wavelet. Anticipating the results of chapter 4 let us suppose for the moment that for some admissible mother wavelet $\psi(t)$ and the dyadic discretization parameters $\eta_0 = 2$ and $\tau_0 = 1$, the corresponding discretized functions $\psi_{mn}(t) \equiv 2^{-m/2} \psi(2^{-m}t - n)$, $m, n \in \mathbb{Z}$, happen to form a complete and orthonormal basis of $L_2(\mathbb{R})$ (the Haar wavelet is one such example). Thus, for a function $x(t)$ we assume that we have the wavelet expansion

$$x(t) = \sum_{m,n=-\infty}^{\infty} d_{mn} \psi_{mn}(t), \quad d_{mn} = \langle \psi_{mn} \mid x \rangle. \tag{3.87}$$

The discrete nature of the dilation and shifts with the given parameters together with the orthonormality requirement set rather stringent conditions on the mother wavelet, in addition to the admissibility requirement of equation (3.55). We will describe these conditions in more detail later but for now show that if an admissible mother wavelet exists whose discretized versions (with $\tau_0 = 1$ and $\eta_0 = 2$) satisfy all the requirements (as yet unspecified) for an orthonormal basis then the transform coefficients of a given function under the continuous wavelet transform, namely, $X_{\tau\eta}$, and the transform coefficients $d_{mn}$, constituting the discrete wavelet transform, are related by a simple formula. Substituting $\psi_{mn}(t)$ for $\psi_{\tau\eta}(t)$ in equation (3.45) and using $\tau = 2^m n$ and $\eta = 2^m$ we find $X_{2^m n, 2^m} = d_{mn}$. That is, given a mother wavelet $\psi(t)$, the discrete wavelet transform coefficients are obtained from the continuous wavelet transform coefficients by sampling the latter on a dyadic grid. In addition, we have

$$X_{\tau\eta} = \langle \psi_{\tau\eta} \mid x \rangle = \sum_{m,n=-\infty}^{\infty} d_{mn} \, K\left(\tau, \eta; n\tau_0, m\eta_0\right), \qquad (3.88)$$

where $K\left(\tau, \eta; n\tau_0, m\eta_0\right)$ is the CWT reproducing kernel function of equation (3.67).

## 3.12   Algorithm A' Trous

The algorithm *A' Trous* (*with holes*) refers to interpolation of a discrete time function by inserting zeros between the function samples. There are two important uses of the algorithm described in this section [43, 44].

Consider a square-integrable function $x(t)$ and its continuous wavelet transform coefficients evaluated on an irregular dyadic grid (see (3.81)) defined by $\eta_0 = 2$ and $\tau_0 = 1$,

$$X_{2^m n, 2^m} = 2^{-m/2} \int_{-\infty}^{\infty} \psi^*\left(2^{-m}t - n\right)x\left(t\right)dt, \qquad (3.89)$$

which can be approximated by the discrete values $t = kT_0$, $k \in \mathbb{Z}$,

$$X_{2^m n, 2^m} \approx T_0 2^{-m/2} \sum_{k=-\infty}^{\infty} x(kT_0)\psi^*(2^{-m}kT_0 - n). \qquad (3.90)$$

If we were using the Morlet wavelet we could evaluate and use samples of the wavelet function on any grid because the mother wavelet is given in terms of known functions. In the general case when an analytic formula for the mother wavelet is not available and setting $T_0 = 1$ (without loss of generality) it is evident from the above equations that, although the data values are only needed at the integers, the wavelet function $\psi(t)$ values are needed at $2^{-m}k$, $k \in \mathbb{Z}$. The required values can be computed by interpolation of the wavelet function values at integers $k$ using a discrete time interpolation function $f[n]$,

$$\psi_{\text{int}}(2^{-m}n) = \sum_{j=-\infty}^{\infty} f[n - 2j]\,\psi(2^{-(m-1)}j), \quad m = 1, 2, \dots. \qquad (3.91)$$

Clearly when $n = 2k$ all but one term on the right hand side must vanish and so we have the A' Trous condition

$$f[2k - 2j] = \delta_{jk} \Leftrightarrow f[2k] = \delta_{0k}, \qquad (3.92)$$

Thus, the values of the interpolated function at the integers $\psi_{\text{int}}(n)$ are exactly the same as $\psi(n)$ and so the effect of the filter $f[n]$ is to first insert zeros between the given samples (the A' Trous condition) and then fill in the zeros ("holes") by interpolation. For instance, when $m = 1$, and starting with the initial wavelet function values at the integers $\psi(j)$, zeros are inserted at half integers to be filled in by interpolation. This is shown in figure 3.15 where the "holes" at half integers and indicated by crosses will be filled in by interpolation of the values at the integers. Finally, the algorithm proceeds

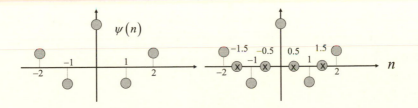

Figure 3.15: The A' Trous filter for $m = 1$.

to use the interpolated function $\psi_{\text{int}}$ in equation (3.90), namely,

$$
\begin{aligned}
X_{2^m n, 2^m} &\approx 2^{-m/2} \sum_{k=-\infty}^{\infty} x(k)\,\psi_{\text{int}}^*(2^{-m}k - n) \\
&= 2^{-m/2} \sum_{k=-\infty}^{\infty} x(k) \sum_{j=-\infty}^{\infty} f[k - 2j - 2^m n]\,\psi^*\left(2^{-(m-1)}j\right) \qquad (3.93) \\
&= 2^{-m/2} \sum_{j=-\infty}^{\infty} \psi^*\left(2^{-(m-1)}j\right) \sum_{k=-\infty}^{\infty} f[k - 2j - 2^m n]\,x(k).
\end{aligned}
$$

For instance, when $m = 1$ we have

$$X_{2n,2} \approx 2^{-1/2} \sum_{k=-\infty}^{\infty} \psi^*(k - n) \sum_{j=-\infty}^{\infty} f[j - 2k]\,x(j). \qquad (3.94)$$

The sum over $j$ on the right hand side is a discrete correlation between $f[j]$ and $x(j)$ (equivalently, it is a convolution between $f[-j]$ and $x(j)$) followed by taking the even numbered samples (down-sampling by a factor of 2 as shown in equation (2.4)). To find a recursive method to compute the transform coefficients we look at the case $m = 2$. Using (3.91) we find

$$
\begin{aligned}
X_{2^2 n, 2^2} &\approx 2^{-1} \sum_{k=-\infty}^{\infty} x(k)\, \psi_{\text{int}}^* \left(2^{-2}k - n\right) \\
&= 2^{-1} \sum_{k=-\infty}^{\infty} x(k) \sum_{j=-\infty}^{\infty} f[k - 2j - 4n]\, \psi^*(j/2) \qquad (3.95) \\
&= 2^{-1} \sum_{j=-\infty}^{\infty} \psi^*(j - n) \sum_{l=-\infty}^{\infty} f[l - 2j] \sum_{k=-\infty}^{\infty} f[k - 2l]\, x(k).
\end{aligned}
$$

Defining the quantities $r_j$, $j = 0, 1, 2, \ldots$,

$$
r_0(l) \equiv \sum_{k=-\infty}^{\infty} f[k - 2l]\, x(k), \quad r_j(l) = \sum_{k=-\infty}^{\infty} f[k - 2l]\, r_{j-1}(k), j \geq 1, \quad (3.96)
$$

we arrive at the result

$$
X_{2^m n, 2^m} \approx 2^{-m/2} \sum_{k=-\infty}^{\infty} \psi^*(k - n) r_{m-1}(k), \quad m \geq 1. \qquad (3.97)
$$

Thus, given the samples of a mother wavelet at the integers $\psi(n)$ and an A' Trous interpolation filter $f[n]$, the above two equations can be used to compute approximate values of the CWT on the irregular grid $\tau = 2^m n, \eta = 2^m$, $m \geq 1$.

An important class of A' Trous interpolation filters is the Lagrange interpolator of a given degree, say, $2N - 1$. The filter is defined so as to produce exact results for polynomials of degree $\leq 2N - 1$, i.e., if $p_{2N-1}(t)$ is such a polynomial, then

$$
p_{2N-1}(n/2) = \sum_{k=-\infty}^{\infty} f_L[n - 2k]\, p_{2N-1}(n), \qquad (3.98)
$$

where $f_L[n]$ denotes the Lagrange A' Trous filter of order $2N - 1$. The following theorem establishes the form of the Lagrange A' Trous filters of finite length [44].

**Theorem 54.** *Let $f_L[n]$, $|n| \leq 2N - 1$, be a finite length A' Trous filter satisfying the A' Trous condition (3.92). Furthermore, assume that the filter coefficients at odd indices are given by*

$$
f_L[2k - 1] = L_{1-k}^{2N-1}(0.5), \quad f_L[1 - 2k] = L_k^{2N-1}(0.5), \quad 1 \leq k \leq N, \qquad (3.99)
$$

where $L_k^{2N-1}(t)$ are Lagrange polynomials of degree $2N-1$ given by

$$L_k^{2N-1}(t) \equiv \frac{\prod\limits_{j \neq k} (t - j)}{\prod\limits_{j \neq k} (k - j)}, \quad j, k = -N + 1, \dots, N. \qquad (3.100)$$

Then $f_L[n]$ is the unique real and symmetric Lagrange A' Trous filter satisfying (3.97) whose support is limited to the interval $[-2N + 1, 2N - 1]$.

The Lagrange A' Trous filter of length $2N - 1$ is closely connected with the Daubechies orthonormal maximally flat filters of even length $N$ that are discussed in chapters 5 and 6. The relationship is the result of the following theorem [44].

**Theorem 55.** *Let $f_L[n]$, $|n| \leq 2N - 1$, be the Lagrange A' Trous filter of theorem 54. Let $h_0^{(N)}$ denote the orthogonal maximally flat Daubechies low-pass filter of even length $N$ (e.g., $N = 2, 4, 6, \dots$). Then,*

$$f_L[n] = \sum_{m=-\infty}^{\infty} h_0^{(N)}[m] h_0^{(N)}[n - m], \quad 0 \leq |n| \leq 2N - 1. \qquad (3.101)$$

The second application of the A' Trous algorithm is to compute approximations to the continuous wavelet transform coefficients on a semiregular dyadic grid specified by $\tau = n$ and $\eta = 2^m$ (see equation (3.83)). If the underlying wavelet system is orthogonal the resulting coefficients constitute a shift-invariant discrete wavelet transform, also known as the undecimated discrete wavelet transform, or the redundant discrete wavelet transform. Its implementation, generally similar to the orthogonal discrete wavelet transform (DWT), has an important difference: whereas in the latter the data is decimated (down-sampled) by a factor of 2 at each stage of the transform, in the former the filters are up-sampled by a factor of 2 at each stage. The up-sampling is performed A' Trous, i.e., as in equation (3.92) by inserting zeros. The shift-invariant UDWT is described in section 5.4.

## 3.13 The Morlet Scalogram

Here we present a simple way to compute approximate continuous wavelet transform coefficients for discrete signals of finite length [41]. Since the Morlet wavelet is defined by an analytic equation, its sample values at any time can be computed exactly without the need to use an interpolating A'

Trous filter. In equation (3.71) we discretize the variables $t$ and $\tau$ by choosing $t_k = k\ \Delta T$ and $\tau_m = m\ \Delta T$, where $\Delta T = f_s^{-1}$ and $f_s$ is the sampling frequency of the discrete data, and find the following approximation to the Morlet CWT coefficients

$$X_{m\Delta T, \eta} = \frac{a}{f_s \sqrt{|\eta|}} \sum_{k=-\infty}^{\infty} \left( e^{i\omega_0 (m-k)/\eta f_s} - e^{-\omega_0^2 \sigma^2 / 2} \right) e^{-(m-k)^2 / 2\sigma^2 \eta^2 f_s^2} x\,[k] \ ,$$

(3.102)

where the data samples $x\,[k\Delta T]$ are denoted by $x\,[k]$. For a fixed scale $\eta$ the right hand side is a convolution and can be implemented either in the time or the frequency domain.

We discretize the scale on a dyadic grid $\eta_n = 2^{n/L}$ where $L$ denotes the number of voices in each octave, $0 \leq n \leq KL-1$, and $K$ denotes the number of octave bands. The initial Morlet mother wavelet is computed so as to be centered at a given frequency $f_c$ by choosing $\omega_0 = 2\pi f_c$. The period of each oscillation $P_c$ and the number of samples for each oscillation $N_s$ determine the sampling frequency of the mother wavelet to be $N_s/P_c$. The effective time-width of the wavelet (at half the maximum) can be chosen as one-quarter of the total time for the waveform which will fix $\sigma$ at approximately one-eighth of the total time-span of the oscillations. The Morlet CWT coefficients (3.72) are proportional to the square root of the corresponding scale and so to preserve amplitudes of sinusoids at different scales in a scalogram the coefficients should be divided by $|\eta|^{1/2}$ before being displayed.

Figure 3.16 shows the IDL widget XMorlet[6] (used with the bowhead whale sound shown earlier in figure 3.1) used on a bat chirp. The Morlet mother wavelet was centered at 0.5 Hz (sampling frequency is 1 Hz) with 32 full cycles. Four chirped components of the signal are observable within octaves 3, 4, and 5. The horizontal axis in the scalogram shows all the octave bands with corresponding frequencies increasing to the left. Figure 3.17 shows the same bat chirp result in the time-frequency plane with frequencies ranging from 15.5 to 64 mHz.

---

[6]Xmorlet is a tool for the design of appropriate Morlet wavelets and their application to study scalograms of arbitrary signals [45]. The program and a users' guide are available from the author najmi@jhuapl.edu.

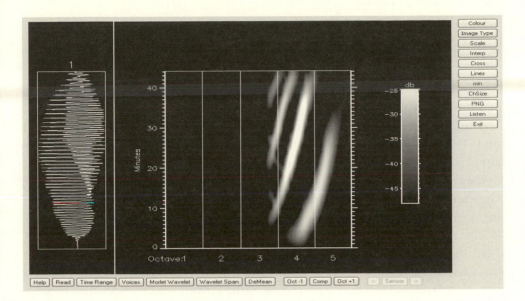

Figure 3.16: XMorlet: an IDL widget to design Morlet wavelets and compute CWT coefficients of arbitrary signals.

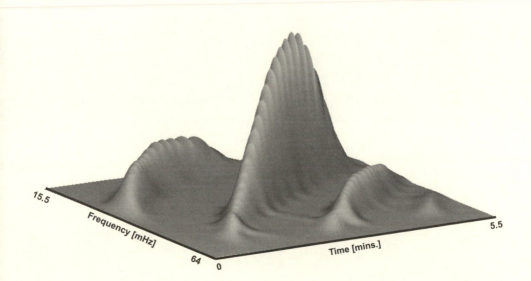

Figure 3.17: CWT coefficients of bat chirp data on the time-frequency plane.

## 3.14    Exercises

**1**   For a Gaussian window function $e^{-t^2/4\sigma^2}$ and using the definitions in equation (3.41), show that $\sigma_f \sigma_t = (4\pi)^{-1}$.

**2**   Reproduce results similar to figure 3.2 using the following data. We assume the sampling frequency is $f_s = 1$ Hz. Define $f_0 = 0.05$ and $f_1 = 0.15$ Hz, and construct two real sinusoids by $s_j[n] = \sin[2\pi f_j n]$ for $j = 0, 1$ and $0 \leq n \leq 250$. Construct the real linearly chirped signal by $c[n] = \sin[2\pi n (f_{\min} + 0.5\beta n)]$ where the chirp rate is $\beta = (f_{\max} - f_{\min})/N_{\max}$ defined for $f_{\min} = 0.2$, $f_{\max} = 0.3$, $0 \leq n \leq 1023$, and $N_{\max} = 1023$. Construct the final signal by adding the two sinusoids to two different starting locations on the chirp, e.g., $y[n] = c[n]$, $n = 0, 99$ or $n = 751, 1023$, $y[n] = c[n] + s_0[n]$, $n = 100, 350$, and $y[n] = c[n] + s_1[n]$, $n = 500, 750$. Now divide the data into 128-point sections, compute a 128-point FFT on each section (you can use a window function for each section), and display the magnitude squared of the results in db (transformation to the db scale is performed by $10 \times \log_{10} |S|^2$).

**3**   Consider the Haar wavelet and the discretization $\tau_0 = 1$ and $\eta_0 = 2$. As we see in chapter 4, the associated discrete wavelet functions are an orthonormal basis in $L_2(\mathbb{R})$. Thus, they form a tight frame with $A = B = 1$. Show that

$$\int_0^\infty \frac{|\Psi(\omega)|^2}{\omega} d\omega = \int_{-\infty}^0 \frac{|\Psi(\omega)|^2}{\omega} d\omega = \frac{\ln 2}{2\pi}$$

where $\Psi(\omega)$ is the Fourier transform of the Haar mother wavelet. Check that this is consistent with the result of theorem 53.

**4**   Prove the final equality of equation (3.95). Calculate $X_{2^3 n, 2^3}$ and show that it is consistent with the general result (3.97).

**5**   Use the method outlined in section 3.13 to compute the approximate wavelet transform coefficients for a linearly chirped signal (see problem 2 for the description of the signal).

# CHAPTER 4

# The Haar and Shannon Wavelets

## 4.1 Introduction

In chapter 3 we found that in order to produce a wavelet frame the discretization parameters $\eta_0$ and $\tau_0$ must be close to 1 and 0, respectively. Even then, the actual computation of the coefficients is cumbersome. The corresponding discrete wavelet frames are, of course, necessarily highly redundant. We saw in section 3.12 a method to compute approximations to CWT coefficients on an irregular dyadic grid. The Mallat approach [13, 46] is computationally superior and is based on orthogonal and biorthogonal bases. This approach, known as multi-resolution analysis (MRA) reduces the problem of computing wavelet coefficients on a dyadic grid to a series of orthogonal projections which can be implemented using finite impulse response (FIR) filters followed by taking even numbered samples (down-sampling by a factor of 2) without ever having to compute or use a scaling function or a mother wavelet for the entire operation. The key to this approach is the idea of successive orthogonal projections into a set of multi-resolution approximation subspaces defined in section 1.15 through equations (1.29)–(1.32). We illustrate the concept in this chapter for the Haar and the Shannon wavelets.

The Haar wavelet of equation (3.77) has the following important property [42]: when discretized on a dyadic grid defined by $\eta_0 = 2$, and $\tau_0 = 1$, it produces a complete and orthonormal basis for $L_2(\mathbb{R})$. Figure 4.1 shows four example functions $\psi(t)$, $\psi(2t)$, $\psi(t/2)$, and $\psi(2t-1)$. The inner product of any two of these examples is zero since either the two functions have no overlap or the complete cycle of one is contained within a half-cycle of the other. Orthogonality across scales and shifts follows similar arguments.

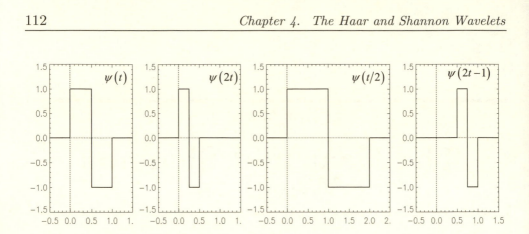

Figure 4.1: Four examples of the Haar wavelet functions.

## 4.2   Haar Multi-Resolution Analysis Subspaces

In this section we show how to arrive at the discretized wavelet functions on a dyadic grid through orthogonal projections onto the Haar multi-resolution analysis subspaces defined in section 1.15. The MRA is the set of subspaces $\mathcal{V}_m$, consisting of functions that are constant on intervals of length $2^m$ defined by $[2^m n, 2^m (n+1)]$, $m, n \in \mathbb{Z}$. We will now proceed to construct an orthonormal basis for each of these subspaces. Consider the functions

$$\phi_{mn}(t) = 2^{-m/2}, \quad 2^m n \le t \le 2^m (n+1), \quad m, n \in \mathbb{Z}, \tag{4.1}$$

whose value is 0 for $t < 2^m n$ or $t > 2^m (n+1)$, as depicted in figure 4.2. For

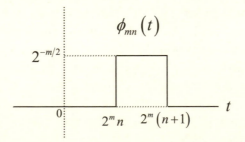

Figure 4.2: The Haar function $\phi_{mn}(t)$.

a fixed integer $m$ these functions are orthonormal (those with different values of the integer shift $n$ are nonzero on disjoint intervals and hence orthogonal), i.e.,

$$\int_{-\infty}^{\infty} \phi_{mn}(t)\,\phi_{mk}(t)\,dt = \delta_{nk}. \tag{4.2}$$

and they span the subspace $\mathscr{V}_m$. For instance, consider a function $x_0(t)$ that is constant on all intervals of unit length $[n, n+1]$, $n \in \mathbb{Z}$, and that has finite norm. Using the functions $\phi_{0n}(t)$ (which by their definition (4.1) are equal to 1 on all intervals of unit length $[n, n+1]$, and span the subspace $\mathscr{V}_0$) we may write

$$x_0(t) = \sum_{n=-\infty}^{\infty} c_{0n}\,\phi_{0n}(t), \tag{4.3}$$

where

$$\|\underline{c}_0\|^2 = \sum_{n=-\infty}^{\infty} |c_{0n}|^2 = \|x_0\|^2 < \infty. \tag{4.4}$$

Now consider the subspace $\mathscr{V}_1$, the space of all functions in $L_2(\mathbb{R})$ that are constant on intervals $[2n, 2n+2]$, $n \in \mathbb{Z}$. These intervals are of length 2 and so any function in this subspace can be written as

$$x_1(t) = \sum_{n=-\infty}^{\infty} c_{1n}\,\phi_{1n}(t). \tag{4.5}$$

Similarly consider the subspace $\mathscr{V}_{-1}$, consisting of all functions in $L_2(\mathbb{R})$ that are constant on intervals $[n, n+1]/2$ of length $1/2$. Clearly $\mathscr{V}_1 \subset \mathscr{V}_0 \subset \mathscr{V}_{-1}$ and

$$x_{-1}(t) = \sum_{n=-\infty}^{\infty} c_{-1,n}\,\phi_{-1,n}(t). \tag{4.6}$$

We note that the orthonormal basis functions of $\mathscr{V}_0$, i.e., $\phi_{0n}(t)$, are simply shifted versions of the Haar scaling function defined by

$$\phi(t) = \begin{cases} 1, & 0 \le t \le 1, \\ 0, & \text{otherwise.} \end{cases} \tag{4.7}$$

Thus $\phi_{0n}(t) = \phi(t - n)$, and the basis functions clearly satisfy the orthonormality relation (4.2) for $m = 0$. $\phi(t)$ is named a scaling function because it satisfies a scaling equation, namely,

$$\phi(t) = \phi(2t) + \phi(2t - 1), \tag{4.8}$$

as is evident in figure 4.3. Equation (4.8) is a special case of a general scaling

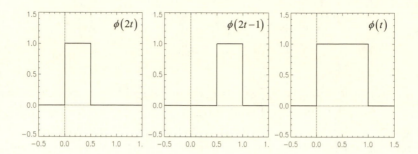

Figure 4.3: The Haar scaling function.

equation that includes an infinite number of integer time shifts [14], namely,

$$\phi(t) = \sqrt{2} \sum_{n=-\infty}^{\infty} h_0[n] \, \phi(2t - n), \qquad (4.9)$$

where the coefficients $h_0[n]$ are known as the low-pass filter coefficients. The Haar scaling equation has only two terms with the corresponding coefficients given by

$$h_0[n] = 2^{-1/2} (\delta_{0n} + \delta_{1n}). \qquad (4.10)$$

We identify a given resolution with the integer $m$ whose associated subspace $\mathscr{V}_m$ is spanned by orthonormal functions $\phi_{mn}(t)$. Using definitions 47 and 48 for the translation and dilation operators we have

$$\phi_{mn}(t) = 2^{-m/2} \phi(2^{-m}t - n) \equiv \mathscr{D}_{2^m} \mathscr{T}_n \phi(t). \qquad (4.11)$$

A simple transformation of the independent variable in the scaling equation (4.8) can be used to write the following scaling equation satisfied by $\phi_{mn}(t)$ (the original scaling equation (4.8) is a special case ($m = 0, n = 0$))

$$\phi_{mn}(t) = \sum_{k=-\infty}^{\infty} h_0[k] \, \phi_{m-1,2n+k}(t). \qquad (4.12)$$

For any function $x(t) \in L_2(\mathbb{R})$, an expansion of the form

$$x_m(t) \equiv \sum_{n=-\infty}^{\infty} c_{mn} \, \phi_{mn}(t) \qquad (4.13)$$

is an approximation to the original function $x(t)$ at resolution (scale) $m$ and the coefficients $c_{mn}$ are found using the orthonormality relations (4.2)

$$c_{mn} = \langle \phi_{mn}, x \rangle = 2^{-m/2} \int_{2^m n}^{2^m (n+1)} x(t) \, dt. \qquad (4.14)$$

This approximation can be viewed as an orthogonal projection of the original function onto the subspace spanned by the basis functions $\phi_{mn}(t)$, $m, n \in \mathbb{Z}$. We define the projection by $\mathbf{P}_m : L_2(\mathbb{R}) \to \mathscr{V}_m$ where $\mathbf{P}_m x \equiv x_m(t)$ as shown in equation (4.13); $\mathscr{V}_m$ is the approximation subspace and the projection is orthogonal. To see that $\mathbf{P}_m$ is an orthogonal projection we need only show that the difference between the function and the approximation to it is orthogonal to any of the basis functions that span $\mathscr{V}_m$, i.e., we must have

$$\left\langle \phi_{mk}, x - \sum_{n=-\infty}^{\infty} c_{mn} \phi_{mn} \right\rangle = 0, \qquad (4.15)$$

which is equivalent to

$$\langle \phi_{mk}, x \rangle = \sum_{n=-\infty}^{\infty} c_{mn} \langle \phi_{mk}, \phi_{mn} \rangle. \qquad (4.16)$$

Using the orthonormality relation $\langle \phi_{mk}, \phi_{mn} \rangle = \delta_{kn}$, equation (4.16) is the same as $c_{mn} = \langle \phi_{mn}, x \rangle$. But the latter is precisely expressed in equation (4.14) and so the projection is orthogonal.

As an example consider the function $x(t) = t \exp(-t^2)$ defined for $t \geq 0$, and 0 for $t < 0$. Figure 4.4 shows the approximations to this function performed at two resolutions $m = -1$ and $m = -3$. The spaces $\mathscr{V}_m$ considered here are examples of the multi-resolution approximation (analysis) spaces described in section 1.15.

The approximation error (detail) at scale $m$ is in the orthogonal complement of $\mathscr{V}_m$ in $\mathscr{V}_{m-1}$ denoted by $\mathscr{V}_m^\perp$. Thus we have

$$\mathscr{V}_{m-1} = \mathscr{V}_m \oplus \mathscr{V}_m^\perp. \qquad (4.17)$$

The space $\mathscr{V}_m^\perp$ is characterized by the error in the approximation at resolution (scale) $m$, i.e., the difference between the projections of the given functions onto the subspaces $\mathscr{V}_{m-1}$ and $\mathscr{V}_m$,

$$\mathbf{E}_m x \equiv \mathbf{P}_{m-1} x - \mathbf{P}_m x. \qquad (4.18)$$

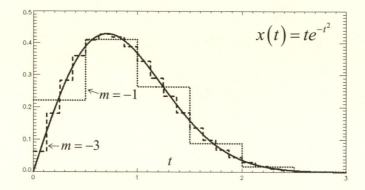

Figure 4.4: The Haar scaling function approximations to $t \exp(-t^2)$, $t \geq 0$ at levels $m = -1$ and $m = -3$.

This can be rewritten as

$$\mathbf{E}_m x \equiv \sum_{n=-\infty}^{\infty} c_{m-1,n} \phi_{m-1,n}(t) - \sum_{n=-\infty}^{\infty} c_{mn} \phi_{mn}(t). \qquad (4.19)$$

The error subspace $\mathscr{V}_m^{\perp}$ is spanned by the discretized Haar wavelet functions on a dyadic grid. To show this we first use equation (4.11) to write

$$\phi_{m-1,2n}(t) = 2^{-(m-1)/2}, \quad 2^{-m}n \leq t \leq 2^{-m}, \qquad (4.20)$$

and

$$\phi_{m-1,2n+1}(t) = 2^{-(m-1)/2} \quad 2^{-m}(n+1/2) \leq t \leq 2^{-m}(n+1), \qquad (4.21)$$

where the functions are zero outside the indicated ranges. Hence,

$$\phi_{m-1,2n}(t) + \phi_{m-1,2n+1}(t) = 2^{-(m-1)/2}, \quad 2^{-m}n \leq t \leq 2^{-m}(n+1). \qquad (4.22)$$

Comparing this with equation (4.11) we conclude that

$$\phi_{m-1,2n}(t) + \phi_{m-1,2n+1}(t) = \sqrt{2}\phi_{mn}(t). \qquad (4.23)$$

The first sum in equation (4.19) can be broken into even indices $2n$ and odd indices $2n+1$, to obtain

$$\sum_{n=-\infty}^{\infty} c_{m-1,n} \phi_{m-1,n}(t) = \sum_{n=-\infty}^{\infty} c_{m-1,2n} \phi_{m-1,2n}(t)$$

$$+ \sum_{n=-\infty}^{\infty} c_{m-1,2n+1} \phi_{m-1,2n+1}(t). \qquad (4.24)$$

As for the expansion coefficients $c_{mn}$, using equations (4.13) and (4.23) we find

$$
\begin{aligned}
c_{mn} = \langle \phi_{mn}, x \rangle &= 2^{-1/2} \langle \phi_{m-1,2n} + \phi_{m-1,2n+1}, x \rangle \\
&= 2^{-1/2} \left( c_{m-1,2n} + c_{m-1,2n+1} \right).
\end{aligned}
\tag{4.25}
$$

Using equations (4.23) and (4.24), the projection at level $m$ (the approximation at resolution $m$), namely, $\mathbf{P}_m x$, can now be written as

$$
\begin{aligned}
\mathbf{P}_m x &\equiv \sum_{n=-\infty}^{\infty} c_{mn}\, \phi_{mn}(t) \\
&= 2^{-1} \sum_{n=-\infty}^{\infty} \left( c_{m-1,2n} + c_{m-1,2n+1} \right) \left[ \phi_{m-1,2n}(t) + \phi_{m-1,2n+1}(t) \right].
\end{aligned}
\tag{4.26}
$$

Finally we compute the approximation error (between resolution levels $m-1$ and $m$) defined in equation (4.18). Using equation (4.24) for $\mathbf{P}_{m-1} x$ and (4.26) for $\mathbf{P}_m x$ we find

$$
\begin{aligned}
\mathbf{E}_m x &= \sum_{n=-\infty}^{\infty} c_{m-1,2n}\phi_{m-1,2n}(t) + \sum_{n=-\infty}^{\infty} c_{m-1,2n+1}\phi_{m-1,2n+1}(t) - \\
&\quad - \frac{1}{2} \sum_{n=-\infty}^{\infty} \left( c_{m-1,2n} + c_{m-1,2n+1} \right) \left[ \phi_{m-1,2n}(t) + \phi_{m-1,2n+1}(t) \right] = \\
&\quad \sum_{n=-\infty}^{\infty} \phi_{m-1,2n}(t) \left\{ c_{m-1,2n} - \tfrac{1}{2} \left( c_{m-1,2n} + c_{m-1,2n+1} \right) \right\} + \\
&\quad \sum_{n=-\infty}^{\infty} \phi_{m-1,2n+1}(t) \left\{ c_{m-1,2n+1} - \tfrac{1}{2} \left( c_{m-1,2n} + c_{m-1,2n+1} \right) \right\} = \\
&\quad \frac{1}{2} \sum_{n=-\infty}^{\infty} \left\{ \phi_{m-1,2n}(t) - \phi_{m-1,2n+1}(t) \right\} \left\{ c_{m-1,2n} - c_{m-1,2n+1} \right\}.
\end{aligned}
\tag{4.27}
$$

The last equality of equation (4.27) can be rewritten in terms of two newly defined quantities: the Haar wavelet functions $\psi_{mn}(t)$,

$$
\psi_{mn}(t) \equiv 2^{-1/2} \left\{ \phi_{m-1,2n}(t) - \phi_{m-1,2n+1}(t) \right\},
\tag{4.28}
$$

and the wavelet coefficients $d_{mn}$,

$$
d_{mn} \equiv 2^{-1/2} \left\{ c_{m-1,2n} - c_{m-1,2n+1} \right\}.
\tag{4.29}
$$

Equation (4.27) then becomes

$$
\mathbf{E}_m x = \sum_{n=-\infty}^{\infty} d_{mn}\psi_{mn}(t),
\tag{4.30}
$$

which is valid for any function $x \in L_2(\mathbb{R})$. Since $\mathbf{E}_m x \in \mathscr{V}_m^\perp$ (orthogonal complement of $\mathscr{V}_m$ in $\mathscr{V}_{m-1}$) and $\mathscr{V}_{m-1} = \mathscr{V}_m \oplus \mathscr{V}_m^\perp$, we conclude that $\mathscr{V}_m^\perp$ is spanned by $\psi_{mn}(t)$. Consider the wavelet function

$$\psi_{00}(t) = 2^{-1/2}\left\{\phi_{-1,0}(t) - \phi_{-1,1}(t)\right\} \tag{4.31}$$

obtained from equation (4.28) and shown in figure 4.1. All the wavelet basis functions $\psi_{mn}(t)$ are found from dilations and time shifts (in that order) of the single function $\psi_{00}(t)$ which we denote by $\psi(t)$: the mother wavelet function. Thus,

$$\psi_{mn}(t) = 2^{-m/2}\psi\left(2^{-m}t - n\right), \tag{4.32}$$

are precisely the discretized Haar wavelet functions on a dyadic grid.

While the Haar scaling function was characterized by the low-pass filter coefficients $h_0[n]$ in equation (4.9), figures 4.1 and 4.3 show that a similar equation, the wavelet equation, holds for the Haar wavelet function

$$\psi(t) = \phi(2t) - \phi(2t - 1). \tag{4.33}$$

This is a special form (series with two terms) of the more general wavelet equation [15, 14]

$$\psi(t) = \sqrt{2}\sum_{n=-\infty}^{\infty} h_1[n]\phi(2t - n). \tag{4.34}$$

In the present example, the Haar high-pass filter coefficients $h_1[n]$ have only two nonzero values and are given by

$$h_1[n] = 2^{-1/2}(\delta_{0n} - \delta_{1n}). \tag{4.35}$$

The definition of the Haar wavelets in terms of the Haar scaling function ensures that the wavelets form an orthonormal set at all scales, as well as being orthogonal to the scaling functions at all scales:

$$\int_{-\infty}^{\infty} \psi_{mk}(t)\psi_{ln}(t)dt = \delta_{ml}\delta_{kl}, \quad \int_{-\infty}^{\infty} \psi_{mk}(t)\phi_{ln}(t)dt = 0. \tag{4.36}$$

Equations (4.18) and (4.19), when rewritten in the form

$$\mathbf{P}_{m-1}x = \sum_{n=-\infty}^{\infty} c_{mn}\phi_{mn}(t) + \sum_{n=-\infty}^{\infty} d_{mn}\psi_{mn}(t), \tag{4.37}$$

have a simple interpretation. The approximation $\mathbf{P}_{m-1}x$ to a function $x(t) \in L_2(\mathbb{R})$ that is obtained by projecting it orthogonally onto the subspace $\mathscr{V}_{m-1}$ of a multi-resolution analysis space is the sum of two terms: one is the approximation to the function obtained by orthogonal projection onto the next coarser scale $\mathscr{V}_m$, represented by the term $\mathbf{P}_m x = \sum\limits_{n=-\infty}^{\infty} c_{mn}\phi_{mn}(t)$, and the other is the difference between the two approximations represented by the term $\mathbf{E}_m x = \sum\limits_{n=-\infty}^{\infty} d_{mn}\psi_{mn}(t)$. The latter quantity is the detail that is lost in going from the finer scale $m-1$ to the coarser scale $m$.

Equation (4.18) and its equivalent form $\mathbf{P}_{m-1}x = \mathbf{P}_m x + \mathbf{E}_m x$, when recursively continued give the result

$$\mathbf{P}_{m-1}x = \sum_{k=m}^{\infty} \mathbf{E}_k x = \sum_{k=m}^{\infty}\sum_{n=-\infty}^{\infty} d_{kn}\psi_{kn}. \tag{4.38}$$

Thus, each approximation subspace $\mathscr{V}_m$ is the direct sum of all the detail subspaces at lower resolutions (scales),

$$\mathscr{V}_{m-1} = \mathscr{V}_m^{\perp} \oplus \mathscr{V}_{m+1}^{\perp} \oplus \cdots = \overset{\infty}{\underset{k=m}{\oplus}} \mathscr{V}_k^{\perp}. \tag{4.39}$$

In addition, taking the limit as $m \to -\infty$ and using $\lim_{m\to-\infty}\mathbf{P}_m x(t) = x(t)$, we have the wavelet expansion equation, or the wavelet representation,

$$x(t) = \sum_{m=-\infty}^{\infty}\sum_{n=-\infty}^{\infty} d_{mn}\psi_{mn}(t), \tag{4.40}$$

where the wavelet transform coefficients $d_{mn}$ are defined in equation (4.29). The wavelet representation is equivalent to the multi-resolution decomposition formula [14]

$$L_2(\mathbb{R}) = \overset{\infty}{\underset{m=-\infty}{\oplus}} \mathscr{V}_m^{\perp}. \tag{4.41}$$

## 4.3   Summary and Generalization of Results

It is clear from the foregoing that the subspaces $\mathscr{V}_m$, $m \in \mathbb{Z}$ form a nested and complete family of subspaces as defined in section 1.15, i.e.,

$$\mathscr{V}_m \subset \mathscr{V}_{m-1},\ m \in \mathbb{Z},\quad \bigcup_{m\in\mathbb{Z}} \mathscr{V}_m = L_2(\mathbb{R}),\quad \bigcap_{m\in\mathbb{Z}} \mathscr{V}_m = \{0\}. \tag{4.42}$$

This family of subspaces satisfies the multi-resolution property, definition 38. The Haar scaling function $\phi(t)$ satisfies a scaling equation for a particular set of low-pass filter coefficients $h_0[n]$ with only two nonzero values, which can be written in the general form

$$\phi(t) = \sqrt{2} \sum_{n=-\infty}^{\infty} h_0[n] \phi(2t - n). \tag{4.43}$$

Each subspace $\mathscr{V}_m$ is spanned by orthonormal basis functions

$$\phi_{mn}(t) = 2^{-m/2} \phi\left(2^{-m}t - n\right) \equiv \mathscr{D}_{2^m} \mathscr{T}_n \phi(t), \quad m, n \in \mathbb{Z}, \tag{4.44}$$

and the orthogonal complement of $\mathscr{V}_m$ inside $\mathscr{V}_{m-1}$, i.e., $\mathscr{V}_m^\perp$, is spanned by an orthonormal basis of Haar wavelet functions that are all constructed from a single mother wavelet $\psi(t)$,

$$\psi_{mn}(t) = 2^{-m/2} \psi\left(2^{-m}t - n\right) \equiv \mathscr{D}_{2^m} \mathscr{T}_n \psi(t), \quad m, n \in \mathbb{Z}. \tag{4.45}$$

The mother wavelet function $\psi(t)$ satisfies the wavelet equation for a particular set of high-pass filter coefficients $h_1[n]$ with only two nonzero values, which can be written in the general form

$$\psi(t) = \sqrt{2} \sum_{n=-\infty}^{\infty} h_1[n] \phi(2t - n). \tag{4.46}$$

The scaling and wavelet functions satisfy the orthonormality and orthogonality conditions

$$\int_{-\infty}^{\infty} \psi_{mk}(t)\psi_{ln}(t)dt = \delta_{ml}\delta_{kl}, \quad \int_{-\infty}^{\infty} \phi_{mk}(t)\phi_{mn}(t)dt = \delta_{kn},$$

$$\int_{-\infty}^{\infty} \psi_{mk}(t)\phi_{ln}(t)dt = 0, \tag{4.47}$$

which will provide constraints on the low-pass filter $h_0[n]$ and the high-pass filter $h_1[n]$. The scaling and the wavelet equations can be written at all scales in the following general forms (cf. (4.43) and (4.46)):

$$\phi_{mn}(t) = \sum_{k=-\infty}^{\infty} h_0[k] \phi_{m-1,2n+k}(t), \tag{4.48}$$

and

$$\psi_{mn}(t) = \sum_{k=-\infty}^{\infty} h_1[k]\,\phi_{m-1,2n+k}(t). \qquad (4.49)$$

The scaling coefficients and the wavelet coefficients, equations (4.25) and (4.29), can be written in terms of the low-pass and high-pass filters in the following forms

$$c_{mn} = \frac{1}{\sqrt{2}} \sum_k (\delta_{0,k-2n} + \delta_{1,k-2n})\,c_{m-1,k} = \sum_{k=-\infty}^{\infty} h_0[k-2n]\,c_{m-1,k},$$

$$(4.50)$$

$$d_{mn} = \frac{1}{\sqrt{2}} \sum_k (\delta_{0,k-2n} - \delta_{1,k-2n})\,c_{m-1,k} = \sum_{k=-\infty}^{\infty} h_1[k-2n]\,c_{m-1,k}.$$

Linear filter theory interpretations of the above equations are as follows. The scaling coefficients $c_{mn}$ at scale $m$ are obtained from the coefficients $c_{m-1,n}$ at the previous finer scale $m-1$ by correlating them with the low-pass filter coefficients $h_0[n]$ (or equivalently, convolving with the time-reversed version $h_0[-n]$) and then taking all the even numbered samples $2n$. The act of taking all even numbered samples in this context is also known as decimation or down-sampling by a factor of 2: decimating or down-sampling a discrete time sequence $x[n]$, $n \in \mathbb{Z}$, by a factor of 2 is given by $y[n] = x[2n]$ as depicted in figure 4.5. A similar interpretation exists for the wavelet

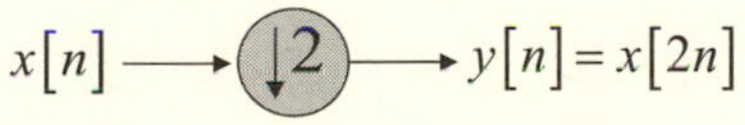

$$x[n] \longrightarrow \boxed{\downarrow 2} \longrightarrow y[n] = x[2n]$$

Figure 4.5: Down-sampling (decimation) by a factor of 2.

coefficients except that the correlation is performed with the coefficients $h_1[n]$ (or equivalently, convolution with the time-reversed version $h_1[-n]$), followed by taking only the even numbered samples (down-sampling by a factor of 2). Figure 4.6 summarizes the results expressed in equation (4.50); it turns out to be applicable to all orthogonal wavelets and not just the Haar, as we show in chapter 5. In addition, the inverse operation to figure 4.6 will be derived in chapter 5 (see figure 5.1).

$$c_{m-1,n} \rightarrow \boxed{h_0[-n]} \rightarrow \downarrow 2 \rightarrow c_{mn}$$
$$\boxed{h_1[-n]} \rightarrow \downarrow 2 \rightarrow d_{mn}$$

Figure 4.6: Discrete convolution interpretation for the scaling and wavelet coefficients.

## 4.4   The Spectra of the Haar Filter Coefficients

It is useful for later developments to look at the spectra (square magnitudes of Fourier transforms) of the filter coefficients for the Haar functions. Given the discrete sequences $h_0[n]$ and $h_1[n]$, we use equation (1.72) to define $H_0(\omega)$ and $H_1(\omega)$. Using (4.10) and (4.35) we find

$$
\begin{aligned}
H_0(\omega) &= 2^{-1/2}\left(1 + e^{-i\omega}\right) = \sqrt{2}e^{-i\omega/2}\cos(\omega/2), \\
H_1(\omega) &= 2^{-1/2}\left(1 - e^{-i\omega}\right) = \sqrt{2}ie^{-i\omega/2}\sin(\omega/2).
\end{aligned}
\tag{4.51}
$$

The spectra of these two functions, defined by the square of the magnitude of the Fourier transforms, are shown in figure 4.7. The horizontal axis is the linear (normalized) frequency $f = (2\pi)^{-1}\omega$.[1] The functions $H_0(\omega)$ and

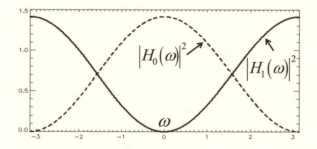

Figure 4.7: Spectra of the Haar filter coefficients.

$H_1(\omega)$ satisfy the PR-QMF equations (2.35) of section 2.7, viz.,

$$
H_1(\omega) = -e^{-i\omega}H_0^*(\omega + \pi),
\tag{4.52a}
$$

$$
|H_0(\omega)|^2 + |H_0(\omega + \pi)|^2 = 2,
\tag{4.52b}
$$

$$
|H_1(\omega)|^2 + |H_1(\omega + \pi)|^2 = 2,
\tag{4.52c}
$$

---

[1]The square of the magnitude of the discrete time Fourier transform of a real time series is a symmetric function of the frequency $\omega \in [-\pi, +\pi]$.

which define the two functions as power complementary half-band PR-QMF filters, although neither is an ideal filter.[2] The Haar filter coefficients illustrate that a finite number of filter coefficients results in nonideal filter response. The half-band low-pass and high-pass properties of the spectra of the Haar filter coefficients are, as we shall see, general properties of all orthogonal wavelet systems. The complementary half-band property of the filter coefficients applies, equivalently, to the time-reversed sequences $h_0[-n]$ and $h_1[-n]$. Since the convolution of two time sequences is equivalent to the product of their Fourier transforms in the frequency domain, figure 4.6 has the frequency domain implication shown in figure 4.8. Assuming that

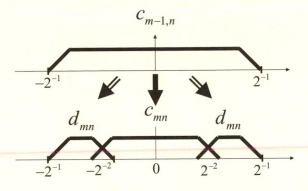

Figure 4.8: Frequency band implication of figure 4.6.

the discrete real sequence $c_{m-1,n}$, for a fixed $m$, has a spectrum contained in the normalized linear frequency interval $[-0.5, 0.5]$,[3] then the spectrum of the $c_{mn}$ will be contained in the middle half of the band, i.e., $[-0.25, 0.25]$, while the spectrum of the $d_{mn}$ will be contained in the upper half-band, i.e., $[-0.5, -0.25] \cup [0.25, 0.5]$. A further decomposition of the $c_{mn}$ coefficients will result in the sequences $c_{m+1,n}$, whose spectrum is contained in the normalized interval $[-0.125, 0.125]$, and $d_{m+1,n}$, whose spectrum is contained in the normalized interval $[-0.25, -0.125] \cup [0.125, 0.25]$, and so on.

The Fourier transforms of the scaling and wavelet functions are given by

$$\Phi(\omega) = e^{-i\omega/2}\sin(\omega/2)/(\omega/2),$$
$$\Psi(\omega) = ie^{-i\omega/2}\sin^2(\omega/4)/(\omega/4). \tag{4.53}$$

---

[2]An ideal half-band low-pass filter would have a spectrum that is exactly 1 in the (normalized) frequency range $[-0.25, +0.25]$ Hz, and zero elsewhere, whereas an ideal half-band high-pass filter's spectrum would be 1 in the interval $[-0.5, -0.25] \cup [0.25, 0.5]$ and zero elsewhere.

[3]If the sampling time for the sequence $c_{mn}$, fixed $m$, is denoted by $\Delta T$, then the sampling frequency is $f_s = 1/\Delta T$ and the normalized linear frequency is defined by $f/f_s$.

and the corresponding spectra are shown in figure 4.9. Clearly the scaling function is a low-pass function, whereas the wavelet is a high-pass function with zero DC ($\omega = 0$) component (see the admissibility condition (3.55)). The latter properties too will be true for all discrete orthogonal wavelets.

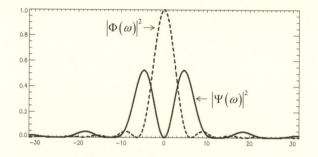

Figure 4.9: Normalized spectra of the Haar scaling and wavelet functions.

## 4.5   Half-Band Finite Impulse Response Filters

Equations (4.52b) and (4.52c) define low-pass and high-pass half-band filters. It is useful here to discuss some general properties of half-band FIR filters and then derive the Haar filter coefficients as a special case. Given a finite length $N$ sequence (FIR filter) $h_n, n = 0, \ldots, N - 1$, its spectrum $S(\omega)$, defined by equation (2.14), when substituted into the half-band equation (4.52b) gives

$$\sum_{k=-(N-1)}^{N-1} \left( s_k + (-1)^k s_k \right) e^{-ik\omega} = 2.$$

This equation can be solved for the coefficients that have even indices, namely, $s_{2l} = \delta_{0l}$, while coefficients $s_{2l+1}$ with odd indices remain undetermined. We now use the half-band property together with spectral factorization, theorem 43, to derive the Haar filter sequences. Let $N = 2$ and consider the half-band low-pass filter equation (4.52b). The associated spectrum $S_0(\omega)$ satisfies the equation

$$S_0(\omega) + S_0(\omega + \pi) = 2.$$

To find a low-pass filter with real coefficients we consider real $s_k$. The low-pass property implies that $S_0(\omega = \pi) = 0$ which together with the half-band equation leads to $S_0(\omega = 0) = 2$. These two conditions together with $s_0 = 1$

(even index solution) and the fact that $s_{-1} = s_1$ (equation (2.16) applied to real coefficients) give $s_{-1} = s_1 = 0.5$. Hence we find

$$S_0(\omega) = s_{-1}e^{i\omega} + s_0 + s_1 e^{-i\omega} = 1 + \cos(\omega)$$

and the associated polynomial factorization

$$S_0(z) = 1 + \frac{z + z^{-1}}{2} = \frac{1}{2}\left(1 + z\right)\left(1 + z^{-1}\right) \ ,$$

which result in the Haar function $H_0(z) = \left(1 + z^{-1}\right)/\sqrt{2}$ that corresponds to the same coefficients given in (4.10), viz., $h_0[0] = h_0[1] = 1/\sqrt{2}$. The half-band high-pass filter equation (4.52c) can be similarly solved to give $s_{-1} = s_1 = -0.5$ and $H_1(z) = \left(1 - z^{-1}\right)/\sqrt{2}$ that corresponds to the coefficients (4.35), $h_1[0] = -h_1[1] = 1/\sqrt{2}$. The following definition and theorem establish that the Haar low-pass filter $h_0$ of equation (4.10) is the only symmetric solution to the half-band equation (4.52a).

**Definition 49.** A finite impulse response (FIR) filter of length $N$ defined by the sequence of complex coefficients $h_n$, $n = 0, \ldots, N - 1$, is said to be conjugate symmetric if

$$h_n = h^*_{N-1-n}, \quad n = 0, \ldots, N - 1. \tag{4.54}$$

**Theorem 56.** *The conjugate symmetric real solution of the half-band filter equation (4.52a) is of length 2 and it is the Haar filter sequence $h_0$ shown in equation (4.10).*

## 4.6 The Shannon Scaling Function

The Haar multi-resolution approximation subspaces corresponded to a set of wavelet functions that are orthonormal within and across all scales but are discontinuous functions of the independent variable $t$. Band-limited square integrable functions $x_\Omega(t)$ whose Fourier transforms $X_\Omega(\omega)$ have compact support in the frequency domain have excellent frequency resolving properties, but are spread out in the $t$ variable and so have poor time resolution.

We now study the Shannon multi-resolution approximation subspaces $\mathscr{V}_m$ consisting of functions whose band is limited to the interval $2^{-m}[-\pi, +\pi]$. Thus, for an infinite band function $x(t) \in L_2(\mathbb{R})$, we define an approximation by the projection $\mathbf{P}_m x \in \mathscr{V}_m$, $\mathbf{P}_m^2 = \mathbf{P}_m$. To show that $\mathbf{P}_m$ is an orthogonal projection we use the general Parseval relation (1.60). If $x_1(t) \in \mathscr{V}_m$ and

$x_2(t)$ is in the complement of $\mathscr{V}_m$, then the Fourier transforms $X_1(\omega)$ and $X_2(\omega)$ have no overlapping region of support: $X_1(\omega)$ (by definition) is zero outside the interval $2^{-m}[-\pi, +\pi]$ while $X_2(\omega)$ is zero in this interval. We conclude that $x_1(t)$ and $x_2(t)$ are orthogonal and so $\mathbf{P}_m$ is an orthogonal projection. As in the Haar example we can now proceed to characterize the multi-resolution approximation subspaces associated with the above orthogonal projection in terms of a scaling function and a wavelet function.

The subspace $\mathscr{V}_0$ includes functions whose Fourier transforms have support in the interval $[-\pi, \pi]$,

$$\mathscr{V}_0 = \{x(t) \in L_2(\mathbb{R}) : X(\omega) = 0, \text{ when } \omega < -\pi \text{ or } \omega > \pi\}, \qquad (4.55)$$

and the function $\phi(t) = \sin(\pi t)/(\pi t)$ is clearly in this space (see chapter 1 exercise 1.5). An orthonormal basis for $\mathscr{V}_0$ is provided by the functions

$$\phi_{0n} = \frac{\sin[\pi(t-n)]}{\pi(t-n)}, \quad n \in \mathbb{Z}. \qquad (4.56)$$

Next, let $\mathscr{V}_{-1}$ denote the space of all functions in $L_2(\mathbb{R})$ whose Fourier transform is identically zero outside the interval $[-2\pi, +2\pi]$. Clearly, $\mathscr{V}_0 \subset \mathscr{V}_{-1}$ and an orthonormal basis for $\mathscr{V}_{-1}$ is given by

$$\phi_{-1,n} = \sqrt{2}\,\frac{\sin[\pi(2t-n)]}{\pi(2t-n)}, \quad n \in \mathbb{Z}. \qquad (4.57)$$

The orthogonal complement of $\mathscr{V}_0$ in $\mathscr{V}_{-1}$ is $\mathscr{V}_0^{\perp}$ and it is spanned by all functions whose band is limited to $[-2\pi, -\pi] \cup [\pi, 2\pi]$. Figure 4.10 shows the frequency bands defining the functions belonging to the three subspaces in the relation $\mathscr{V}_{-1} = \mathscr{V}_0 \oplus \mathscr{V}_0^{\perp}$. Next we apply the sampling theorem 33,

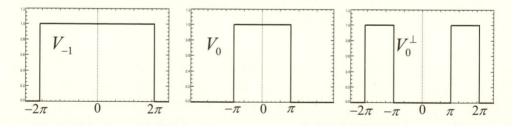

Figure 4.10: Support of the spectra for functions in $\mathscr{V}_{-1}, \mathscr{V}_0, \mathscr{V}_0^{\perp}$.

equation (1.95) with $B = 1$, to the sinc function itself,

$$\frac{\sin(\pi t)}{\pi t} = \sum_{n=-\infty}^{\infty} c_n \frac{\sin[2\pi(t - n/2)]}{2\pi(t - n/2)}, \qquad (4.58)$$

where the expansion coefficients are given by

$$c_n = \phi\left(n/2\right) = \frac{\sin\left(n\pi/2\right)}{n\pi/2}. \tag{4.59}$$

Using these coefficients in equation (4.58) we find

$$\phi\left(t\right) = \sum_{n=-\infty}^{\infty} \frac{\sin\left(n\pi/2\right)}{n\pi/2}\phi\left(2t - n\right), \tag{4.60}$$

which is a scaling equation of the same form as that shown in equation (4.43) with the filter coefficients given by

$$h_0\left[n\right] = \sqrt{2}\,\frac{\sin\left(n\pi/2\right)}{n\pi}. \tag{4.61}$$

To complete our example and the analogy to the Haar scaling and wavelet functions we must obtain the orthonormal basis for the detail (wavelet) subspace $\mathscr{V}_0^\perp$. Let $\psi\left(t\right)$ denote the mother wavelet. Then by analogy to the Haar case (see equation (4.45) an orthonormal basis of $\mathscr{V}_m^\perp$ is given by $\psi_{mn}\left(t\right)$, i.e.,

$$\mathscr{V}_m^\perp = \text{span}\left\{\psi_{mn}\left(t\right) \equiv 2^{-m/2}\psi\left(2^{-m}t - n\right), \quad m, n \in \mathbb{Z}\right\}, \tag{4.62}$$

and, in particular, an orthonormal basis for $V_0^\perp$ is provided by $\psi_{0n}\left(t\right)$.

$$\mathscr{V}_0^\perp = \text{span}\left\{\psi_{0n}\left(t\right) \equiv \psi\left(t - n\right), \quad n \in \mathbb{Z}\right\}. \tag{4.63}$$

We now proceed to find the Shannon mother wavelet. Define the Fourier transform of the mother wavelet by $\Psi\left(\omega\right)$, and the Fourier transform of the scaling function by $\Phi\left(\omega\right)$. Since $|\Phi\left(\omega\right)| = 1$ when $\omega \in \left[-\pi, +\pi\right]$, and 0 otherwise, we use the Fourier transform pair relation

$$2\phi\left(2t\right) \leftrightarrow \Phi\left(\omega/2\right), \tag{4.64}$$

to find

$$\Phi\left(\frac{\omega}{2}\right) - \Phi\left(\omega\right) = \begin{cases} 1, & -2\pi \leq \omega \leq -\pi \text{ or } \pi \leq \omega \leq 2\pi, \\ 0, & \text{otherwise.} \end{cases} \tag{4.65}$$

Let $\Gamma\left(\omega\right) \equiv \Phi\left(\frac{\omega}{2}\right) - \Phi\left(\omega\right)$. Since the functions $\Gamma\left(\omega\right)$ and $\Phi\left(\omega\right)$ have compact and nonoverlapping regions of support, we have, for any two integers $k$ and $l$,

$$\int_{-\infty}^{\infty} \Gamma^*\left(\omega\right) e^{i\omega l}\Phi\left(\omega\right) e^{-i\omega k}d\omega = 0, \tag{4.66}$$

The general Parseval relation of equation (1.60) and the two Fourier transform pairs

$$\gamma^* (t - l) \leftrightarrow \Gamma^* (\omega) e^{i\omega l}, \quad \phi (t - k) \leftrightarrow \Phi (\omega) e^{-i\omega k} \tag{4.67}$$

now imply that the corresponding time functions $\gamma^* (t - l)$ and $\phi (t - k)$ are orthogonal, and we may choose the Fourier transform of the mother wavelet $\Psi (\omega)$ to be any multiple of $\Gamma (\omega)$ to ensure that it is orthogonal to the Scaling function. The usual choice is

$$\Psi (\omega) = -e^{-i\omega/2} \left\{ \Phi (\omega/2) - \Phi (\omega) \right\}. \tag{4.68}$$

Since this differs from $\Gamma (\omega)$ by a pure phase, the orthogonality of the mother wavelet and the scaling functions is unaffected. Taking the inverse Fourier transform of equation (4.68) we find

$$\psi (t) = \phi (t - 1/2) - 2\phi (2t - 1) = \text{sinc} (t - 1/2) - 2\text{sinc} (2t - 1). \tag{4.69}$$

Figure 4.11 shows the Shannon scaling and mother wavelet functions. Using the above construction it is now easy to show that all the general orthogonality relations for the wavelet functions are satisfied. The sub-

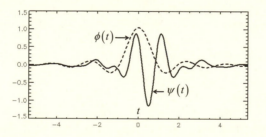

Figure 4.11: The Shannon scaling and mother wavelet functions.

space $V_m$ contains those $L_2$ functions whose band is limited to the interval $[-\pi/2^m, +\pi/2^m]$, and the subspace $V_m^\perp$ contains those $L_2$ functions whose band is limited to the interval $[-\pi/2^{m-1}, -\pi/2^m] \cup [\pi/2^m, \pi/2^{m-1}]$. The latter, for different integers $m$, do not overlap and the general Parseval relation then ensures orthogonality of the wavelet functions $\psi_{mn} (t)$ (obtained from the Shannon mother wavelet by dilation and translation) across all scales.

The Shannon wavelet function of equation (4.69) can be put in the general form of equation (4.46),

$$\psi (t) = \sqrt{2} \sum_{n=-\infty}^{\infty} h_1 [n] \phi (2t - n), \tag{4.70}$$

where the filter coefficients $h_1[n]$ are found to be

$$h_1[n] = (-1)^n h_0[1-n] = (-1)^n \frac{\sqrt{2}}{\pi(1-n)} \cos(n\pi/2). \qquad (4.71)$$

Both the scaling and the wavelet filter coefficients are shown in figure 4.12.

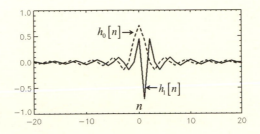

Figure 4.12: The Shannon filter coefficients $h_0[n]$ and $h_1[n]$.

The relation between the Haar scaling and wavelet expansion coefficients expressed in equation (4.50) and figure 4.6 is valid in the present case too. We illustrate this by looking at the expansion coefficients at scales 0 and $-1$. The scaling expansion coefficients at scale 0 for a function $x(t)$ are

$$c_{0n} = \int_{-\infty}^{\infty} x(t)\,\phi_{0n}(t)dt = \int_{-\infty}^{\infty} x(t)\,\frac{\sin[\pi(t-n)]}{\pi(t-n)}dt \qquad (4.72)$$

which on using the scaling equation (4.58) can be written as

$$c_{0n} = \sqrt{2}\sum_{k=-\infty}^{\infty} h_0[k] \int_{-\infty}^{\infty} x(t)\,\frac{\sin[\pi(2t-2n-k)]}{\pi(2t-2n-k)}dt. \qquad (4.73)$$

The scaling expansion coefficients at scale $-1$ are

$$c_{-1,n} = \int_{-\infty}^{\infty} x(t)\,\phi_{-1,n}(t)dt = \sqrt{2}\int_{-\infty}^{\infty} x(t)\,\frac{\sin[\pi(2t-n)]}{\pi(2t-n)}dt. \qquad (4.74)$$

Use of (4.74) in (4.73) gives

$$c_{0n} = \sum_{k=-\infty}^{\infty} h_0[k]\,c_{-1,2n+k}, \qquad (4.75)$$

which on changing the summation index becomes

$$c_{0n} = \sum_{k=-\infty}^{\infty} h_0 \left[ k - 2n \right] c_{-1,k}. \tag{4.76}$$

Equation (4.76), relating the scaling coefficients at scale 0 to those at scale $-1$, is in fact a general equation valid for all scales $m$:

$$c_{mn} = \sum_{k=-\infty}^{\infty} h_0 \left[ k - 2n \right] c_{m-1,k}. \tag{4.77}$$

In addition, the wavelet expansion coefficients $d_{mn}$ defined by

$$\mathbf{E}_m f = \mathbf{P}_{m-1} f - \mathbf{P}_m f \equiv \sum_{n=-\infty}^{\infty} d_{mn} \, \psi_{mn} \left( t \right) \tag{4.78}$$

satisfy the equation (see equation (4.50))

$$d_{mn} = \sum_{k=-\infty}^{\infty} h_1 \left[ k - 2n \right] c_{m-1,k}. \tag{4.79}$$

Equations (4.77) and (4.79) have exactly the same interpretation as in the Haar example shown in figure 4.6.

## 4.7   The Spectrum of the Shannon Filter Coefficients

As in the case of the Haar functions (see equation (4.51)), and dropping the overall factor of $\sqrt{2}$, we find

$$H_0 \left( \omega \right) = \sum_{n=-\infty}^{\infty} \frac{\sin \left( n\pi/2 \right)}{n\pi} e^{-in\omega}, \tag{4.80a}$$

$$H_1 \left( \omega \right) = \sum_{n=-\infty}^{\infty} \left( -1 \right)^n \frac{\cos \left( n\pi/2 \right)}{\left( 1 - n \right) \pi} e^{-in\omega}. \tag{4.80b}$$

The sums on the right hand sides of the above equations can be performed[4] to give,

$$H_0\left(\omega\right) = \begin{cases} 1, & \text{when } \omega \in [-\pi/2, +\pi/2], \\ 0, & \text{otherwise,} \end{cases} \tag{4.81}$$

and

$$H_1\left(\omega\right) = \begin{cases} 0, & \text{when } \omega \in [-\pi/2, +\pi/2], \\ 1, & \text{otherwise.} \end{cases} \tag{4.82}$$

Thus $H_0\left(\omega\right)$ and $H_1\left(\omega\right)$ are ideal low-pass and high-pass half-band filters, as shown in figure 4.13. The Shannon filter coefficients are examples of infinite impulse response (IIR) filters; ideal response is obtained at the expense of an infinity of coefficients (recall that for the Haar filter coefficients, nonideal response was obtained at the expense of a finite number of elements in each of the two sets of filter sequences).

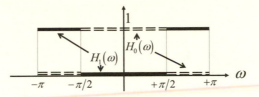

Figure 4.13: Spectra of the Shannon filter coefficients $H_0(\omega)$ and $H_1(\omega)$.

## 4.8  Meyer's Wavelet

The Shannon scaling function $\phi(t)$ is in the subspace $\mathcal{V}_0$ and so its spectrum has infinitely sharp edges at $\omega = \pm\pi$ (see figure 4.10). This results in ripples (known in digital filter design as ringing) in the corresponding time function (equation (4.56) with $n = 0$). Meyer's scaling and wavelet functions are based on smoothing the sharp edges while preserving the orthogonality of the system by preserving the relation (see exercise 4) [36]

$$\sum_{n=-\infty}^{\infty} \left|\Phi\left(\omega + 2n\pi\right)\right|^2 = 1.$$

---

[4]Consider the periodic function $F\left(\omega\right) = 1$ when $\omega \in [-\pi/2, +\pi/2]$, and zero on the rest of the interval $[-\pi, +\pi]$. Using the discrete time Fourier transform relations (1.72) and (1.73) we have $F\left(\omega\right) = \sum_{n=-\infty}^{\infty} f_n e^{-in\omega}$, and $f_n = (2\pi)^{-1} \int_{-\pi/2}^{\pi/2} e^{in\omega} d\omega = (n\pi)^{-1} \sin\left(n\pi/2\right)$. The required sum in equation (4.80a) now follows by substituting the relation $H_1\left(\omega\right) = -e^{-i\omega} H_0^*\left(\omega + \pi\right)$ that leads to equation (4.68). The last relation is explained in chapters 5 and 6.

The solution is for the Fourier transform of the Meyer scaling function to be defined in terms of the modified frequency variable

$$\sigma \equiv \frac{3}{2\pi} |\omega| - 1 \tag{4.83}$$

and by the equation [47, 48, 49]

$$\Phi(\omega) = \begin{cases} 1, & |\omega| \leq 2\pi/3 \Leftrightarrow |\sigma + 1| \leq 1, \\ \cos\left(\frac{\pi}{2} F(\sigma)\right), & 2\pi/3 \leq |\omega| \leq 4\pi/3 \Leftrightarrow 1 \leq |\sigma + 1| \leq 2, \\ 0, & \text{otherwise.} \end{cases} \tag{4.84}$$

where the function $F(\sigma)$ must satisfy the relations

$$F(\sigma) + F(1 - \sigma) = 1, \tag{4.85a}$$
$$F(\sigma) = 0 \ \text{ for } \ \sigma \leq 0, \tag{4.85b}$$
$$F(\sigma) = 1 \ \text{ for } \ \sigma \geq 1. \tag{4.85c}$$

If we require that the first $n$ derivatives of $\Phi(\omega)$ vanish at $\omega = 2\pi/3$, then the function $F(\sigma)$ is proportional to $\sigma^{n+1}$. The same smoothness requirement at $\omega = 4\pi/3$ can then be used to solve for a unique polynomial solution of degree $2n + 1$. Several such polynomials are given in table 4.1. The Meyer

| $F(\sigma)$ ,   $\sigma \equiv 3|\omega|/2\pi - 1$ |
| --- |
| $\sigma$ |
| $\sigma^2 (3 - 2\sigma)$ |
| $\sigma^3 \left(10 - 15\sigma + 6\sigma^2\right)$ |
| $\sigma^4 \left(35 - 84\sigma + 70\sigma^2 - 20\sigma^3\right)$ |
| $\sigma^5 \left(126 - 420\sigma + 540\sigma^2 - 315\sigma^3 + 70\sigma^4\right)$ |

Table 4.1: Polynomial solutions used in the Fourier transform of the Meyer scaling and wavelet functions.

wavelet has the following Fourier transform:

$$\Psi(\omega) = \begin{cases} -e^{-i\omega/2} \sin\left(\frac{\pi}{2} F(\sigma)\right), & \frac{2\pi}{3} \leq |\omega| \leq \frac{4\pi}{3} \Leftrightarrow 1 \leq |\sigma + 1| \leq 2, \\ -e^{-i\omega/2} \cos\left(\frac{\pi}{2} F\left(\frac{\sigma}{2} - \frac{1}{2}\right)\right), & \frac{4\pi}{3} \leq |\omega| \leq \frac{8\pi}{3} \Leftrightarrow 2 \leq |\sigma + 1| \leq 4, \\ 0, & |\omega| \geq \frac{8\pi}{3} \Leftrightarrow |\sigma + 1| \geq 4. \end{cases}$$

Figure 4.14 shows the spectra of the Shannon and the Meyer scaling functions for the fifth-order polynomial function in table 4.1 in addition to the Meyer scaling and the wavelet functions of time.

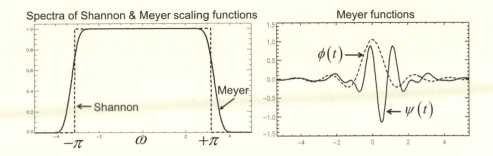

Figure 4.14: Spectra of Meyer and Shannon scaling functions.

## 4.9 Exercises

**1** Reproduce the Haar scaling function approximations (see figure 4.4) for several resolutions $m$. For this example compute the function in the interval $[0, 3]$ using $\Delta t = 0.001$. Then use equation (4.14) (for a fixed $m$) to compute the exact scaling function coefficients $c_{mn}$ for the given function $te^{-t^2}$. The approximation to $x(t)$ at scale $m$ is then given by $x_m(t) = \sum_n c_{mn}\phi_{mn}(t)$, where the sum over $n$ is limited by the range of the original signal (in our example the signal is zero to the left of 0 and negligible to the right of 3).

**2** Prove equations (4.51) and (4.53) and reproduce the associated spectra.

**3** Given the relation $H_1(\omega) = -e^{-i\omega}H_0^*(\omega + \pi)$ and using equations (1.72) and (1.73) show that the corresponding time series are related by $h_1[n] = (-1)^n h_0[1-n]$. This is the result (4.71).

**4** Here we will verify the equation that guarantees the orthogonality of the Meyer wavelet system (the general result for any orthogonal wavelet system is proven in theorem 67 of section 5.7).

Consider the Shannon scaling function $\phi(t)$, whose Fourier transform $\Phi(\omega)$ is depicted in figure 4.14. Then it is easy to see the validity of the equation $\sum_{k=-\infty}^{\infty} |\Phi(\omega + 2k\pi)|^2 = 1$: the terms in the sum have no overlap and each is a box of height 1 on an interval of length $2\pi$. For instance, $|\Phi(\omega + 2\pi)|^2$ is a box of height 1 on the interval $[-3\pi, -\pi]$, whereas $|\Phi(\omega - 2\pi)|^2$ is a box of height 1 on the interval $[\pi, 3\pi]$. Using the definition (4.84) for the Meyer scaling function, show that the Fourier transform of the Meyer scaling functions satisfies the same equation, i.e., $\sum\limits_{k=-\infty}^{\infty} |\Phi(\omega + 2k\pi)|^2 = 1$. Note that the two functions $\Phi(\omega)$ and $\Phi(\omega - 2\pi)$ overlap only in the interval $[2\pi/3, 4\pi/3]$, for which $\omega \leq 2\pi$, and so in this interval $|\omega - 2\pi| = 2\pi - \omega$. Then use equation (4.85a).

**5** Reproduce figure 4.11 using the following procedure. Given the analytic formula for a symmetric $\Phi(\omega)$ first construct an $\omega$ axis with the range $[-\omega_{max}, \omega_{max}]$ and a suitably large number $N$ of points in that interval. Thus, $\Delta\omega = 2\omega_{max}/N$. The range must be chosen so that $\Phi(\omega)$ is negligible outside that range. Then circularly shift the function $\Phi(\omega)$ by $N/2$, compute the $N$-point inverse FFT, reverse the circular shift, and take the real part (recall that we assumed a symmetric $\Phi(\omega)$) to obtain the time series $\phi(t)$. Using the relations

$$\Delta\omega = 2\pi\Delta f = 2\pi f_s/N = 2\pi/(N \ \Delta T),$$

where $f_s$ is the nominal sampling frequency, we find $\Delta T = 2\pi/(N \ \Delta\omega)$ and then construct the time axis to have $N$ points $t_n$ in the range $[-N/2, N/2] \Delta T$. Finally normalize $\phi(t)$ so that it has unit norm, i.e., divide $\phi(t)$ by the square root of $\sum\limits_{n=0}^{N-1} |\phi(t_n)|^2 \Delta T$. Test this procedure for the function $\Phi(\omega) = \sin^2(\omega/2) \big/ (\omega/2)^2$ whose inverse transform must be the triangle function limited in time to the interval $[-1, 1]$ using $\omega_{max} = 100$ and $N = 1024$, as shown in figure 4.15.

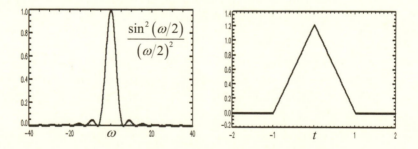

Figure 4.15: The triangle function and its spectrum.

CHAPTER 5

# General Properties of Scaling and Wavelet Functions

## 5.1  Introduction

In section 4.3 we saw the Haar scaling and wavelet functions as basis functions for subspaces that allowed approximations to a function at different resolutions. The associated filters, namely, $h_0$ and $h_1$, satisfied a set of equations that we claimed to be quite general relations for orthogonal wavelets. In addition, the filter spectra satisfied the equations of PR-QMF. In this chapter we discuss the general properties of orthogonal and biorthogonal scaling and wavelet functions, in the time and the frequency domains, and derive equivalent relations for the associated filter functions.

## 5.2  Multi-Resolution Analysis Spaces

Here we formalize the results of the last chapter assuming that the scaling and the wavelet are real functions. Consider a function $x(t) \in L_2(\mathbb{R})$ which is seen as the limit of successive approximations $x_m(t)$ at a different resolution (scale) and which is a "smoothed" version of $x(t)$. From section 1.15 we recall the definition of multi-resolution analysis (MRA) subspaces of $L_2(\mathbb{R})$ denoted by $\{\mathscr{V}_m : m \in \mathbb{Z}\}$ that possess the following properties [15, 14]:

- Nesting or containment:

$$\cdots \mathscr{V}_m \subset \cdots \subset \mathscr{V}_1 \subset \mathscr{V}_0 \subset \mathscr{V}_{-1} \cdots \mathscr{V}_{-m} \subset \cdots .$$

- Completeness (note that $\mathscr{V}_{\pm\infty} \equiv \lim_{m\to\infty} \mathscr{V}_{\pm m}$):

$$\bigcap_{m=-\infty}^{\infty} \mathscr{V}_m = \{0\}, \quad \bigcup_{m=-\infty}^{\infty} \mathscr{V}_m = L_2(\mathbb{R}), \mathscr{V}_\infty = L_2(\mathbb{R}), \mathscr{V}_{-\infty} = \{0\}.$$

- The multi-resolution or scaling property (defining resolution and scale):

$$x\left(t\right) \in \mathscr{V}_m \Leftrightarrow x\left(2t\right) \in \mathscr{V}_{m-1}, \; \forall \, x\left(t\right) \in L_2\left(\mathbb{R}\right).$$

- The basis or frame property: There exists a scaling function $\phi\left(t\right) \in \mathscr{V}_0$ such that all shifted and scaled versions of it,

$$\phi_{mn}(t) \equiv 2^{-m/2} \, \phi\left(2^{-m}t - n\right) \equiv \mathscr{D}_{2^m} \mathscr{T}_n \, \phi(t),$$

form an orthonormal basis for $\mathscr{V}_m$, i.e., using the Dirac notation (no summation on the index $m$ that represents the fixed resolution)

$$\sum_{n=-\infty}^{\infty} |\phi_{mn}\rangle \langle \phi_{mn}| = \mathbf{1}_m, \quad \langle \phi_{mn} \mid \phi_{mk} \rangle = \delta_{nk}, \tag{5.1}$$

where $\mathbf{1}_m$ is the unit operator in $\mathscr{V}_m$.

The scaling and the frame properties imply that $\phi\left(t\right) \in \mathscr{V}_0$ can be expressed as a linear combination of the basis for $V_{-1}$, i.e.,

$$\phi\left(t\right) = \sqrt{2} \sum_{n=-\infty}^{\infty} h_0\left[n\right] \phi\left(2t - n\right), \tag{5.2}$$

defining the low-pass filter coefficients $h_0\left[n\right]$. Note that the scaling equation can be written in the form

$$\phi_{mn}\left(t\right) = \sum_{k=-\infty}^{\infty} h_0\left[k\right] \phi_{m-1,2n+k}\left(t\right), \tag{5.3}$$

valid for all scales. Next we denote the orthogonal complement of $\mathscr{V}_m$ in $\mathscr{V}_{m-1}$ by $\mathscr{V}_m^\perp$. The wavelet function $\psi\left(t\right) \in V_0^\perp$, and its shifted and scaled versions

$$\psi_{mn}(t) \equiv 2^{-m/2} \, \psi(2^{-m}t - n) \equiv \mathscr{D}_{2^m} \mathscr{T}_n \psi\left(t\right), \tag{5.4}$$

form an orthonormal basis for $\mathscr{V}_m^\perp$ in addition to being orthogonal across scales (no summation on the index $m$ that represents the fixed resolution),

$$\sum_{n=-\infty}^{\infty} |\psi_{mn}\rangle \langle \psi_{mn}| = \mathbf{1}_m, \quad \langle \psi_{mn} \mid \psi_{lk} \rangle = \delta_{ml}\delta_{nk}. \tag{5.5}$$

where, for each fixed $m$, $\mathbf{1}_m$ is now the unit operator in $\mathscr{V}_m^\perp$.[1]

---

[1]As we shall see in theorem 57, if we sum over all scales $m$ in addition to the integer shifts $n$ we will have $\sum_{m,n=-\infty}^{\infty} |\psi_{mn}\rangle \langle \psi_{mn}| = \mathbf{1}$ where $\mathbf{1}$ is the unit operator for the entire Hilbert space $L_2(\mathbb{R})$.

The wavelet functions are also orthogonal at each scale to the corresponding scaling functions,[2] i.e.,

$$\langle \psi_{mn} \mid \phi_{mk} \rangle \equiv \int_{-\infty}^{\infty} \psi_{mn}(t) \, \phi_{mk}(t) \; dt = 0. \tag{5.6}$$

The multi-resolution/scaling and the frame properties further imply that since $\psi(t) \in \mathcal{V}_0^{\perp} \subset \mathcal{V}_{-1}$, the wavelet function can be expressed as a linear combination of the basis for $\mathcal{V}_{-1}$, i.e.,

$$\psi(t) = \sqrt{2} \sum_{n=-\infty}^{\infty} h_1[n] \, \phi(2t - n), \tag{5.7}$$

defining the wavelet high-pass coefficients $h_1[n]$. The wavelet equation can also be written as

$$\psi_{mn}(t) = \sum_{k=-\infty}^{\infty} h_1[k] \, \phi_{m-1, 2n+k}(t), \tag{5.8}$$

valid for all scales $m$.

Given a square-integrable signal $x(t)$ and using the completeness and orthonormality of the scaling bases, the scaling coefficients of the signal (at resolution index $m$) are given by

$$c_{mn} = \langle \phi_{mn} \mid x \rangle \equiv \int_{-\infty}^{\infty} \phi_{mn}(t) \, x(t) \; dt, \tag{5.9}$$

while the signal wavelet coefficients are given by

$$d_{mn} = \langle \psi_{mn} \mid x \rangle \equiv \int_{-\infty}^{\infty} \psi_{mn}(t) \, x(t) \; dt. \tag{5.10}$$

When studying the Haar and the Shannon wavelets we found that the corresponding coefficients $c_{mn}$ and $d_{mn}$ satisfied certain recursive relations. The Haar recursions are shown in equation (4.50) while the Shannon recursions are shown in equations (4.77) and (4.79). Linear filter theory interpretations

---

[2]Note that in this chapter we assume all the scaling and wavelet functions of interest to be real functions and so in all integrals involving the scaling and wavelet functions (e.g., inner products) complex conjugates do not appear.

of the recursions were depicted in figure 4.6. To derive the same results in this general setting we begin by expressing the scaling functions $\phi_{mn}(t)$ in terms of $\phi_{m-1,n}(t)$, i.e., equation (5.3). Thus,

$$
\begin{aligned}
\phi\left(2^{-m}t - n\right) &= \sqrt{2} \sum_{k=-\infty}^{\infty} h_0\left[k\right] \phi\left(2 \times \left(2^{-m}t - n\right) - k\right) \\
&= \sqrt{2} \sum_{k=-\infty}^{\infty} h_0\left[k\right] \phi\left(2^{-m+1}t - 2n - k\right),
\end{aligned}
\tag{5.11}
$$

which after a change in the summation index becomes

$$
\phi\left(2^{-m}t - n\right) = \sqrt{2} \sum_{k=-\infty}^{\infty} h_0\left[k - 2n\right] \varphi\left(2^{-m+1}t - k\right).
\tag{5.12}
$$

Equation (5.9) for the coefficients $c_{mn}$ then gives

$$
\begin{aligned}
c_{mn} &= 2^{-m/2} \int_{-\infty}^{\infty} x\left(t\right) \phi\left(2^{-m}t - n\right)\ dt \\
&= 2^{-m/2} \sum_{k} \sqrt{2} h_0\left[k - 2n\right] \int_{-\infty}^{\infty} x\left(t\right) \phi\left(2^{-m+1}t - k\right) dt \\
&= \sum_{k} h_0\left[k - 2n\right] c_{m-1,k}.
\end{aligned}
\tag{5.13}
$$

Similarly,

$$
d_{mn} = 2^{-m/2} \int_{-\infty}^{\infty} x\left(t\right) \psi\left(2^{-m}t - n\right)\ dt = \sum_{k} h_1\left[k - 2n\right] c_{m-1,k}.
\tag{5.14}
$$

To prove the last equality in equation (5.14) consider the approximation to a function $x\left(t\right)$ at scale $m - 1$, i.e.,

$$
x_{m-1}\left(t\right) \equiv \sum_{n=-\infty}^{\infty} c_{m-1,n}\ \phi_{m-1,n}\left(t\right).
\tag{5.15}
$$

Since $\mathcal{V}_{m-1} = \mathcal{V}_m \oplus \mathcal{V}_m^\perp$, we have

$$
x_{m-1}\left(t\right) \equiv \sum_{n=-\infty}^{\infty} c_{mn}\ \phi_{mn}\left(t\right) + \sum_{n=-\infty}^{\infty} d_{mn}\ \psi_{mn}\left(t\right).
$$

Taking inner products of both sides with the wavelet functions $\psi_{mn}(t)$ and using the wavelet equation (5.8) we find

$$
\begin{aligned}
d_{mn} &= \langle x_{m-1}(t), \psi_{mn}(t) \rangle \\
&= \left\langle \sum_{k=-\infty}^{\infty} c_{m-1,k}\, \phi_{m-1,k}(t), \sum_{l=-\infty}^{\infty} h_1[l]\, \phi_{m-1,2n+l}(t) \right\rangle \\
&= \sum_{k=-\infty}^{\infty} \sum_{l=-\infty}^{\infty} c_{m-1,k}\, h_1[l]\, \langle \phi_{m-1,k}(t), \phi_{m-1,2n+l}(t) \rangle \\
&= \sum_{k=-\infty}^{\infty} \sum_{l=-\infty}^{\infty} c_{m-1,k}\, h_1[l]\, \delta_{k,2n+l} = \sum_{k=-\infty}^{\infty} h_1[k-2n]\, c_{m-1,k}.
\end{aligned}
\tag{5.16}
$$

The above recursions show that the scaling and wavelet coefficients of expansion in an orthogonal multi-resolution analysis scheme (as described above) at each scale are found by first correlating the scaling coefficients at the previous finer scale with the scaling and wavelet filter coefficients (or, equivalently, convolving with the time-reversed versions of the same filter coefficients), and then taking the even numbered samples (down-sampling by a factor of 2). This interpretation, depicted in figure 4.6 for the Haar wavelet, is valid for all orthogonal wavelet expansions and forms the basis for a fast algorithm in computing the discrete wavelet transform coefficients of sampled and finite length sequences; calculations that can be performed without ever having to compute the scaling or the wavelet functions and that only use the scaling and the wavelet filter coefficients $h_0[n]$ and $h_1[n]$.

The decomposition

$$
\mathscr{V}_{m-1} = \mathscr{V}_m \oplus \mathscr{V}_m^{\perp} = \left( \mathscr{V}_{m+1} \oplus \mathscr{V}_{m+1}^{\perp} \right) \oplus \mathscr{V}_m^{\perp}
$$

when continued indefinitely together with the completeness property of the subspaces $\mathscr{V}_m$ allows us to write

$$
L_2(\mathbb{R}) = \lim_{m \to -\infty} \mathscr{V}_m = \bigoplus_{m=-\infty}^{\infty} \mathscr{V}_m^{\perp},
\tag{5.17}
$$

leading to the following theorem.

**Theorem 57.** *Any function $x(t) \in L_2(\mathbb{R})$ has the following purely orthogonal wavelet expansion*

$$
x(t) = \sum_{m,n=-\infty}^{\infty} d_{mn} \psi_{mn}(t).
\tag{5.18}
$$

*Thus the orthogonal wavelets satisfy the completeness relation (summation on both repeated indices $m, n$ is implied)*

$$|\psi_{mn}\rangle \langle \psi_{mn}| = \mathbf{1}. \tag{5.19}$$

The purely wavelet expansion requires an infinite number of resolutions as represented by the summation over the index $m$. In practice, we consider using a finite resolution expansion that uses the scaling function as well as the wavelet functions. To this end consider the approximation at scale $M$ to the function $x(t)$, i.e.,

$$x_M(t) = \sum_{n=-\infty}^{\infty} c_{Mn}\, \phi_{Mn}(t). \tag{5.20}$$

Using $\mathcal{V}_M = \mathcal{V}_{M+1} \oplus \mathcal{V}_{M+1}^{\perp}$ we can write

$$x_M(t) = \sum_{n=-\infty}^{\infty} c_{M+1,n}\, \phi_{M+1,n}(t) + \sum_{n=-\infty}^{\infty} d_{M+1,n} \psi_{M+1,n}(t), \tag{5.21}$$

and continue to decompose the "coarse" approximations (the first sum on the right hand side in the above equation) up to scale $M + L$ to obtain

$$x_M(t) = \sum_{n=-\infty}^{\infty} c_{M+L,n}\phi_{M+L,n}(t) + \sum_{l=1}^{L} \sum_{n=-\infty}^{\infty} d_{M+l,n}\psi_{M+l,n}(t). \tag{5.22}$$

Hence the approximation $x_M(t)$ can be represented as the sum of a low-pass approximation at scale $M + L$, and a total of $L$ detail components at scales $M + 1$ through $M + L$. This expansion is the basis for a practical implementation of the discrete wavelet transform.

## 5.3    The Inverse Relations

For a given function $x(t)$ and scaling coefficients $c_{m-1,n}$ we can use equations (5.13) and (5.14) to compute the coefficients at the next coarser scale, namely, $c_{mn}$ and $d_{mn}$ as shown in figure 4.6. Here we derive the inverse relations, i.e., equations that allow us to use the coefficients $\{c_{mn}, d_{mn}\}$ at scale $m$, to construct the set $\{c_{m-1,n}\}$ at scale $m - 1$.

Consider $x_{m-1}(t) \in \mathcal{V}_{m-1}$. Then

$$x_{m-1}(t) = \sum_{n=-\infty}^{\infty} c_{m-1,n}\, \phi_{m-1,n}(t)$$

$$= \sum_{n=-\infty}^{\infty} c_{mn}\, \phi_{mn}(t) + \sum_{n=-\infty}^{\infty} d_{mn}\, \psi_{mn}(t). \tag{5.23}$$

Using the scaling and the wavelet equations (5.3) and (5.8) we have

$$x_{m-1}(t) = \sum_{n=-\infty}^{\infty} c_{mn} \sum_{k=-\infty}^{\infty} h_0[k]\, \phi_{m-1,2n+k}(t) +$$

$$\sum_{n=-\infty}^{\infty} d_{mn} \sum_{k=-\infty}^{\infty} h_1[k]\, \phi_{m-1,2n+k}(t).$$

Defining a new index $k = j - 2n$ in the above equation we have

$$x_{m-1}(t) = \sum_{n=-\infty}^{\infty} c_{mn} \sum_{j=-\infty}^{\infty} h_0[j-2n]\, \phi_{m-1,j}(t) +$$

$$\sum_{n=-\infty}^{\infty} d_{mn} \sum_{j=-\infty}^{\infty} h_1[j-2n]\, \phi_{m-1,j}(t)$$

$$= \sum_{j=-\infty}^{\infty} \phi_{m-1,j}(t) \sum_{n=-\infty}^{\infty} c_{mn} h_0[j-2n] +$$

$$\sum_{j=-\infty}^{\infty} \phi_{m-1,j}(t) \sum_{n=-\infty}^{\infty} d_{mn} h_1[j-2n]. \tag{5.24}$$

Comparing the right hand side to the defining relations for $c_{m-1,n}$ given in equation (5.15) we have

$$c_{m-1,j} = \sum_{n=-\infty}^{\infty} c_{mn}\, h_0[j-2n] + \sum_{n=-\infty}^{\infty} d_{mn}\, h_1[j-2n]. \tag{5.25}$$

The filtering interpretation of this equation is found by defining the sequence $\hat{c}_{mn} = c_{mn}$ when $n$ is even and $\hat{c}_{mn} = 0$ when $n$ is odd. The sequence $\hat{c}_{mn}$ is simply the result of interpolating the sequence $c_{mn}$ with zeros. Defining the sequence $\hat{d}_{mn}$ similarly, i.e., through zero-interpolation of the sequence $d_{mn}$, we state that the sequence $c_{m-1,n}$ is found by first convolving $\hat{c}_{mn}$ with

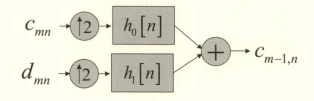

Figure 5.1: Discrete convolution interpretation for the inverse scaling and wavelet coefficients.

$h_0[n]$, and adding the output to the result of convolving the sequence $\hat{d}_{mn}$ with $h_1[n]$. This filtering interpretation is shown in figure 5.1. Figure 5.2 shows the combined forward (figure 4.6) and inverse (figure 5.1) relations for computing the discrete wavelet transform. The forward transform is also known as the analysis operation, while the inverse transform is known as the synthesis operation. Comparison with figure 2.4 of section 2.7 clearly

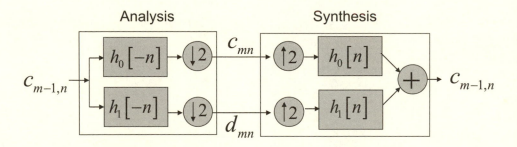

Figure 5.2: Discrete convolution interpretation for computing the discrete wavelet coefficients.

shows that the filter coefficients are equivalent to a perfect reconstruction (PR) filter system if we make the replacements $g_j[n] = h_j[-n]$, $j = 0, 1$. Thus, the PR equation (2.28) is now [14]

$$\sum_k \{h_0[n-2k]\, h_0[l-2k] + h_1[n-2k]\, h_1[l-2k]\} = \delta_{nl}. \qquad (5.26)$$

Before examining all other conditions that are required by scaling and wavelet orthonormality relations we discuss the shift-invariant discrete wavelet transform in the next section.

## 5.4 The Shift-Invariant Discrete Wavelet Transform

As noted earlier (see equation (3.83)) a shift-invariant or redundant discrete wavelet transform can be defined on a semi-regular dyadic grid. The corresponding scaling and wavelet functions are given by

$$\phi_{mn}(t) \equiv D_{2^m} T_n \phi(t) = 2^{-m/2} \phi\left(\frac{t-n}{2^m}\right), \tag{5.27a}$$

$$\psi_{mn}(t) \equiv D_{2^m} T_n \psi(t) = 2^{-m/2} \psi\left(\frac{t-n}{2^m}\right). \tag{5.27b}$$

We define the shift-invariant scaling and wavelet coefficients $c_{mn}$ and $d_{mn}$ in analogy with equations (5.9) and (5.10) except that the above scaling and wavelet functions are used in the integrals. For instance for $m \geq 1$ we have

$$c_{mn} = 2^{-m/2} \int_{-\infty}^{\infty} x(t) \phi\left(\frac{t-n}{2^m}\right) dt$$

$$= 2^{-(m-1)/2} \sum_{k=-\infty}^{\infty} h_0[k] \int_{-\infty}^{\infty} x(t) \phi\left(\frac{t-n}{2^{m-1}} - k\right) dt$$

$$= \sum_{k=-\infty}^{\infty} h_0[k] c_{m-1,n+2^{m-1}k}. \tag{5.28}$$

If we now define the A' Trous interpolated low-pass filter sequence $h_0^{(m)}$, $m \geq 1$, by

$$h_0^{(m)}[2^m k] = h_0[k], m \geq 1, \quad \text{and } 0, m < 1, \tag{5.29}$$

and similarly for the A' Trous interpolated high-pass filter sequence $h_1^{(m)}$ [50], we have the following recursive relationship starting with $h_0^{(0)} = h_0$, $h_1^{(0)} = h_1$, and $n \in \mathbb{Z}$:

$$h_j^{(m)}[2n] = h_j^{(m-1)}[n], \ h_j^{(m)}[2n+1] = 0, \ j = 0, 1, \quad m = 1, 2, 3, \ldots. \tag{5.30}$$

With respect to this equation (5.28) and a similarly derived one for $d_{mn}$ become

$$c_{mn} = \sum_{k=-\infty}^{\infty} h_0^{(m-1)}[k] c_{m-1,n+k}, \quad d_{mn} = \sum_{k=-\infty}^{\infty} h_1^{(m-1)}[k] c_{m-1,n+k}. \tag{5.31}$$

The correlation interpretation of (5.31) is shown in figure 5.3 where the correlations are performed with the A' Trous interpolated filters defined in (5.29). Comparison with the left hand part of figure 5.2 shows the absence of the down-sampling (decimation); hence the undecimated wavelet transform (UDWT) title.[3] To find the inverse UDWT/RDWT consider (5.31) for even

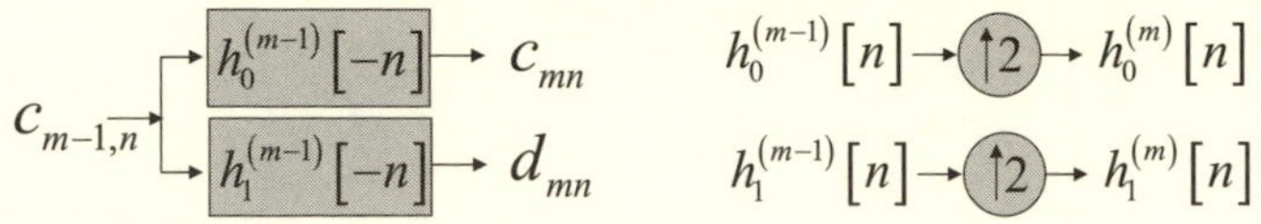

$$c_{m-1,n} \rightarrow \begin{array}{|c|} \hline h_0^{(m-1)}[-n] \\ \hline h_1^{(m-1)}[-n] \\ \hline \end{array} \rightarrow \begin{array}{l} c_{mn} \\ d_{mn} \end{array} \qquad h_0^{(m-1)}[n] \rightarrow \uparrow 2 \rightarrow h_0^{(m)}[n]$$
$$h_1^{(m-1)}[n] \rightarrow \uparrow 2 \rightarrow h_1^{(m)}[n]$$

Figure 5.3: The shift-invariant wavelet transform coefficients' recursion with A' Trous interpolated filter sequences.

index $2n$ written in terms of the original sequences $h_0$ and $h_1$,

$$
\begin{aligned}
c_{m,2n} &= \sum_{k=-\infty}^{\infty} h_0[k]\, c_{m-1,2n+2^{m-1}k}, \\
d_{m,2n} &= \sum_{k=-\infty}^{\infty} h_1[k]\, c_{m-1,2n+2^{m-1}k}.
\end{aligned}
\tag{5.32}
$$

The index $2n$ now allows us to invoke the PR relation (5.26) for the orthogonal pair $\{h_0, h_1\}$ and arrive at the following inverse (note the A' Trous interpolated filters)

$$
c_{m-1,2n} = \sum_{k=-\infty}^{\infty} h_0^{(m-1)}[k]\, c_{m,2n-k} + \sum_{k=-\infty}^{\infty} h_1^{(m-1)}[k]\, d_{m,2n-k},
\tag{5.33}
$$

while the odd index $2n+1$ will, in a similar fashion, reproduce the coefficients $c_{m-1,2n+1}$. The shift-invariant discrete wavelet transform is illustrated in figure 5.4.

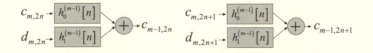

Figure 5.4: The shift-invariant wavelet transform inverse (even and odd indices).

---

[3]We use the acronyms UDWT and RDWT synonymously.

## 5.5  Time Domain Properties

Here we derive some general properties of real scaling and wavelet functions. For instance, the completeness and orthogonality properties of the scaling and the wavelet functions can be used to find a set of equations satisfied by the filter coefficients, which in simple instances and with some additional constraints, can be used to solve for unique solutions.

The filter coefficients can be expressed in terms of the scaling function and wavelet function integrals as follows. Equation (5.3) for $m = 0$ and $n = 0$ is

$$\phi(t) = \sum_{n=-\infty}^{\infty} h_0[n]\,\phi_{-1,n}(t), \qquad (5.34)$$

where $\phi_{00}(t) \equiv \phi(t)$ and $\phi_{-1,n}(t) \equiv 2^{1/2}\phi(2t-n)$. Taking appropriate inner products of both sides we find

$$h_0[n] = \int_{-\infty}^{\infty} \phi_{00}(t)\,\phi_{-1,n}(t)\,dt = \sqrt{2}\int_{-\infty}^{\infty} \phi(t)\,\phi(2t-n)\,dt. \qquad (5.35)$$

Similarly, and using equation (5.8), we find

$$h_1[n] = \int_{-\infty}^{\infty} \psi_{00}(t)\,\phi_{-1,n}(t)\,dt = \sqrt{2}\int_{-\infty}^{\infty} \psi(t)\,\phi(2t-n)\,dt, \qquad (5.36)$$

where $\psi_{00}(t) \equiv \psi(t)$. The last two equations are special cases of the relation

$$\phi_{mk}(t) = \sum_{j=-\infty}^{\infty} h_0[k-2j]\,\phi_{m+1,j}(t) + \sum_{j=-\infty}^{\infty} h_1[k-2j]\,\psi_{m+1,j}(t). \qquad (5.37)$$

An immediate consequence of (5.35) and (5.36) is that if, for instance, $h_0[n]$ is identically zero for $n < 0$ and $n > N$, then the integral on the right hand side of (5.35) must be zero for all $n$ outside the interval $[0, N-1]$. This means that the two functions $\phi(t)$ and $\phi(2t-n)$ must have no intersection for those values of $n$, and thus we have the following theorem.

**Theorem 58.** *If the low-pass filter coefficients $h_0[n]$ are zero for $n < 0$ and $n > N$, then the corresponding scaling function $\phi(t)$ has compact support in the interval $[0, N-1]$. The same statements hold for the coefficients $h_1[n]$ and the corresponding wavelet function $\psi(t)$. The converse is also true: A scaling function $\phi(t)$ that has support in the interval $[0, N-1]$ for some* *integer $N$, the smallest such integer, is associated with a sequence $h_0[n]$ (defined by the scaling equation) that has exactly $N$ nonzero elements. The same result holds for the wavelet function $\psi(t)$ and the associated sequence $h_1[n]$.*

The simplest condition satisfied by the low-pass filter coefficients $h_0[n]$ is found by integrating both sides of the scaling equation (5.34) with respect to time:

$$\int_{-\infty}^{\infty} \phi(t)\, dt = \sqrt{2} \sum_{n=-\infty}^{\infty} h_0[n] \int_{-\infty}^{\infty} \phi(2t-n)\, dt$$

$$= \sqrt{2} \sum_{n=-\infty}^{\infty} h_0[n] \frac{1}{2} \int_{-\infty}^{\infty} \phi(s)\, ds, \tag{5.38}$$

from which we arrive at the following theorem.

**Theorem 59.**

$$\sum_{n=-\infty}^{\infty} h_0[n] = \sqrt{2}\;, \quad provided \;\; \int_{-\infty}^{\infty} \phi(t)\, dt \neq 0. \tag{5.39}$$

The proviso in (5.39) is always satisfied by a low-pass function $\phi(t)$, i.e., one with a nonzero average value. To calculate the average value we square the scaling equation (5.34) and use the result (5.39),

$$\left(\int_{-\infty}^{\infty} \phi(t)\, dt\right)^2 = \sum_{n=-\infty}^{\infty} \sum_{m=-\infty}^{\infty} h_0[n]\, h_0[m] \int_{-\infty}^{\infty} \phi_{-1,n}(t)\, \phi_{-1,m}(s)\, dt\, ds$$

$$= 2^{-1} \left(\sum_{n=-\infty}^{\infty} h_0[n]\right)^2 = 1. \tag{5.40}$$

We, therefore, assume the following normalization for the (real) scaling function:

$$\int_{-\infty}^{\infty} \phi(t)\, dt = 1. \tag{5.41}$$

The normalization is consistent with the orthonormality of the scaling function, which in view of the assumption that $\phi(t)$ is real, is

$$\|\phi\|^2 = \int_{-\infty}^{\infty} \phi^2(t)\, dt = 1. \tag{5.42}$$

On the other hand, equation (5.42) must be consistent with (5.39), since by squaring and integrating the scaling equation (5.34) and using the orthonormality conditions (5.1) we have

$$\int_{-\infty}^{\infty} \phi^2(t)\, dt = \sum_{n=-\infty}^{\infty} (h_0[n])^2. \tag{5.43}$$

The right hand side of the above equation can be computed from (5.39), and the result must agree with (5.42). To see this we begin by substituting the scaling equation (5.34) into the orthonormality relation (put $m = n = 0$ in (5.2))

$$\int_{-\infty}^{\infty} \phi(t)\, \phi(t-k)\, dt = \delta_{0k}, \tag{5.44}$$

to obtain

$$2 \sum_{n=-\infty}^{\infty} \sum_{m=-\infty}^{\infty} h_0[n]\, h_0[m] \int_{-\infty}^{\infty} \phi(2t-n)\, \phi(2t-2k-m)\ dt = \delta_{0k}. \tag{5.45}$$

Changing the variable of integration $2t \to t$ and using the orthonormality of the scaling function we have the following result.

**Theorem 60.**

$$\sum_{n=-\infty}^{\infty} h_0[n]\, h_0[n-2k] = \delta_{0k}. \tag{5.46}$$

Summing both sides of the above over the index $k = -\infty, \ldots, \infty$ we find

$$\sum_{k=-\infty}^{\infty} \sum_{n=-\infty}^{\infty} h_0[n]\, h_0[n-2k] = \sum_{k=-\infty}^{\infty} \sum_{n=-\infty}^{\infty} h_0[n]\, h_0[n+2k] = 1. \tag{5.47}$$

This sum is next broken into even and odd values of the summation index $n$ to find

$$\left( \sum_{n=-\infty}^{\infty} h_0[2n] \right)^2 + \left( \sum_{n=-\infty}^{\infty} h_0[2n+1] \right)^2 = 1. \tag{5.48}$$

Equation (5.39) too can be written as a sum over even and odd indices,

$$\sum_{n=-\infty}^{\infty} h_0[2n] + \sum_{n=-\infty}^{\infty} h_0[2n+1] = \sqrt{2}. \tag{5.49}$$

Solving the two equations (5.48) and (5.49) for the two unknown sums (even index sum and odd index sum) proves the following theorem.

**Theorem 61.**

$$\sum_{n=-\infty}^{\infty} h_0\left[2n\right] = \sum_{n=-\infty}^{\infty} h_0\left[2n+1\right] = 2^{-1/2}. \tag{5.50}$$

This result together with (5.39) and (5.46) are the only requirements on the low-pass filter coefficients $h_0[n]$ imposed by the normalization (5.41) and orthonormality of the real scaling function (5.1). An important property of the scaling function is the "partition of unity:"

**Theorem 62.**

$$\sum_{n=-\infty}^{\infty} \phi\left(t+n\right) = 1. \tag{5.51}$$

Note that the sum on the left hand side is periodic in $t$ with period 1. The theorem can be proven by summing the orthonormality assumption (5.44) over the index $k$ to obtain

$$\int_{-\infty}^{\infty} \phi\left(t\right)\left(A_t - 1\right) dt = 0, \quad A_t \equiv \sum_{k=-\infty}^{\infty} \phi\left(t+k\right). \tag{5.52}$$

Since $\phi(t)$ is orthogonal to all integer translates of itself, this shows that $A_t = 1$ and hence (5.51) holds.[4] Substituting equation (5.51) in the orthogonality relation (5.7) (putting $m = k = 0$ and summing over the index $n$) we obtain the (necessary) wavelet admissibility relation:

**Theorem 63.**

$$\int_{-\infty}^{\infty} \psi\left(t\right) dt = 0. \tag{5.53}$$

Integrating both sides of the wavelet equation (5.8) and using the normalization equation (5.41) and the above we find the counterpart to equation (5.39):

**Theorem 64.**

$$\sum_{n=-\infty}^{\infty} h_1\left[n\right] = 0. \tag{5.54}$$

---

[4]The partition of unity can also be proven using the frequency domain results in section 5.7.

The orthonormality of the wavelet functions (5.5) and the orthogonality of the scaling and wavelet functions (5.6) can be used in a similar manner to the derivation of (5.46), to arrive at the following result.

**Theorem 65.**

$$\sum_{n=-\infty}^{\infty} h_1[n]\, h_1[n-2k] = \delta_{0k}, \tag{5.55a}$$

$$\sum_{n=-\infty}^{\infty} h_1[n]\, h_0[n-2k] = 0. \tag{5.55b}$$

## 5.6 Examples of Finite Length Filter Coefficients

As we saw earlier, scaling functions of compact support lead to a finite set of coefficients $h_0[n]$. Here we give some examples of the use of the above equations to compute the corresponding finite length low-pass filter coefficients $h_0[n]$ (a systematic method of solution is presented in chapter 7). The simplest case is the Haar scaling function. In section 2.7 we derived the Haar PR filter using the PR-QMF equation (2.35b). Let us assume two nonzero coefficients, namely, $h_0[0]$ and $h_0[1]$ (this follows from our earlier result that if $\phi(t)$ is limited to the interval $[0,1]$ then $h_0$ has only two nonzero coefficients). Equations (5.39) and (5.49) become

$$h_0[0] + h_0[1] = \sqrt{2}, \quad (h_0[0])^2 + (h_0[1])^2 = 1, \tag{5.56}$$

whose solution set is $h_0[0] = h_0[1] = 1/\sqrt{2}$ which is precisely the solution found in section 2.7.

Next we assume four nonzero coefficients for $h_0[n]$. Using (5.46) for $k = 0$ and $k = 1$, together with (5.39), we find three equations for the four unknowns:

$$h_0[0] + h_0[1] + h_0[2] + h_0[3] = \sqrt{2}, \tag{5.57a}$$

$$(h_0[0])^2 + (h_0[1])^2 + (h_0[2])^2 + (h_0[3])^2 = 1, \tag{5.57b}$$

$$h_0[0]h_0[2] + h_0[1]h_0[3] = 0, \tag{5.57c}$$

whose solution is a one-parameter family [51, 49]

$$h_0[0] = (1 - \cos\theta + \sin\theta)/2\sqrt{2}, \quad h_0[1] = (1 + \cos\theta + \sin\theta)/2\sqrt{2},$$

$$h_0[2] = (1 + \cos\theta - \sin\theta)/2\sqrt{2}, \quad h_0[3] = (1 - \cos\theta - \sin\theta)/2\sqrt{2}. \tag{5.58}$$

This parametrization allows the construction of an infinite class of orthogonal wavelets of length 4. The case $\theta = \pi/3$ corresponds to the Daubechies compactly supported wavelet of length 4 (DAUB-4), also found in section 2.7 for unitary filter banks,

$$h_0^{D4} = \left\{1 + \sqrt{3},\ 3 + \sqrt{3},\ 3 - \sqrt{3},\ 1 - \sqrt{3}\right\}\Big/4\sqrt{2}. \qquad (5.59)$$

This is found by augmenting the three equations (5.57) with the additional equation

$$\sum_{n=0}^{3} (-1)^n n h_0 [3 - n] = 0. \qquad (5.60)$$

To see the meaning of this equation we note that for an orthogonal compactly supported wavelet whose filter coefficients are of length $N$ we have

$$\frac{d}{d\omega} H_1 (\omega) = \sum_{n=0}^{N-1} (-in) h_1 [n] e^{-in\omega},$$

which upon setting $N = 4$ and $\omega = 0$ produces equation (5.60) provided that

$$h_1 [n] = (-1)^n h_0 [N - 1 - n]. \qquad (5.61)$$

Thus, if the "alternating flip" relation (the QMF condition of unitary filter banks in section 2.7) between $h_0$ and $h_1$ holds, then the additional constraint (5.60) is equivalent to

$$\sum_{n=0}^{3} n h_1 [n] = 0, \qquad (5.62)$$

which is interpreted as the vanishing of the first moment of the sequence $h_1 [n]$.[5] We discuss equation (5.61), relating the $h_1$ and the $h_0$ filter sequences of compactly supported wavelets, at the end of the next section and in the context of the orthogonality properties of the scaling and the wavelet functions in the frequency domain (see equation (5.88)). The vanishing moment condition, equation (5.62), is discussed in chapter 7.

## 5.7   Frequency Domain Relations

In this section we study the frequency domain results that follow from the scaling and the wavelet equations, (5.2) and (5.7), and the orthonormality

---

[5]Given a sequence $h[n]$, its $k$th moment is defined by $\sum_{n} n^k h[n]$. This is defined in equation (7.10) of chapter 7.

and orthogonality conditions, (5.1), (5.5), and (5.6). Taking Fourier transforms of the scaling equation (5.2) we find

$$\Phi(\omega) = 2^{-1/2} H_0(\omega/2) \Phi(\omega/2), \tag{5.63}$$

where we define $\Phi(\omega)$ as the Fourier transform of $\phi(t)$, and $H_0(\omega)$ as the Fourier transform of $h_0[n]$ . Changing $t$ to $2t$ in the scaling equation (5.2) gives

$$\phi(2t) = \sqrt{2} \sum_{n=-\infty}^{\infty} h_0[n] \phi(4t - n), \tag{5.64}$$

which after taking Fourier transforms yields

$$\Phi(\omega/2) = 2^{-1/2} H_0(\omega/4) \Phi(\omega/4). \tag{5.65}$$

This process can be repeated indefinitely to prove the following theorem.

**Theorem 66.** *The Fourier transform of the scaling function is the infinite resolution product of the discrete time Fourier transform of the low-pass filter sequence $h_0[n]$, i.e.,*

$$\Phi(\omega) = \Phi(0) \prod_{n=1}^{\infty} \left( 2^{-1/2} H_0(\omega/2^n) \right) = \prod_{n=1}^{\infty} \left( 2^{-1/2} H_0(\omega/2^n) \right). \tag{5.66}$$

Note that we have used the normalization of the scaling function (5.41), $\Phi(0) = 1$. Each factor $H_0(\omega/2^n)$ in the infinite product is periodic with period $2^n \pi$, $n = 1, 2, \ldots$, while $\Phi(\omega)$ is, of course, not a periodic function of $\omega$.

An important result on the orthonormality of the scaling functions $\phi(t - n)$ is found by using the Poisson summation formula in theorem 32 and the functions

$$X(\omega) \equiv |\Phi(\omega)|^2, \quad Y(\omega) \equiv \sum_{k=-\infty}^{\infty} X(\omega + 2k\pi).$$

Now $Y(\omega)$ is a periodic function of $\omega$ with period $2\pi$ and has a Fourier series representation

$$Y(\omega) = \sum_{n=-\infty}^{\infty} y_n e^{in\omega}$$

whose coefficients $y_n$ are found using the summation formula of theorem 32

$$y_n = (2\pi)^{-1} x(-n),$$

where $x(t)$ and $X(\omega)$ form a Fourier transform pair. Next we use theorem 27 to write

$$\langle \phi_{0m} \mid \phi_{0n} \rangle = \frac{1}{2\pi} \int\limits_{-\infty}^{\infty} e^{-i(m-n)\omega} X(\omega)\, d\omega = \frac{1}{2\pi} x(n-m) = y_{m-n},$$

where $\phi_{0n} = \phi(t-n)$ denotes the shifted scaling function. Thus,

$$\langle \phi_{0m} \mid \phi_{0n} \rangle \Leftrightarrow y_n = \delta_{0n} \Leftrightarrow Y(\omega) = 1,$$

whcih leads to the following theorem.

**Theorem 67.** *The orthonormal basis property of the scaling functions $\phi_{mn}(t)$ is equivalent to the equation*

$$\sum_{n=-\infty}^{\infty} |\Phi(\omega + 2n\pi)|^2 = 1. \tag{5.67}$$

An important use of equation (5.67) is in orthogonalization of an otherwise independent basis set using the following theorem [15] (an example is presented in the next section).

**Theorem 68.** *Let $\chi(t-n)$, $n \in \mathbb{Z}$ be a linearly independent set of functions that are a basis for $\mathcal{V}_0$, and let $\chi(t) \leftrightarrow X(\omega)$ be a Fourier transform pair. Define the function $\phi(t)$ by its Fourier transform*

$$\Phi(\omega) = \left( \sum_{k=-\infty}^{\infty} |X(\omega + 2k\pi)|^2 \right)^{-1/2} X(\omega). \tag{5.68}$$

*Then the integer translates $\phi(t-n)$, $n \in \mathbb{Z}$, form an orthogonal basis of $\mathcal{V}_0$.*

The results expressed in (5.66) and (5.67) have their counterparts if we use the wavelet equation. Taking Fourier transforms of the wavelet equation (5.7) we find

$$\Psi(\omega) = 2^{-1/2} H_1(\omega/2)\, \Phi(\omega/2), \tag{5.69}$$

where we have defined the Fourier transform pairs $\psi(t) \leftrightarrow \Psi(\omega)$ and $h_1[n] \leftrightarrow H_1(\omega)$. Changing $t$ to $2t$ in the wavelet equation (5.7) and taking Fourier transforms, and repeating the process indefinitely, now leads to the following theorem.

**Theorem 69.** *The Fourier transform of the wavelet function is the infinite resolution product of the discrete time Fourier transform of the high-pass filter sequence $h_1[n]$, i.e.,*

$$\Psi(\omega) = \prod_{n=1}^{\infty} \left(2^{-1/2} H_1(\omega/2^n)\right). \tag{5.70}$$

Again, we have used the normalization of the scaling function (5.41), $\Phi(0) = 1$. Each factor $H_0(\omega/2^n)$ in the infinite product is periodic with period $2^n \pi$, $n = 1, 2, \ldots$, while $\Psi(\omega)$ is, of course, not a periodic function of $\omega$. The following theorem, a complement to theorem 67, is proven using the Poisson summation formula of theorem 32 and arguments similar to those preceding that theorem.

**Theorem 70.** *The orthonormal basis property of the wavelet functions $\psi_{mn}(t)$ and their orthogonality to all the scaling functions, (5.5) and (5.6) are equivalent to the following relations*

$$\sum_{n=-\infty}^{\infty} |\Psi(\omega + 2n\pi)|^2 = 1, \tag{5.71}$$

$$\sum_{n=-\infty}^{\infty} \Psi(\omega + 2n\pi) \Phi^*(\omega + 2n\pi) = 0. \tag{5.72}$$

Equations (5.67), (5.71), and (5.72) are direct results of the orthonormality and orthogonality of the scaling and wavelet functions. We will now investigate their consequences on the filter functions $H_0(\omega)$ and $H_1(\omega)$. This will enable us to solve for the latter in terms of the former and eventually help us implement the orthogonal discrete wavelet transform, figure 5.2, with a knowledge of the filter coefficients $h_0[n]$ alone. We begin by rewriting equations (5.63) and (5.67) for the variable $2\omega$,

$$\Phi(2\omega) = 2^{-1/2} H_0(\omega) \Phi(\omega), \quad \sum_{n=-\infty}^{\infty} |\Phi(2\omega + 2n\pi)|^2 = 1, \tag{5.73}$$

which together imply

$$\sum_{n=-\infty}^{\infty} |\Phi(\omega + n\pi)|^2 |H_0(\omega + n\pi)|^2 = 2. \tag{5.74}$$

The above summation is next split into even and odd indices, which using the $2\pi$ periodicity of $H_0(\omega)$ becomes

$$|H_0(\omega)|^2 \sum_{n=-\infty}^{\infty} |\Phi(\omega + 2n\pi)|^2 + |H_0(\omega + \pi)|^2 \sum_{n=-\infty}^{\infty} |\Phi(\omega + \pi + 2n\pi)|^2 = 2.$$

Now we use equation (5.67) for the first summation and the same equation applied to the variable $(\omega + \pi)$ for the second summation. This together with similar arguments applied to the orthonormal wavelet function, and orthogonal scaling and wavelet functions, equations (5.71) and (5.72), lead to the following result [15, 14].

**Theorem 71.** *Necessary conditions for the orthonormality of the scaling functions $\phi_{mn}(t)$, and the orthonormality of the wavelet functions $\psi_{mn}(t)$, and their mutual orthogonality are the following relations*

$$|H_0(\omega)|^2 + |H_0(\omega + \pi)|^2 = 2, \qquad (5.75a)$$

$$|H_1(\omega)|^2 + |H_1(\omega + \pi)|^2 = 2, \qquad (5.75b)$$

$$H_1(\omega)H_0^*(\omega) + H_1(\omega + \pi)H_0^*(\omega + \pi) = 0. \qquad (5.75c)$$

Note that equation (5.75c) is complex and so its complex conjugate is also a valid condition. All the above conditions can be summarized in matrix form

$$\mathbf{H}^+(\omega) \cdot \mathbf{H}(\omega) = \mathbf{1}, \qquad (5.76)$$

where $\mathbf{1}$ is the $2 \times 2$ identity matrix,

$$\mathbf{H}(\omega) \equiv \frac{1}{\sqrt{2}} \left[ \begin{array}{cc} H_0(\omega) & H_0(\omega + \pi) \\ H_1(\omega) & H_1(\omega + \pi) \end{array} \right] \qquad (5.77)$$

and the superscript "+" indicates Hermitian conjugation (complex conjugation and transposition of the matrix).[6] Equations (5.75a) and (5.75b) show both $H_0(\omega)$ and $H_1(\omega)$ as half-band filter functions. Equation (5.75c) can be rewritten in the form

$$H_1(\omega)/H_0^*(\omega + \pi) = -H_1(\omega + \pi)/H_0^*(\omega) \equiv \eta(\omega),$$

defining the function $\eta(\omega)$. Since $H_0$ and $H_1$ are both periodic functions of $\omega$ with period $2\pi$ and in view of the matrix equation (5.76)) the following conditions must be satisfied

$$\eta(\omega) = \eta(\omega + 2\pi), \quad \eta(\omega) = -\eta(\omega + \pi), \quad |\eta(\omega)| = 1, \qquad (5.78)$$

which result in the following solution.

---

[6]In the parlance of multirate signal processing the matrix $\mathbf{H}(\omega)$ is known as the paraunitary matrix of the two-band perfect reconstruction quadrature mirror filter bank (PR-QMF) which is discussed in section 2.7.

**Theorem 72.** *The general solution to* (5.78) *is of the form*

$$\eta(\omega) = P(2\omega) e^{\pm i\omega}, \quad P(\omega) = P(\omega + 2k\pi), \quad k \in \mathbb{Z}. \tag{5.79}$$

The most common choice is

$$\eta(\omega) = -e^{-i\omega}. \tag{5.80}$$

Thus,

$$H_1(\omega) \equiv -e^{-i\omega} H_0^*(\omega + \pi), \tag{5.81}$$

which after an inverse Fourier transform and using equations (1.72) and (1.73) yields

$$h_1[n] = (-1)^n h_0[1 - n], \quad n \in \mathbb{Z}. \tag{5.82}$$

These are the same QMF conditions of unitary filter banks discussed in section 2.7. The Shannon wavelet filter coefficients (4.71) provide an example of the use of equations (5.81) and (5.82). The discrete analog of the relation (5.81) is

$$H_1(z) = -z^{-1} H_0(-1/z). \tag{5.83}$$

Using this equation, the results of theorem 71 in discrete time can be stated as follows.

**Theorem 73.** *Necessary conditions for the orthonormality of the scaling functions* $\phi_{mn}(t)$ *and the orthonormality of the wavelet functions* $\psi_{mn}(t)$, *and their mutual orthogonality, are equation* (5.83) *and the relation*

$$H_0(z) H_0(1/z) + H_0(-z) H_0(-1/z) = 2. \tag{5.84}$$

Note that theorem 60 can be deduced from the above relation using the results

$$H_0(z) H_0(1/z) = \sum_{n=-\infty}^{\infty} \sum_{k=-\infty}^{\infty} h_0[n] h_0[n-k] z^{-k}, \tag{5.85a}$$

$$H_0(-z) H_0(-1/z) = \sum_{n=-\infty}^{\infty} \sum_{k=-\infty}^{\infty} h_0[n] h_0[n-k] (-z)^{-k}. \tag{5.85b}$$

When the number of elements of the filter coefficients is finite (a finite impulse response filter), say $N$, then $h_0[n]$ and $h_1[n]$ are nonzero only for $n = 0, \ldots, N - 1$ and we have a compactly supported orthogonal wavelet system. As we saw in section 2.7 on unitary filter banks, $N$ must be an even integer. If the $h_0[n]$ indices are assumed to lie in $[0, N - 1]$ and in order that indices for $h_1[n]$ lie in the same interval the solution shown in (5.80) can be changed to (see the discussion after equations (2.29) and (2.30))

$$\eta_N\left(\omega\right) = -e^{-i(N-1)\omega},\tag{5.86}$$

and equation (5.83) becomes

$$H_1\left(z\right) = \left(-z\right)^{-(N-1)} H_0\left(-1/z\right).\tag{5.87}$$

This is equivalent to the time domain relation

$$h_1\left[n\right] = \left(-1\right)^n h_0\left[N-1-n\right], \quad n = 0, \ldots, N-1,\tag{5.88}$$

which is the alternating flip [36] relation expressed by the QMF condition for finite length unitary filter banks (2.32).

The characterization of orthogonal wavelets in terms of the functions $H_0(z)$ and $H_1(z)$ (and the relation (5.87)), is particularly useful in finding suitable wavelet families of compact support; for then simple algebraic methods can be used to solve equation (5.84) and any additional moment equations (such as (5.60) used in solving for the $N = 4$ Daubechies wavelet). The results of this section can be summarized in the following theorem [15].

**Theorem 74.** *Given the function $H_0(e^{i\omega})$ satisfying the equation (5.75a), viz.,*

$$\left|H_0\left(\omega\right)\right|^2 + \left|H_0\left(\omega + \pi\right)\right|^2 = 2,$$

*and the normalization $H_0\left(1\right) = \sqrt{2}$, and the function $H_1(e^{i\omega})$ related to $H_0(e^{i\omega})$ by (5.81), viz., $H_1(\omega) \equiv -e^{-i\omega} H_0^*(\omega + \pi)$, let us define $\Phi(\omega)$ and $\Psi(\omega)$ by (5.66) and (5.70), viz.,*

$$\Phi\left(\omega\right) = \prod_{n=1}^{\infty} \left(2^{-1/2} H_0\left(\omega/2^n\right)\right), \quad \Psi\left(\omega\right) = \prod_{n=1}^{\infty} \left(2^{-1/2} H_1\left(\omega/2^n\right)\right).$$

*Then, the inverse Fourier transforms $\phi(t)$ and $\psi(t)$ are in $L_2(\mathbb{R})$ and they satisfy the relations (5.2) and (5.7), viz.,*

$$\phi\left(t\right) = \sqrt{2} \sum_{n=-\infty}^{\infty} h_0\left[n\right]\phi\left(2t-n\right), \quad \psi\left(t\right) = \sqrt{2} \sum_{n=-\infty}^{\infty} h_1\left[n\right]\phi\left(2t-n\right),$$

*where $h_0$ and $h_1$ are the inverse Fourier transforms of the functions $H_0$ and $H_1$. Moreover, the functions defined in (5.4), viz.,*

$$\psi_{mn}\left(t\right) \equiv 2^{-m/2}\,\psi(2^{-m}t-n) \equiv \mathscr{D}_{2^m}\,\mathscr{T}_n\psi\left(t\right),$$

*form an orthonormal basis of $L_2(\mathbb{R})$ if, and only if,*

$$\left| H_0 \left( e^{i\omega/2^k} \right) \right| > 0, \quad \forall \omega, k.$$

*If $h_0$ (and therefore $h_1$) has a finite number of nonzero terms then $\phi(t)$ and $\psi(t)$ have compact support.*

## 5.8 Orthogonalization of a Basis Set: $b_1$ Spline Wavelet

As mentioned in theorem 68 of the last section, equation (5.67) can be used to orthogonalize an otherwise independent basis set. An example of the use of this theorem is provided by the construction of the $b_1$ spline orthogonal scaling and wavelet functions. Let us define the symmetric $b_1$ spline function (spline function of order 1) by the formula[7]

$$b_1(t) = \begin{cases} 1 - |t|, & -1 \le t \le 1, \\ 0, & \text{otherwise.} \end{cases} \tag{5.89}$$

It is easy to show that $b_1(t)$ satisfies a scaling equation

$$b_1(t) = 0.5\, b_1(2t - 1) + b_1(2t) + 0.5\, b_1(2t + 1). \tag{5.90}$$

The space of all polynomials of a given order is spanned by the b-spline of that order and all its integer translates. For instance, the space $V_0^{(b_1)}$ denoting the space of all square integrable functions that are at most polynomials of degree 1 in any interval $[n, n+1]$, $n \in \mathbb{Z}$, is spanned by $b_1(t)$ and all its integer translates $b_1(t - n)$. The translates, although linearly independent, are not orthogonal, i.e.,

$$\int_{-\infty}^{\infty} b_1(t - m)b_1(t - n)dt \ne 0, \quad m, n \in \mathbb{Z}, \ m \ne n.$$

The Fourier transform of $b_1(t)$ is

$$B_1(\omega) = \left[ \sin(\omega/2)/(\omega/2) \right]^2, \tag{5.91}$$

---

[7]General definitions and **Z** transform properties of splines of all order are discussed in section 7.7.

and (see problem 3 of this chapter's exercises)

$$\sum_{n=-\infty}^{\infty} |B_1(\omega + 2n\pi)|^2 = (2\sin(\omega/2))^4 \sum_{n=-\infty}^{\infty} \frac{1}{|\omega + 2n\pi|^4}$$

$$= (2 + \cos\omega)/48. \tag{5.92}$$

The orthogonalized functions $\phi(t)$ are then given by the inverse Fourier transform of

$$\Phi(\omega) = \frac{4\sqrt{3}}{\sqrt{2 + \cos\omega}} \left(\frac{\sin(\omega/2)}{(\omega/2)}\right)^2. \tag{5.93}$$

The corresponding low-pass filter sequence $h_0[n]$ is the inverse Fourier transform of $H_0(\omega)$. Writing equation (5.63) with $\omega$ replaced with $2\omega$ we have

$$H_0(\omega) = \sqrt{2}\, \Phi(2\omega)/\Phi(\omega) = \sqrt{2}\cos^2(\omega/2)\sqrt{\frac{2 + \cos(\omega)}{2 + \cos(2\omega)}}. \tag{5.94}$$

The high-pass filter sequence $h_1[n]$ is the inverse Fourier transform of $H_1(\omega)$ which is (see equation (5.81))

$$H_1(\omega) = -e^{-i\omega} H_0^*(\omega + \pi) = -\sqrt{2}e^{-i\omega}\sin^2(\omega/2)\sqrt{\frac{2 - \cos(\omega)}{2 + \cos(2\omega)}}. \tag{5.95}$$

Finally, the wavelet function $\psi(t)$ is found using equation (5.69) as the inverse Fourier transform of

$$\Psi(\omega) = -4\sqrt{3}e^{-i\omega/2}\frac{\sin^4(\omega/4)}{(\omega/4)^2}\sqrt{\frac{2 - \cos(\omega/2)}{(2 + \cos\omega)(2 + \cos(\omega/2))}}. \tag{5.96}$$

Although the original $b_1(t)$ spline satisfies a three-term scaling equation, the orthogonalized scaling functions satisfy a scaling equation with an infinite number of coefficients (the inverse Fourier transform of (5.94)). Figure 5.5 shows the orthogonalized $b_1$ spline scaling and wavelet functions, obtained by taking inverse Fourier transforms of the expressions (5.93) and (5.96).

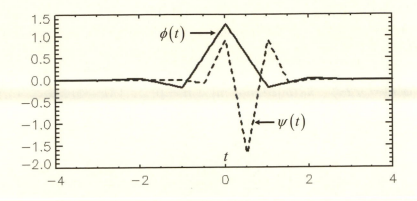

Figure 5.5: $b_1$ spline orthogonal scaling and wavelet functions.

## 5.9 The Cascade Algorithm

The scaling equation (5.2) can be used to numerically calculate and plot a compactly supported scaling function using an iterative scheme given by the equation

$$\phi^{[l]}(t) = \sqrt{2} \sum_{m=0}^{M-1} h_0[m]\, \phi^{[l-1]}(2t - m), \ l = 1, 2, \ldots, \tag{5.97}$$

which is known as the Cascade algorithm [15]. Here we describe a method to compute the values of the scaling and the wavelet functions of compact support on a dyadic grid $n/2^l$ for $n = 0, \ldots, 2^l(M-1)$, at each iteration $l$, assuming known values for the filter coefficients $h_0[m]$ and $h_1[m]$, $m = 0, \ldots, M-1$, and $h_0[0] \neq 0$ and $h_0[M-1] \neq 0$.

Let us assume that the scaling function values $\phi[n/2^l]$ are known at iteration $l > 0$, for $n = 0, \ldots, 2^l(M-1)$. Then at iteration $l+1$ and for even indices $2n$ we have

$$\phi\left[\frac{2n}{2^{l+1}}\right] = \phi\left[\frac{n}{2^l}\right], \tag{5.98}$$

and so they are the known values of the scaling function at iteration $l$. To compute the function at odd values $2n+1$ we use

$$\phi\left[\frac{2n+1}{2^{l+1}}\right] = \sqrt{2} \sum_{m=0}^{M-1} h_0[m]\, \phi\left[\frac{2n+1}{2^l} - m\right]. \tag{5.99}$$

The right hand side of (5.99) is just a linear combination of known values of the scaling function at iteration $l$.

To start the iteration we need the $l = 0$ values, i.e., the scaling function at the integers. This can be computed from the scaling equation on the integers $t = n$,

$$\phi[n] = \sqrt{2} \sum_{m=0}^{M-1} h_0[m]\phi[2n - m], \quad n = 0, \ldots, (M-1), \tag{5.100}$$

together with the equation

$$\sum_{m=0}^{M-1} \phi[m] = 1, \tag{5.101}$$

which follows from the partition of unity (equation (5.51) with $t = 0$), and the relations

$$\phi[0] = \phi[M-1] = 0, \tag{5.102}$$

which follow from the scaling equation (5.100) written for $n = 0$ and $n = M-1$,

$$\phi[0] = \sqrt{2}h_0[0]\phi[0], \text{ and } \phi[M-1] = \sqrt{2}h_0[M-1]\phi[M-1]. \tag{5.103}$$

The values of the scaling function outside the interval $[0, M-1]$ are, of course, zero and so we need only solve equations (5.100), which with a change of summation index can be written as

$$\phi[n] = \sqrt{2} \sum_{m=2n-(M-1)}^{2n} h_0[2n - m]\,\phi[m], \quad n = 1, \ldots, (M-3), \tag{5.104}$$

and

$$\sum_{n=1}^{M-2} \phi[n] = 1. \tag{5.105}$$

Noting that the actual summation indices in equation (5.104) are

$$m_{\min} = \text{Max}\,[1, 2n - (M-1)], \quad m_{\max} = \text{Min}\,[M-2, 2n],$$

equations (5.104) and (5.105) constitute a set of $M-2$ equations in the unknowns $\phi[n]$, $n = 1, \ldots, M-2$. Once the initial scaling function values on the integers 0 through $M-1$ have been found, the wavelet function values at integers are computed from

$$\psi[n] = \sqrt{2} \sum_{m=0}^{M-1} h_1[m]\,\phi[2n - m]. \tag{5.106}$$

All values outside the range $[0, M - 1]$ are, of course, zero. The wavelet function values at a given iteration $l > 0$ are found using equations similar to (5.98) and (5.99),

$$\psi\left[\frac{2n}{2^{l+1}}\right] = \psi\left[\frac{n}{2^l}\right], \tag{5.107a}$$

$$\psi\left[\frac{2n+1}{2^{l+1}}\right] = \sqrt{2}\sum_{m=0}^{M-1} h_1[m]\,\phi\left[\frac{2n+1}{2^l} - m\right]. \tag{5.107b}$$

Figure 5.6 shows the Daubechies scaling and wavelet functions of compact support for $M = 4, 8$. Cases $M = 12, 16$ are shown in 5.7.

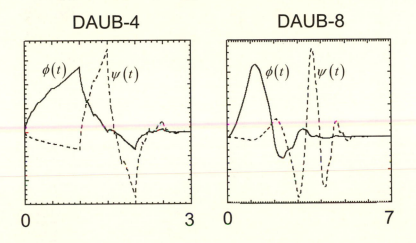

Figure 5.6: Daubechies scaling and wavelet functions of compact support for $M = 4, 8$ after eight iterations.

Table 5.1 lists the $h_0$ coefficients for $M = 4, 8, 12, 16$ Daubechies orthogonal wavelet systems [15]. The corresponding $h_1$ coefficients are found by using equation (5.88). In chapter 7 we discuss the derivation of the Daubechies filter coefficients.

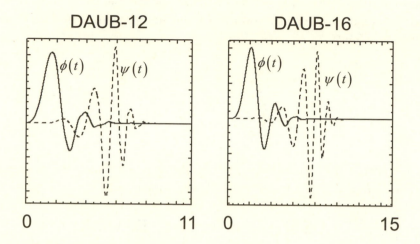

Figure 5.7: Daubechies scaling and wavelet functions of compact support for $M = 12, 16$ after eight iterations.

| D4 | D8 | D12 | D16 |
|---|---|---|---|
| .4829629131445341 | .2303778133088964 | .1115407433501095 | .0544158422431072 |
| .8365163037378077 | .7148465705529154 | .4946238903984533 | .3128715909143166 |
| .2241438680420134 | .6308807679398587 | .7511339080210959 | .6756307362973195 |
| −.1294095225512603 | −.0279837694168599 | .3152503517091982 | .5853546836542159 |
| | −.1870348117190931 | −.2262646939654400 | −.0158291052563823 |
| | .0308413818355607 | −.1297668675672625 | −.2840155429615824 |
| | .0328830116668852 | .0975016055873225 | .0004724845739124 |
| | −.0105974017850690 | .0275228655303053 | .1287474266204893 |
| | | −.0315820393174862 | −.0173693010018090 |
| | | .0005538422011614 | −.0440882539307971 |
| | | .0047772575109455 | .0139810279174001 |
| | | −.0010773010853085 | .0087460940474065 |
| | | | −.0048703529934520 |
| | | | −.0003917403733770 |
| | | | .0006754494064506 |
| | | | −.0001174767841248 |

Table 5.1: $h_0$ coefficients for DAUB-4, DAUB-8, DAUB-12, and DAUB-16 Daubechies compactly supported wavelets of orders 4, 8, 12, and 16.

## 5.10 Biorthogonal Wavelets

The filter sequence pair $\{h_0, h_1\}$ of the Haar wavelet system is the only example of a pair of finite impulse response (FIR) quadrature mirror filters (QMFs) satisfying the relation (5.88) resulting from an orthogonal multi-resolution analysis that have linear phase. Image processing applications often require symmetric FIR filters which for a discrete filter $h[n]$ of finite length $N$ is equivalent to the requirement

$$h\,[n] = h\,[N - n - 1]\,, \ n = 0, \ldots, N - 1.$$

The Fourier transform of a symmetric sequence is then either

$$2\,e^{-iM\omega} \left( h_M/2 + \sum_{m=0}^{M-1} h_m \cos\left( \left( \frac{N-1}{2} - m \right) \omega \right) \right), \quad N = 2M + 1,$$

or

$$2\,e^{-i(M-1/2)\omega} \sum_{m=0}^{M-1} h_m \cos\left( \left( \frac{N-1}{2} - m \right) \omega \right), \quad N = 2M,$$

clearly exhibiting the linear phase property.

Symmetric filter coefficients can be obtained by abandoning the requirement of orthogonality in favour of biorthogonality and, as we shall see later, the resulting filter pairs $\{h_0, h_1, h_0', h_1'\}$ can be of even or odd lengths. We use the first pair $\{h_0', h_1'\}$ for analysis (forward transform) and the second pair $\{h_0, h_1\}$ for synthesis (inverse transform).

The concept of biorthogonality was introduced in section 1.24. Biorthogonal scaling functions form a pair of scaling functions $\phi(t)$ and $\phi'(t)$ satisfying a pair of scaling equations [52]

$$\phi(t) = \sqrt{2} \sum_{n=-\infty}^{\infty} h_0\,[n] \phi\,(2t - n)\,, \tag{5.108a}$$

$$\phi'(t) = \sqrt{2} \sum_{n=-\infty}^{\infty} h_0'\,[n] \phi'\,(2t - n)\,, \tag{5.108b}$$

where the filter coefficients $h_0$ and $h_0'$ are said to be dual to each other. Both scaling functions can be taken to have the normalization (5.41) and so we have

$$\sum_{n=-\infty}^{\infty} h_0\,[n] = \sum_{n=-\infty}^{\infty} h_0'\,[n] = \sqrt{2}. \tag{5.109}$$

These results can also be proven using the frequency domain relations developed below (equation (5.117)). Dilated and translated families are defined as

$$\phi_{mn}(t) = 2^{-m/2}\phi\left(2^{-m}t - n\right), \quad \phi'_{mn}(t) = 2^{-m/2}\phi'\left(2^{-m}t - n\right), \quad (5.110)$$

with scaling equations

$$\phi_{mn}(t) = \sum_{k=-\infty}^{\infty} h_0\,[k]\phi_{m-1,2n+k}(t), \qquad (5.111a)$$

$$\phi'_{mn}(t) = \sum_{k=-\infty}^{\infty} h'_0\,[k]\phi'_{m-1,2n+k}(t). \qquad (5.111b)$$

There is also a pair of wavelet functions $\{\psi(t), \psi'(t)\}$ whose dilated and translated families,

$$\psi_{mn}(t) = 2^{-m/2}\psi\left(2^{-m}t - n\right), \quad \psi'_{mn}(t) = 2^{-m/2}\psi'\left(2^{-m}t - n\right), \quad (5.112)$$

satisfy the wavelet equations

$$\psi_{mn}(t) = \sum_{k=-\infty}^{\infty} h_1\,[k]\phi_{m-1,2n+k}(t), \qquad (5.113a)$$

$$\psi'_{mn}(t) = \sum_{k=-\infty}^{\infty} h'_1\,[k]\phi'_{m-1,2n+k}(t). \qquad (5.113b)$$

The high-pass nature of the wavelet functions (zero DC component) imply

$$\sum_{n=-\infty}^{\infty} h_1\,[n] = \sum_{n=-\infty}^{\infty} h'_1\,[n] = 0. \qquad (5.114)$$

Biorthogonality relations (replacing the orthogonality relations (5.1) and (5.5) and (5.6)), are

$$\int_{-\infty}^{\infty} \phi_{mk}(t)\,\phi'_{mn}(t)dt = \delta_{kn}, \int_{-\infty}^{\infty} \psi_{mk}(t)\,\psi'_{ln}(t)dt = \delta_{ml}\delta_{kn}, \qquad (5.115a)$$

$$\int_{-\infty}^{\infty} \phi_{mk}(t)\,\psi'_{mn}(t)dt = 0, \int_{-\infty}^{\infty} \phi'_{mk}(t)\,\psi_{mn}(t)dt = 0. \qquad (5.115b)$$

Consider the special case of the biorthogonality relation between $\phi_{00}(t) \equiv \phi(t)$ and $\phi'_{0n}(t) \equiv \phi'(t-n)$ in which we use the scaling equations (5.111) for the corresponding scaling functions to obtain

$$\sum_{m=-\infty}^{\infty} h_0\left[2n+m\right] h'_0\left[m\right] = \delta_{0n}. \tag{5.116}$$

Using equations (5.140) and (5.141) of problem 2 of the exercises at the end of this chapter we find the Fourier transform relation

$$H_0^*(\omega)H_0'(\omega) + H_0^*(\omega+\pi)H_0'(\omega+\pi) = 2. \tag{5.117}$$

Similarly, using the biorthogonality relation between $\psi_{00}(t) \equiv \psi(t)$ and $\psi'_{0n}(t) \equiv \psi'(t-n)$ we find the time domain relation

$$\sum_{m=-\infty}^{\infty} h_1\left[2n+m\right] h'_1\left[m\right] = \delta_{0n} \tag{5.118}$$

and the equivalent frequency domain result

$$H_1^*(\omega)H_1'(\omega) + H_1^*(\omega+\pi)H_1'(\omega+\pi) = 2. \tag{5.119}$$

The orthogonality relation between $\phi_{00}(t) \equiv \phi(t)$ and $\psi'_{0n}(t) \equiv \psi'(t-n)$, and that between $\phi'_{00}(t) \equiv \phi'(t)$ and $\psi_{0n}(t) \equiv \psi(t-n)$, on the other hand, lead to the time domain relations

$$\sum_{m=-\infty}^{\infty} h_0\left[2n+m\right] h'_1\left[m\right] = 0, \tag{5.120a}$$

$$\sum_{m=-\infty}^{\infty} h'_0\left[2n+m\right] h_1\left[m\right] = 0, \tag{5.120b}$$

and the equivalent frequency domain results

$$H_0^*\left(\omega\right) H_1'\left(\omega\right) + H_0^*\left(\omega+\pi\right) H_1'\left(\omega+\pi\right) = 0, \tag{5.121a}$$

$$H_1^*\left(\omega\right) H_0'\left(\omega\right) + H_1^*\left(\omega+\pi\right) H_0'\left(\omega+\pi\right) = 0. \tag{5.121b}$$

These together with (5.117) and (5.119) can be written in matrix form. Defining

$$\mathbf{U}\left(\omega\right) \equiv \frac{1}{\sqrt{2}} \left[ \begin{array}{cc} H_0\left(\omega\right) & H_1\left(\omega\right) \\ H_0\left(\omega+\pi\right) & H_1\left(\omega+\pi\right) \end{array} \right], \tag{5.122a}$$

$$\mathbf{U}'\left(\omega\right) \equiv \frac{1}{\sqrt{2}} \left[ \begin{array}{cc} H_0'\left(\omega\right) & H_1'\left(\omega\right) \\ H_0'\left(\omega+\pi\right) & H_1'\left(\omega+\pi\right) \end{array} \right], \tag{5.122b}$$

we have

$$\mathbf{U}^{+}\left(\omega\right)\mathbf{U}'\left(\omega\right)=\mathbf{1}. \tag{5.123}$$

The two equations corresponding to the zeros, (5.121), are identically satisfied if we choose (cf. the QMF condition of orthogonal filters (5.81))

$$H_1\left(\omega\right)\equiv-e^{-i\omega}H_0'^{\,*}\left(\omega+\pi\right),\ H_1'\left(\omega\right)\equiv-e^{-i\omega}H_0^*\left(\omega+\pi\right), \tag{5.124}$$

thus reducing (5.123) to a single equation relating $H_0$ and $H_0'$, namely, equation (5.117). The choices in equation (5.124) imply the time domain relations

$$h_1[n]\equiv(-1)^n h_0'[1-n],\ h_1'[n]\equiv(-1)^n h_0[1-n], \tag{5.125}$$

which are the analogs to equation (5.81) for orthogonal wavelets. In addition, these relations show that the determinants of the matrices $\mathbf{U}(\omega)$ and $\mathbf{U}'(\omega)$ are nonzero and so equation (5.123) can also be written as

$$\mathbf{U}'\left(\omega\right)\mathbf{U}^{+}\left(\omega\right)=\mathbf{1}. \tag{5.126}$$

The analogs of the Fourier transforms of orthogonal scaling and wavelet functions, equations (5.63) and (5.69), now become

$$\Phi\left(\omega\right)=2^{-1/2}H_0\left(\omega/2\right)\Phi\left(\omega/2\right), \tag{5.127a}$$

$$\Phi'\left(\omega\right)=2^{-1/2}H_0'\left(\omega/2\right)\Phi'\left(\omega/2\right), \tag{5.127b}$$

$$\Psi\left(\omega\right)=2^{-1/2}H_1\left(\omega/2\right)\Phi\left(\omega/2\right), \tag{5.127c}$$

$$\Psi'\left(\omega\right)=2^{-1/2}H_1'\left(\omega/2\right)\Phi'\left(\omega/2\right), \tag{5.127d}$$

while the biorthogonality relations in the frequency domain (cf. equations (5.67), (5.71) and (5.72) for orthogonal wavelets) are

$$\sum_{n=-\infty}^{\infty}\Phi^*\left(\omega+2n\pi\right)\Phi'\left(\omega+2n\pi\right)=1, \tag{5.128a}$$

$$\sum_{n=-\infty}^{\infty}\Psi^*\left(\omega+2n\pi\right)\Psi'\left(\omega+2n\pi\right)=1, \tag{5.128b}$$

$$\sum_{n=-\infty}^{\infty}\Phi^*\left(\omega+2n\pi\right)\Psi'\left(\omega+2n\pi\right)=0, \tag{5.128c}$$

$$\sum_{n=-\infty}^{\infty}\Phi'\left(\omega+2n\pi\right)\Psi^*\left(\omega+2n\pi\right)=0. \tag{5.128d}$$

The high-pass nature of the mother wavelets (zero DC components) requires that $H_1(0) = H'_1(0) = 0$ which using the QMF conditions (5.124) implies $H_0(\pi) = H'_0(\pi) = 0$. Setting $\omega = 0$ in equation (5.117) gives $H^*_0(\omega = 0)H'_0(\omega = 0) = 2$. In terms of the **Z** transform functions this is equivalent to $H^*_0(z = 1)H'_0(z = 1) = 2$. Thus, we may take the normalization $H_0(z = 1) = H'_0(z = 1) = \sqrt{2}$ which implies the normalizations in equation (5.109) [52].

The derivation of some biorthogonal wavelets of compact support is presented in chapter 7. Once a biorthogonal pair $\{h_0, h_1\}$ and $\{h'_0, h'_1\}$ has been determined, the Cascade algorithm of section 5.9 can be used to compute the associated scaling and wavelet functions, as, for instance, shown in figures 7.4, 7.12, and 7.13.

## 5.11 Multi-Resolution Analysis Using Biorthogonal Wavelets

Multi-resolution analysis using orthogonal wavelets and computation of the orthogonal discrete wavelet transform coefficients, and the associated inverse, are summarized in the basic forward and inverse relations depicted in figure 5.2. Our task in this section is to show that when using biorthogonal wavelets, the same basic structure is preserved except that the forward transform, or the analysis stage, is now accomplished using the filters $h'_0[-n]$ and $h'_1[-n]$, whereas the inverse, or the synthesis stage, is performed using the pair $h_0[n]$ and $h_1[n]$ [14]. This structure is shown in figure 5.8 which when compared to figure 2.4 of section 2.7 clearly shows a perfect reconstruction filter system if we make the replacements $g_j[n] = h'_j[-n]$, $j = 0, 1$. Thus, the PR equation (2.28) is now

$$\sum_k \left\{ h_0[n - 2k] h'_0[l - 2k] + h_1[n - 2k] h'_1[l - 2k] \right\} = \delta_{nl}. \qquad (5.129)$$

Consider the function $x(t) \in L_2(\mathbb{R})$ and its approximation $x_m(t)$ at scale $m \in \mathbb{Z}$,

$$x_m(t) = \sum_{n=-\infty}^{\infty} c_{mn} \, \phi_{mn}(t), \quad c_{mn} = \int_{-\infty}^{\infty} x(t) \phi'_{mn} dt. \qquad (5.130)$$

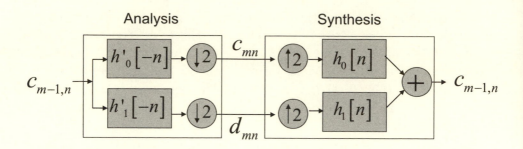

Figure 5.8: Single-stage biorthogonal wavelet analysis and synthesis.

Using the scaling equation (5.111) for $\phi'_{mn}$ in the above we find

$$
c_{mn} = \sum_{k=-\infty}^{\infty} c_{m-1,k} \int_{-\infty}^{\infty} \phi_{m-1,k}(t)\phi'_{m,n}dt
$$

$$
= \sum_{k=-\infty}^{\infty} c_{m-1,k} \int_{-\infty}^{\infty} \phi_{m-1,k}(t) \sum_{l=-\infty}^{\infty} h'_0[l]\ \phi'_{m-1,2n+l}dt
$$

$$
= \sum_{k=-\infty}^{\infty} h'_0[k - 2n]c_{m-1,k}, \tag{5.131}
$$

where the biorthogonality relations (5.115) between $\phi$ and $\phi'$ have been used. In other words, the approximation coefficients at scale $m$ are obtained from those at the previous finer scale $m - 1$, namely, $c_{m-1,k}$, by convolving the latter with the sequence $h'_0 [-k]$,[8] and down-sampling the output by a factor of 2. This is shown on the top left hand portion of figure 5.8.

Next, we expand the detail $x_{m-1}(t) - x_m(t)$ in terms of the corresponding wavelet function $\psi_{mn}(t)$,

$$
x_{m-1}(t) - x_m(t) = \sum_{n=-\infty}^{\infty} d_{mn}\ \psi_{mn}(t), \tag{5.132}
$$

where

$$
d_{mn} = \int_{-\infty}^{\infty} (x_{m-1}(t) - x_m(t))\ \psi'_{mn}dt. \tag{5.133}
$$

---

[8]Recall that convolution with $h'_0 [-k]$ is equivalent to correlation with the sequence $h'_0 [k]$.

Now using the scaling function expansion of equation (5.74) for $x_m(t)$ in the equation for $d_{mn}$ above, and also the orthogonality of the $\phi$ and $\psi'$ functions, the second term on the right hand side of (5.75c) vanishes and we have

$$d_{mn} = \int_{-\infty}^{\infty} x_{m-1}(t)\, \psi'_{mn} dt. \tag{5.134}$$

Using the wavelet expansion for $\psi'_{mn}$ from equation (5.113) and the definition of the scaling function coefficients in equation (5.74) we obtain

$$d_{mn} = \sum_{l=-\infty}^{\infty} h'_1[l] \int_{-\infty}^{\infty} x_{m-1}(t)\phi'_{m-1,2n+l}(t)dt$$

$$= \sum_{l=-\infty}^{\infty} h'_1[l]c_{m-1,2n+l} = \sum_{k=-\infty}^{\infty} h'_1[k-2n]c_{m-1,k}. \tag{5.135}$$

Thus, the detail coefficients in the approximation at scale $m$ are obtained from the approximation coefficients at the previous finer scale $m-1$, namely, $c_{m-1,n}$, by convolving the latter with the sequence $h'_1[-l]$ and down-sampling the output by a factor of 2. This is represented by the bottom left hand part of figure 5.8.

Having established the forward transform, or the analysis stage shown on the left hand part of figure 5.8, in terms of the pair of filters $h'_0[-l]$ and $h'_1[-l]$, we now turn to the right hand part of the same figure and the synthesis (the inverse transform) stage. We use the following identity for $x_{m-1}(t)$

$$x_{m-1}(t) \equiv x_m(t) + (x_{m-1}(t) - x_m(t))$$

$$= \sum_{n=-\infty}^{\infty} c_{mn}\phi_{mn}(t) + \sum_{n=-\infty}^{\infty} d_{mn}\psi_{mn}(t). \tag{5.136}$$

On the other hand, we use the scaling function expansion (5.74) for $x_{m-1}(t)$,

$$x_{m-1}(t) = \sum_{n=-\infty}^{\infty} c_{m-1,n}\phi_{m-1,n}(t). \tag{5.137}$$

Equating the right hand sides of the above two equations, (5.81) and (5.82), and substituting for $\phi_{mn}(t)$ and $\psi_{mn}(t)$ from the scaling equation (5.111) and the wavelet equation (5.113) we obtain

$$
\sum_{n=-\infty}^{\infty} c_{m-1,n}\phi_{m-1,n}(t) = \sum_{n=-\infty}^{\infty} c_{mn} \sum_{l=-\infty}^{\infty} h_0[l]\,\phi_{m-1,2n+l}(t)+
$$
$$
\sum_{n=-\infty}^{\infty} d_{mn} \sum_{l=-\infty}^{\infty} h_1[l]\,\phi_{m-1,2n+l}(t) \tag{5.138}
$$
$$
= \sum_{k=-\infty}^{\infty} \left\{ \sum_{n=-\infty}^{\infty} (h_0[k-2n]\,c_{mn} + h_1[k-2n]\,d_{mn}) \right\} \phi_{m-1,k}(t),
$$

where the last line was obtained from the previous line by a change in the summation index from $l$ to $k \equiv 2n + l$. Comparing the first and the last parts of the above equation we conclude that

$$
c_{m-1,k} = \sum_{n=-\infty}^{\infty} \{h_0[k-2n]c_{mn} + h_1[k-2n]d_{mn}\}. \tag{5.139}
$$

The filtering interpretation of this equation is exactly the same as that following equation (5.25) for orthogonal wavelets. Thus, the coefficients $\{c_{mn}, d_{mn}\}$ at scale $m$ are used to produce the approximation coefficients $c_{m-1,k}$ at the finer scale $m-1$ by the following steps (for a fixed $m$):

1. Up-sample the set $c_{mn}$ by a factor of 2 by inserting one zero between every two samples, and then convolve with the sequence $h_0[n]$.

2. Up-sample the set $d_{mn}$ by a factor of 2 by inserting one zero between every two samples, and then convolve with the sequence $h_1[n]$.

3. Add the results of steps 1 and 2.

This is shown on the right hand portion (synthesis stage) of figure 5.8.

## 5.12   Exercises

1   Prove equation (5.37) by writing

$$
\phi_{mk}(t) = \sum_{j=-\infty}^{\infty} a_{kj}\phi_{m+1,j}(t) + \sum_{j=-\infty}^{\infty} b_{kj}\psi_{m+1,j}(t),
$$

and then computing the coefficients $a$ and $b$ from

$$
a_{kj} = \int_{-\infty}^{\infty} \phi_{mk}(t)\,\phi_{m+1,j}(t)\,dt, \quad b_{kj} = \int_{-\infty}^{\infty} \phi_{mk}(t)\,\psi_{m+1,j}(t)\,dt.
$$

Use the scaling and wavelet equations (5.3) and (5.8) in the equivalent forms

$$\phi_{m+1,j}(t) = \sum_{l=-\infty}^{\infty} h_0\left[l - 2j\right]\phi_{ml}(t), \quad \psi_{m+1,j}(t) = \sum_{l=-\infty}^{\infty} h_1\left[l - 2j\right]\phi_{ml}(t),$$

to find $a_{kj} = h_0[k - 2j]$ and $b_{kj} = h_1[k - 2j]$. Equation (5.37) specifies the exact form of the relation $\mathcal{V}_m = \mathcal{V}_{m+1} \oplus \mathcal{V}^{\perp}_{m+1}$.

**2** Two discrete sequences $h_n$ and $h'_n$, $n \in \mathbb{Z}$ satisfy the relation (cf. equation (5.46))

$$\sum_{n=-\infty}^{\infty} h_n h'_{n+2k} = \delta_{0k}, \quad k \in \mathbb{Z}. \tag{5.140}$$

Multiply both sides of the above with $\exp\left(-2i\omega k\right)$ and sum over $k \in \mathbb{Z}$. Now let $q = 2k$ and argue that the latter can be written as a sum over all $q \in \mathbb{Z}$, rather than even values $q = 2k$, in the form

$$\sum_{n=-\infty}^{\infty} \sum_{q=-\infty}^{\infty} h_n h'_{n+q} e^{-iq\omega}\left(1 + (-1)^q\right) = 2.$$

Define a new summation variable $m = n + q$ and rewrite the above equation in the form

$$\sum_{n=-\infty}^{\infty} \sum_{m=-\infty}^{\infty} h_n h'_m e^{i(n-m)\omega}\left(1 + (-1)^{m-n}\right) = 2.$$

Using $e^{\pm in\pi} = (-1)^{\pm n}$ show that

$$H^*\left(\omega\right) H'\left(\omega\right) + H^*\left(\omega + \pi\right) H'\left(\omega + \pi\right) = 2, \tag{5.141}$$

where $H(\omega)$ and $H'(\omega)$ are the Fourier transforms of the sequences $h_n$ and $h'_n$, respectively, defined by equation (1.72). This is the proof of equation (5.117).

**3** Here we derive a result used in completing equation (5.92) when orthogonalizing the $b_1$ spline scaling function. Use the Haar scaling function's Fourier transform as given in equation (4.53), the orthonormality condition (5.67), and the identity

$$\sin\left(k\pi + \omega/2\right) = (-1)^k \sin\left(\omega/2\right),$$

to obtain the relation

$$\sum_{k=-\infty}^{\infty} \frac{1}{(\omega + 2k\pi)^2} = \frac{1}{4\sin^2(\omega/2)}.$$

Differentiating both sides twice with respect to $\omega$ will give the required result.

**4**  Verify that (5.58) satisfies equations (5.57) and (5.60).

**5**  Prove equations (5.71) and (5.72) using the scaling and wavelet orthonormality and orthogonality relations (5.1), (5.5), and (5.6).

**6**  Show that equations (5.87) and (5.88) are equivalent.

**7**  Verify that the parametrized low-pass filter coefficients $h_0$ given below

$$h_0[0] = [(1 + \cos\theta_1 + \sin\theta_1)(1 - \cos\theta_2 - \sin\theta_2) + 2\cos\theta_1\sin\theta_2]/4\sqrt{2},$$
$$h_0[1] = [(1 - \cos\theta_1 + \sin\theta_1)(1 + \cos\theta_2 - \sin\theta_2) - 2\cos\theta_1\sin\theta_2]/4\sqrt{2},$$
$$h_0[2] = [1 + \cos(\theta_1 - \theta_2) + \sin(\theta_1 - \theta_2)]/2\sqrt{2},$$
$$h_0[3] = [1 + \cos(\theta_1 - \theta_2) - \sin(\theta_1 - \theta_2)]/2\sqrt{2},$$
$$h_0[4] = 1/\sqrt{2} - h_0[0] - h_0[2],$$
$$h_0[5] = 1/\sqrt{2} - h_0[1] - h_0[3].$$

satisfy equations (5.39), (5.46), and (5.50). Prove that when $\theta_1 = \theta_2$, the only nonzero coefficients are the last two, i.e., $h_0[4]$ and $h_0[5]$, which correspond to the Haar low-pass filter sequence. Show further that when $\theta_1 = 0$ and $\theta_2 = \pi/3$, the only surviving four coefficients constitute the DAUB-4 example (5.59). The DAUB-6 compactly supported wavelets of length 6 correspond to the values $\theta_1 = 1.35980373244182$ and $\theta_2 = -0.78210638474440$. This parametric solution is an example of a general result by Pollen for all Daubechies orthogonal wavelets of compact support with even number of coefficients [51, 49].

# CHAPTER 6

# Discrete Wavelet Transform of Discrete Time Signals

## 6.1 Introduction

The multi-resolution analysis framework discussed in the last two chapters dealt exclusively with continuous time signals. In practice, we deal with discretized signals and so it is essential that we specify the appropriate implementation of the discrete wavelet transform to this class of signals.

Our starting point is figure 5.2 and an assumption that the coefficients $c_{Mn}$ should be interpreted as the discrete signal values for some scale $M$. The actual value of the scale is irrelevant since this is our starting scale and our assumption is that the discrete signal values can be taken to be the approximation to the original continuous time signal for a multi-resolution analysis. We begin with the Shannon scaling function

$$\phi^S(t) \equiv \sin(\pi t)/\pi t, \qquad (6.1)$$

choose $\Delta T = 2^M T_0$, and set $T_0 = 1$, without loss of generality, for some integer $M$, and consider the (normalized) shifted and dilated functions

$$\phi_{Mn}^S(t) \equiv \mathscr{T}_{n\Delta T} \mathscr{D}_{2^M} \phi^S(t) = \mathscr{D}_{2^M} \mathscr{T}_n \phi^S(t)$$

$$= (\Delta T)^{-1/2} \sin(\pi(t/\Delta T - n))/\pi(t/\Delta T - n) \qquad (6.2)$$

which form an orthonormal basis of the subspace $\mathscr{V}_M$ of the Shannon multi-resolution analysis spaces. Equation (1.98) of the sampling theorem can now be written in terms of the above functions in the form

$$x_\Omega(t) = \sum_{n=-\infty}^{\infty} (\Delta T)^{1/2} x[n\Delta T] \phi_{Mn}^S(t). \qquad (6.3)$$

Thus, $x_\Omega(t)$ is the orthogonal projection of the signal $x(t)$ onto the subspace $\mathscr{V}_M$ and $(\Delta T)^{1/2} x[n\Delta T]$ are the $c_{Mn}^S$ expansion coefficients in terms of the shifted and dilated versions of the Shannon scaling function $\phi^S(t)$.

To use the discrete time signal values in figure 5.2, the filter coefficients $h_0$ and $h_1$ must indeed be those associated with the Shannon wavelet, as given in equations (4.61) and (4.71). What we intend to do, however, is to apply other filters associated with other orthogonal (or more generally biorthogonal) scaling and wavelet functions. We explore the justifications and required conditions in the next section. It is important to note, however, that a justification is necessary only if we wish to approximate functions. If, on the other hand, our interest is in a discrete transform and its inverse, no justification is necessary: discrete data is transformed by an orthogonal (or biorthogonal) matrix and its inverse is applied after some nonlinear processing of the coefficients, without any need to justify where the actual discrete data came from or whether it represents an approximation, at some level, to some continuous time function.

## 6.2   Discrete Time Data and Scaling Function Expansions

Consider a band-limited signal $x(t)$ whose band is $[-\Omega, +\Omega]$. Let us assume that this functions has two orthogonal projections $x_M^S(t)$ and $x_M(t)$: the first projection has an expansion in terms of the Shannon family of scaling functions $\phi_{Mn}^S(t)$ with expansion coefficients denoted by $c_{Mn}^S$, while the second projection has an expansion in terms of an arbitrary and orthogonal family of scaling function $\phi_{Mn}(t)$ with expansion coefficients $c_{Mn}$. The scale $M$ is defined by $\Delta T = 2^M$ ($T_0 = 1$ as before) and $\Delta T \leq 1/2\Omega$ as required by the sampling theorem 33. The two projections and expansions are

$$x_M^S(t) = \sum_{n=-\infty}^{\infty} c_{Mn}^S \phi_{Mn}^S(t), \quad c_{Mn}^S \equiv \langle \phi_{Mn}^S \mid x \rangle \equiv (\Delta T)^{1/2} x[n\Delta T], \quad (6.4a)$$

$$x_M(t) = \sum_{n=-\infty}^{\infty} c_{Mn} \phi_{Mn}(t), \quad c_{Mn} \equiv \langle \phi_{Mn} \mid x \rangle. \tag{6.4b}$$

The function $\phi_{Mn}(t)$ can also be expanded in terms of the Shannon family of scaling functions

$$\phi_{Mn}(t) = \sum_{k=-\infty}^{\infty} p_{nMk} \phi_{Mk}^S(t), \quad p_{nMk} = \langle \phi_{Mk}^S \mid \phi_{Mn} \rangle. \tag{6.5}$$

We now substitute (6.5) in (6.4b) to find

$$x_M(t) = \sum_{k=-\infty}^{\infty} \sum_{n=-\infty}^{\infty} c_{Mn} p_{nMk} \phi_{Mk}^S(t). \tag{6.6}$$

The assumed equivalence between the two expansions in (6.4a) and (6.4b) then demands the following equality:

$$\sum_{n=-\infty}^{\infty} c_{Mn} p_{nMk} = c_{Mk}^S = (\Delta T)^{1/2} x[n\Delta T]. \tag{6.7}$$

Our original intention to use the discrete time signal values with filters $h_0$ and $h_1$ associated with the scaling function $\phi(t)$ and wavelet function $\psi(t)$, instead of the Shannon filter coefficients, corresponds to the requirement that the two sets of expansion coefficients $c_{Mn}^S$ and $c_{Mn}$ be identical. Thus,

$$c_{Mn} = c_{Mk}^S \iff p_{nMk} = \delta_{nk}. \tag{6.8}$$

Now

$$p_{nMk} = \langle \phi_{Mk}^S \mid \phi_{Mn} \rangle = \int_{-\infty}^{\infty} \phi_{Mn} \frac{\sin\left[\pi \left(2^{-M}t - n\right)\right]}{2^{M/2}\pi \left(2^{-M}t - n\right)} dt. \tag{6.9}$$

It is easy to see that (6.8) is satisfied provided that

$$\phi_{Mn}(t) = 2^{M/2}\delta\left(t - 2^M n\right), \tag{6.10}$$

i.e., $M$ must be sufficiently large and negative so that $\phi_{Mn}(t)$ is nearly a Dirac delta centered at $t = 2^M n$, since then

$$p_{nMk} = \frac{\sin\left[\pi\left(k - n\right)\right]}{\pi\left(k - n\right)} = \delta_{nk}. \tag{6.11}$$

Clear $t = 2^M n$ is precisely the shift that has to be applied to $\phi_{M0}(t)$ to produce the function $\phi_{Mn}(t)$, i.e.,

$$\phi_{Mn}(t) = \mathscr{T}_{2^M n}\, \phi_{M0}(t). \tag{6.12}$$

Thus, $\phi_{Mn}(t)$ is nearly a Dirac delta centered at $t = 2^M n$ provided that $\phi_{M0}(t)$ is a Dirac delta centered at $t = 0$, which would be true if $\phi_{00}(t)$ is a Dirac delta centered at $t = 0$. Let us define the mean time location $\bar{t}_{00}$, the first moment,[1] of the scaling function $\phi(t) = \phi_{00}(t)$ in a similar manner to equation (3.42), i.e.,

$$\bar{t}_{00} \equiv \int\limits_{-\infty}^{\infty} t\phi(t)dt. \tag{6.13}$$

Then the mean time location $\bar{t}_{Mn}$ for $\phi_{Mn}(t)$ is given by

$$\bar{t}_{Mn} = 2^{-M} \int\limits_{-\infty}^{\infty} t\,\phi\left(2^{-M}t - n\right) dt = 2^{M}\left(\bar{t}_{00} + n\right). \tag{6.14}$$

The requirement (6.10) is then equivalent to choosing the scaling function $\phi(t)$ to have a zero first moment, i.e., $\bar{t}_{00} = 0$ which implies $\bar{t}_{Mn} = 2^{M}n$. Equation (6.8) is then (approximately) valid so long as we use a scaling function with at least one zero moment. In that case

$$x[n] \equiv x[n\,\Delta T] \approx 2^{-M/2}c_{Mn} \tag{6.15}$$

and our discrete data can be interpreted as the coefficients in an orthogonal scaling function approximation (with a regular scaling function that has at least one vanishing moment)

$$x_M(t) = \sum_{n=-\infty}^{\infty} c_{Mn}\phi_{Mn}(t) = 2^{M/2} \sum_{n=-\infty}^{\infty} x[n]\,\phi_{Mn}(t) \tag{6.16}$$

at a scale $M$ which is determined by the sampling time interval.

Many useful wavelet systems, in particular the Daubechies orthogonal wavelets, have scaling functions that have nonzero first moments and as such do not satisfy the requirements discussed above. If we denote such a scaling function by $\phi(t)$, the discrete time data samples $x[n]$ cannot theoretically be used for function approximation using scaling function expansions, although they can be used as starting input in figure 5.2 without any further justification. To find the theoretically correct data samples $\hat{x}[k]$ for scaling function approximations we note that the projection of the original data $x(t)$ onto the subspace $\mathcal{V}_0$ spanned by $\phi_{0k}(t)$ is

$$x_0(t) = \sum_{k=-\infty}^{\infty} \hat{x}[k]\,\phi_{0k}(t). \tag{6.17}$$

---

[1]Note that although $\phi(t)$ satisfies the normalization condition (5.41), it is not a strictly positive function and cannot be considered a density in the statistical sense.

If we evaluate the above equation for a finite number $N$ of sampled values we have a set of equations

$$x_0[n] = \sum_{k=-\infty}^{\infty} \hat{x}[k] \phi_{0k}[n], \quad 0 \le n \le N-1 \tag{6.18}$$

which must be solved for the correct input samples $\hat{x}[k]$ using the available data samples $x_0[n]$ and the scaling function values at the integers $0, \ldots, N-1$, namely, $\phi_{0k}[n] = \phi[n-k]$. The solutions, namely, $\hat{x}[k]$ can then be used in scaling function approximations [36] whereas the original sample values $x[n]$ need no processing before being used together with the filters $h_0$ and $h_1$ of figure 5.2 .

To compute the scaling coefficients $c_{M+L,n}$ in the finite resolution approximation equation

$$x_M(t) = \sum_{n=-\infty}^{\infty} c_{M+L,n} \, \phi_{M+L,n}(t) + \sum_{j=1}^{L} \sum_{n=-\infty}^{\infty} d_{M+j,n} \, \psi_{M+j,n}(t), \tag{6.19}$$

we use (6.10) at scale $M + L$, i.e.,

$$\phi_{M+L,n}(t) \approx 2^{(M+L)/2} \delta\left(t - 2^{(M+L)}n\right), \tag{6.20}$$

to find

$$c_{M+L,n} = \int_{-\infty}^{\infty} \phi_{M+L,n}(t) \, x_M(t) \, dt$$

$$= 2^{(M+L)/2} \int_{-\infty}^{\infty} \delta\left(t - 2^{(M+L)}n\right) x_M(t) \, dt$$

$$= 2^{(M+L)/2} x_M\left[2^{(M+L)}n\right]. \tag{6.21}$$

which in view of the relation $\Delta T = 2^M$ becomes

$$c_{M+L,n} = (\Delta T)^{1/2} \, 2^{L/2} x_M\left[2^L n \Delta T\right]. \tag{6.22}$$

Thus, starting with the available discrete data $x[n]$, we compute the wavelet coefficients through $L$ stages, after which data values $x_M\left[2^L n \Delta T\right]$ remain together with the wavelet coefficients $d_{M+j,n}$, for $j = 1, \ldots, L$. The former are precisely the scaling function coefficients that together with the latter wavelet coefficients will suffice to reconstruct the (approximate) signal $x_M(t)$ as expressed in the finite resolution wavelet and scaling function expansion (6.19).

In summary, a given discrete sequence $x[n]$ can be thought of as the coefficients in a multi-resolution approximation scheme for some scale $M$ and some scaling function $\phi(t)$. Although we can never improve the approximation by going to smaller values of $M$ (indeed the actual value is unknown), we can certainly compute the approximation coefficients for higher scale values $M+1, M+2, \ldots$, and in view of figure 5.2 the process is invertible. This multi-stage forward and inverse transform is illustrated for three stages in figure 6.1. The original signal samples $x_M[n]$ are taken as the ap-

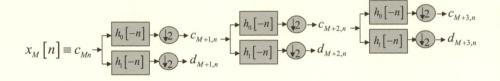

Figure 6.1: Three-stage finite resolution wavelet transform and its inverse.

proximation values $c_{Mn}$ for a starting resolution $M$. Next, these values are correlated with $h_0[n]$ and $h_1[n]$ (alternatively, convolved with their time-reversed versions) and down-sampled (decimated) by a factor of 2 to produce the scaling coefficients $c_{M+1,n}$ and the wavelet coefficients $d_{M+1,n}$. The wavelet coefficients are stored, and the scaling coefficients are run through the same procedure. The process is stopped after $L$ steps at which point the sequences $\{d_{M+1,n}, d_{M+2,n}, \ldots, d_{M+L,n}\}$ together with the last scaling coefficients $\{c_{M+L,n}\}$ are retained. Together, they form the finite resolution wavelet transform coefficients of the signal $x_M[n]$. To recover the original signal we begin with the transform coefficients $c_{M+L,n}$, up-sampled by a factor of 2 and convolved with the sequence $h_0[n]$ and combined (added together) with the output of the convolution of the up-sampled (by a factor of 2) coefficients $d_{M+L,n}$ with the sequence $h_1[n]$ to produce the sequence $c_{M+L-1,n}$. The latter and the corresponding transform coefficients $d_{M+L-1,n}$ are run through the same procedure again, and the process is continued for

$L$ steps, at the end of which the original sequence $x_M[n]$ is recovered exactly.

The implementation of the forward and the inverse processes described above, using finite impulse response filters (finite length sequences $h_0$ and $h_1$) corresponding to scaling and wavelet functions of compact support is discussed in the next section.

## 6.3 Implementing the DWT for Even Length $h_0$ Filters

Here we show an implementation of the finite resolution multi-stage orthogonal discrete wavelet transform, as depicted in figure 6.1, using the results of PR-QMF filter banks discussed in section 2.7. We always assume signal lengths that are powers of 2 to avoid any issues with down- and up-sampling by a factor of 2.

A matrix implementation of correlation (and convolution) is based on the illustration following equation (2.8). This can be modified to include two simultaneous correlations as well as down-sampling by a factor of 2. We illustrate the procedure for $N = 4$ and data samples $x_0$ through $x_7$. We denote the four nonzero low-pass filter coefficients $h_0[n]$ by $h_0, h_1, h_2, h_3$. The first stage of the forward transform producing the scaling function coefficients $c_0$ through $c_3$ and the wavelet function coefficients $d_0$ through $d_3$ are shown below [53]:

$$
\begin{bmatrix}
h_0 & h_1 & h_2 & h_3 & 0 & 0 & 0 & 0 \\
h_3 & -h_2 & h_1 & -h_0 & 0 & 0 & 0 & 0 \\
0 & 0 & h_0 & h_1 & h_2 & h_3 & 0 & 0 \\
0 & 0 & h_3 & -h_2 & h_1 & -h_0 & 0 & 0 \\
0 & 0 & 0 & 0 & h_0 & h_1 & h_2 & h_3 \\
0 & 0 & 0 & 0 & h_3 & -h_2 & h_1 & -h_0 \\
h_2 & h_3 & 0 & 0 & 0 & 0 & h_0 & h_1 \\
h_1 & -h_0 & 0 & 0 & 0 & 0 & h_3 & -h_2
\end{bmatrix}
\begin{bmatrix}
x_0 \\ x_1 \\ x_2 \\ x_3 \\ x_4 \\ x_5 \\ x_6 \\ x_7
\end{bmatrix}
=
\begin{bmatrix}
c_0 \\ d_0 \\ c_1 \\ d_1 \\ c_2 \\ d_2 \\ c_3 \\ d_3
\end{bmatrix} . \quad (6.23)
$$

Note that the last two rows are cyclic translations of the previous two rows necessitated by the requirement that our matrix implementation be orthogonal.[2] Note that even rows $0, 2, 4, 6$ perform both the data correlation with $h_0[n]$ (convolution with the time-reversed version) and the decimation by a factor of 2, resulting in the scaling function coefficients $c_0, c_1, c_2, c_3$. The

---

[2]See reference [54] for the periodic wavelet transform to compute the discrete wavelet transform coefficients of finite length discrete data. Our treatment using orthogonal matrices does not require the periodic extensions for a signal of finite length.

odd rows $(1, 3, 5, 7)$ perform both the correlation with $h_1[n]$ (convolution with the time-reversed version) and the decimation by a factor of 2, resulting in the wavelet function coefficients $d_0, d_1, d_2, d_3$. The latter are stored and the former are processed similarly except that the new matrix is now $4 \times 4$:

$$[c_0, d_0, c_1, d_1, c_2, d_2, c_3, d_3]^T \to [d_0, d_1, d_2, d_3, c_0, c_1, c_2, c_3]^T \ , \qquad (6.24)$$

with superscript $T$ denoting transposition, and thus

$$\begin{bmatrix} h_0 & h_1 & h_2 & h_3 \\ h_3 & -h_2 & h_1 & -h_0 \\ h_2 & h_3 & h_0 & h_1 \\ h_1 & -h_0 & h_3 & -h_2 \end{bmatrix} \begin{bmatrix} c_0 \\ c_1 \\ c_2 \\ c_3 \end{bmatrix} = \begin{bmatrix} c_0' \\ d_0' \\ c_1' \\ d_1' \end{bmatrix} \to \begin{bmatrix} d_0' \\ d_1' \\ c_0' \\ c_1' \end{bmatrix} . \qquad (6.25)$$

This is the last stage for our specific example, after which the two scaling coefficients $c_0', c_1'$ together with the six wavelet coefficients $d_0, d_1, d_2, d_3, d_0', d_1'$ comprise the final transform. Thus, each stage of the transform is performed by first multiplying on the left with an orthogonal matrix of the form shown in equation (6.23) followed by another orthogonal transformation to put the coefficients $c_j$ and $d_j$ in the correct order, as shown in equation (6.25), before the next stage of the transform. The $N \times N$ matrix ($N$ is a power of 2) used to carry out the reordering has the form

$$\begin{bmatrix} 0 & 1 & 0 & & & & & 0 \\ 0 & 0 & 0 & 1 & 0 & & & 0 \\ \vdots & & & & & & & \vdots \\ 0 & & \cdots & & & 0 & 1 & \\ 1 & 0 & & & \cdots & & & 0 \\ 0 & 0 & 1 & 0 & & & & 0 \\ \vdots & & & & & & & \vdots \\ 0 & & & & & 0 & 1 & 0 \end{bmatrix} , \qquad (6.26)$$

which clearly shows it to be an orthogonal matrix. Denoting the initial filter matrix by $\mathbf{H}_N^0$ and the above reordering matrix by $\mathbf{R}_N^0$ we can write an $M$-level transform $\mathbf{y}$ of an initial data vector $\mathbf{x}$ in the form

$$\mathbf{y} = \mathbf{R}_{2^{-M}N}^M \mathbf{H}_{2^{-M}N}^M \cdots \mathbf{R}_{2^{-1}N}^1 \mathbf{H}_{2^{-1}N}^1 \mathbf{R}_N^0 \mathbf{H}_N^0 \mathbf{x} \ , \qquad (6.27)$$

where $\mathbf{H}_{2^{-m}N}^m$ is a block matrix with the block on the bottom right equal to a filter matrix of size $2^{-m}N \times 2^{-m}N$ and the complementary block on the top left equal to an identity matrix of size $(N - 2^{-m}N) \times (N - 2^{-m}N)$

$$\mathbf{H}^m_{2^{-m}N} = \left[ \begin{array}{cc} \mathbf{1}_{N-2^{-m}N} & 0 \\ 0 & \mathbf{H}^0_{2^{-M}N} \end{array} \right], \qquad (6.28)$$

and a similar definition for the reordering matrices $\mathbf{R}^m_{2^{-m}N}$. Thus, the transform (6.27) is equivalent to a single product of the initial data vector, which we assume to be real, with an orthogonal matrix $\mathbf{W}_N$,

$$\mathbf{y} = \mathbf{W}_N \mathbf{x}, \qquad (6.29)$$

whose rows $\mathbf{w}_n$, $0 \le n \le N - 1$, form an orthogonal basis of $\mathbb{R}^N$ (for real data). Every element of the transform vector is then found by taking the dot product of the initial data with one of the rows of the transform matrix,

$$\mathbf{y}_n = \mathbf{w}_n \cdot \mathbf{x}, \quad 0 \le n \le N - 1. \qquad (6.30)$$

The rows $\mathbf{w}_n$ are found by transforming the usual orthonormal basis of $\mathbb{R}^N$: vectors $\mathbf{e}_n$ of length $N$ with 1 in location $n$ and all $N - 1$ remaining entries equal to 0.

We implement the inverse transform using transposes of the forward transform matrices since all matrices in the forward transform are orthogonal. The orthogonality of the forward transform matrices follows from theorems 60 and 64, equations (5.46) and (5.54), and the QMF (alternating flip) condition (5.82), which imply that any two rows are orthogonal to each other (including the cyclically shifted ones) and the sum of the squares of elements of each row is 1. For instance, in the above example and starting with the final set of coefficients $c'_0, c'_1, d'_0, d'_1$, in that order, the last $4 \times 4$ matrix in the forward transform is the first in the inverse transform and it is an orthogonal matrix provided $h_0^2 + h_1^2 + h_2^2 + h_3^2 = 1$ and $h_0 h_2 + h_1 h_3 = 0$. The last two equations are required by the orthogonality relations (5.57). Thus, all the matrices are invertible through transposition and starting with (6.25) we have

$$\left[ \begin{array}{cccc} h_0 & h_3 & h_2 & h_1 \\ h_1 & -h_2 & h_3 & -h_0 \\ h_2 & h_1 & h_0 & h_3 \\ h_3 & -h_0 & h_1 & -h_2 \end{array} \right] \left[ \begin{array}{c} c'_0 \\ d'_0 \\ c'_1 \\ d'_1 \end{array} \right] = \left[ \begin{array}{c} c_0 \\ c_1 \\ c_2 \\ c_3 \end{array} \right]. \qquad (6.31)$$

The next stage is to intersperse the scaling coefficients $c_0, c_1, c_2, c_3$ with the wavelet coefficients $d_0, d_1, d_2, d_3$, and then to compute

$$
\begin{bmatrix}
h_0 & h_3 & 0 & 0 & 0 & 0 & h_2 & h_1 \\
h_1 & -h_2 & 0 & 0 & 0 & 0 & h_3 & h_0 \\
h_2 & h_1 & h_0 & h_3 & 0 & 0 & 0 & 0 \\
h_3 & -h_0 & h_1 & -h_2 & 0 & 0 & 0 & 0 \\
0 & 0 & h_2 & h_1 & h_0 & h_3 & 0 & 0 \\
0 & 0 & h_3 & -h_0 & h_1 & -h_2 & 0 & 0 \\
0 & 0 & 0 & 0 & h_2 & h_1 & h_0 & h_3 \\
0 & 0 & 0 & 0 & h_3 & -h_0 & h_1 & -h_2
\end{bmatrix}
\begin{bmatrix}
c_0 \\ d_0 \\ c_1 \\ d_1 \\ c_2 \\ d_2 \\ c_3 \\ d_3
\end{bmatrix}
=
\begin{bmatrix}
x_0 \\ x_1 \\ x_2 \\ x_3 \\ x_4 \\ x_5 \\ x_6 \\ x_7
\end{bmatrix}. \qquad (6.32)
$$

Figure 6.2 shows 1024 points of a linearly chirped signal

$$x[n] = \sin[2\pi n(0.01 + 0.00005n)], \quad n = 0, \ldots, 1023, \qquad (6.33)$$

with a sampling frequency of 1 Hz, and its DWT coefficients, through six stages of DAUB-4 (5.59) wavelet transform. The original signal and the

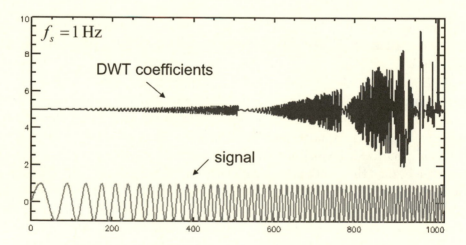

Figure 6.2: Linearly chirped signal and its DWT coefficients (six stages).

reconstruction error (between the original and the inverse transform of the DWT coefficients) are shown in figure 6.3.

With a direct implementation using time series correlation and convolution together with down- and up-sampling by 2 the following observations are useful. According to equation (2.7) when convolving (or correlating) a time series of length $N$ (a power of 2) with a filter of length $M$ (usually $M \ll N$), the output sequence has $N + M - 1$ points. To keep data lengths that are powers of 2 at every stage of the forward transform, therefore, we

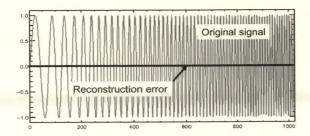

Figure 6.3: Signal and reconstruction error (six-stage DAUB-4 transform).

must discard $M - 1$ points from every output sequence before decimating by a factor of 2. We choose to discard the first $M - 1$ points. In the inverse transform, each input sequence is first up-sampled by a factor of 2 by zero insertion and then convolved with filters of length $M$, and again $M - 1$ points must be discarded, which we now choose to be the last $M - 1$ points. Using the same linearly chirped signal of figure 6.3 but with an additional constant offset of 5, i.e.,

$$x[n] = \sin[2\pi n(0.01 + 0.00005n)] + 5, \quad n = 0, \dots, 1023,$$

we find that the reconstruction (using a six-stage DAUB-4 transform) is exact except for some initial points (in our specific example the first 120 samples of the reconstruction are erroneous), as illustrated in figure 6.4. An

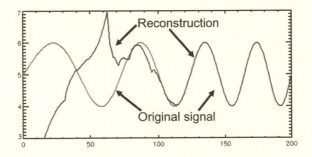

Figure 6.4: Signal and reconstruction (six-stage DAUB-4 transform) using direct convolution method.

exact reconstruction can be performed by doubling the signal size, appending 1024 points with constant value of 5 to the beginning of the signal (avoiding any discontinuous edges), and performing a seven-stage transform on this new signal of length 2048 as shown in figure 6.1 by repeatedly correlating with $h_0[m]$ and $h_1[m]$ coefficients $m = 0, \ldots, M-1$ and each time discarding the first $M-1$ points from the resulting $N+M-1$ outputs before decimating by a factor of 2. The inverse transform is obtained by applying the second part of figure 6.1 in which up-sampling by 2 is achieved by inserting zeros between original samples, and then convolving the corresponding $c$ and $d$ coefficients with $h_0[m]$ and $h_1[m]$, each time discarding the last $M-1$ points from the outputs, before adding the results. Finally, the second segment of length 1024 is extracted from the length-2048 result which will then have zero (to within machine precision limit) reconstruction error.

It is instructive to verify that the matrix implementation of the inverse transform is equivalent to the direct method of up-sampling by a factor of 2, followed by adding the two convolutions. We will limit our discussion to the Haar wavelet; generalization to other orthogonal wavelets is straightforward. Figure 5.2 shows a one-stage forward and inverse orthogonal wavelet transform. Consider the forward transform of four data points, which is implemented by the matrix equation

$$\begin{bmatrix} h_0[0] & h_0[1] & 0 & 0 \\ h_1[0] & h_1[1] & 0 & 0 \\ 0 & 0 & h_0[0] & h_0[1] \\ 0 & 0 & h_1[0] & h_1[1] \end{bmatrix} \begin{bmatrix} x_0 \\ x_1 \\ x_2 \\ x_3 \end{bmatrix} = \begin{bmatrix} c_0 \\ d_0 \\ c_1 \\ d_1 \end{bmatrix}, \qquad (6.34)$$

where $h_0[0] = h_0[1] = h_1[0] = -h_1[1] = 1/\sqrt{2}$ are the Haar filters. The matrix product below performs the grouping of $c$ and $d$ coefficients followed by up-sampling both sets of coefficients by a factor of 2 (inserting zeros between adjacent samples).

$$\begin{bmatrix} 1 & 0 & 0 & 0 \\ 0 & 0 & 0 & 0 \\ 0 & 1 & 0 & 0 \\ 0 & 0 & 0 & 0 \\ 0 & 0 & 1 & 0 \\ 0 & 0 & 0 & 0 \\ 0 & 0 & 0 & 1 \\ 0 & 0 & 0 & 0 \end{bmatrix} \begin{bmatrix} 1 & 0 & 0 & 0 \\ 0 & 0 & 1 & 0 \\ 0 & 1 & 0 & 0 \\ 0 & 0 & 0 & 1 \end{bmatrix} \begin{bmatrix} c_0 \\ d_0 \\ c_1 \\ d_1 \end{bmatrix} = \begin{bmatrix} c_0 \\ 0 \\ c_1 \\ 0 \\ d_0 \\ 0 \\ d_1 \\ 0 \end{bmatrix}. \qquad (6.35)$$

To write the matrix representing the inverse transform, that is, filtering $c$ coefficients with $h_0$ and filtering $d$ coefficients with $h_1$ and summing the results (the rightmost part of figure 5.2), we assume that the first term of the reconstruction is $h_0[0]c_0 + h_1[0]d_0$ which leads to the matrix

$$
\begin{bmatrix}
h_0[0] & 0 & 0 & h_1[0] & 0 & 0 \\
h_0[1] & h_0[0] & 0 & h_1[1] & h_1[0] & 0 \\
0 & h_0[1] & h_0[0] & 0 & h_1[1] & h_1[0] \\
0 & 0 & h_0[1] & 0 & 0 & h_1[1]
\end{bmatrix}. \tag{6.36}
$$

The first three columns perform the convolution of $c$ coefficients with $h_0$ (hence $h_0[1]$ precedes $h_0[0]$) and the next three columns perform the convolution of $d$ coefficients with $h_1$ (hence $h_1[1]$ precedes $h_1[0]$). We note that the inserted zeros on the right hand side of (6.35) (up-sampling by a factor of 2) enable us to replace the values in columns 2 and 5 by zeros and so we arrive at the following matrix to represent the rightmost part of figure 5.2 (convolutions and addition):

$$
\begin{bmatrix}
h_0[0] & 0 & 0 & h_1[0] & 0 & 0 \\
h_0[1] & 0 & 0 & h_1[1] & 0 & 0 \\
0 & 0 & h_0[0] & 0 & 0 & h_1[0] \\
0 & 0 & h_0[1] & 0 & 0 & h_1[1]
\end{bmatrix}. \tag{6.37}
$$

The (left) multiplication of this matrix and the two matrices on the left hand side of (6.35) produce the transpose of the orthogonal forward transform (the correlation and down-sampling) matrix of equation (6.34). The forms of the matrix representations in this section follow from orthonormality relations shown in equations (5.46), (5.55a), and (5.55b).

## 6.4  Denoising and Thresholding

The wavelet transform can be used to denoise signals through an examination of the coefficients at different scales [55, 56]. We illustrate the concept by the following simple example of a 256-point binary signal with added white Gaussian noise with 0 mean and variance 1, shown in figure 6.5. We apply a six-level transform using the Haar wavelet. Figure 6.6 shows the transform coefficients for both the original and the noise-corrupted signals. Clearly the noise coefficients occupy levels 1–4 while the signal is in levels 4–7. In order not to affect the signal we zero out the coefficients in the first three levels and then perform an inverse Haar transform. Figure 6.7 is a plot of the recovered signal together with the original, showing excellent noise cancellation: the original signal levels are clearly derivable from the reconstructed signal.

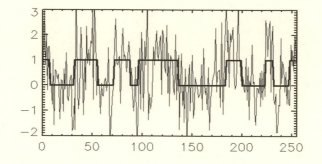

Figure 6.5: Binary signal: original and noise corrupted.

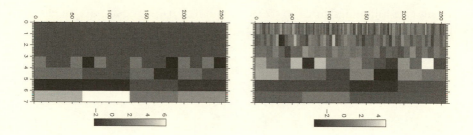

Figure 6.6: Six-level Haar transform of signals in figure 6.5.

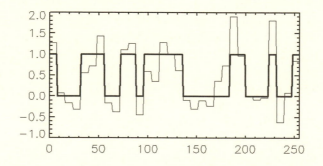

Figure 6.7: Binary signal: original and denoised.

This example illustrates a general concept: additive noise is a low-scale (high-frequency) phenomenon. Thus, suppressing lower-scale coefficients should, in general, produce a less noisy signal. However, when applied to image data (see chapter 9) far more care must be taken in analyzing the lower-scale coefficients with large magnitudes because these are also the same locations at which edge information resides.[3] So instead of a simple suppression of all coefficients in lower scales a thresholding strategy must be used. The threshold values in each scale must be chosen carefully, assuming that coefficients with edge information exceed the calculated thresholds. A simple method to calculate a threshold in a scale is to find the variance $\sigma^2$ of that scale's wavelet coefficients and then set the threshold for that scale to equal $\alpha\sigma$ for some positive constant $\alpha$, which may vary for different scales. More sophisticated procedures are based on fitting a statistical distribution function to the observed wavelet coefficients and then choose thresholds based on percentiles of the fitted distribution [49]. Thresholding techniques can also be used to detect edges in an image, as opposed to local noise at a given scale, using the fact that edge information is present at the same location of the image at all scales. Once a threshold has been determined, it can be applied either as a hard or a soft policy illustrated in figure 6.8.

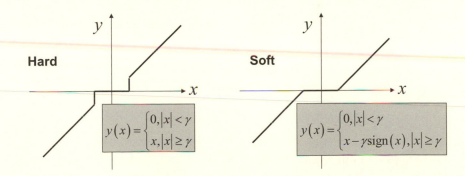

Figure 6.8: Hard and soft thresholding: threshold $= \gamma$.

## 6.5 Biorthogonal Wavelets of Compact Support

We use the Fourier and **Z** transform correspondences of table 2.2 in equation (5.126) to write

$$\begin{bmatrix} H_0'(z) & H_1'(z) \\ H_0'(-z) & H_1'(-z) \end{bmatrix} \begin{bmatrix} H_0(1/z) & H_0(-1/z) \\ H_1(1/z) & H_1(-1/z) \end{bmatrix} = \begin{bmatrix} 2 & 0 \\ 0 & 2 \end{bmatrix}. \qquad (6.38)$$

---

[3]The disadvantage of using the Fourier transform in detecting edges in an image is that the basis functions are not localized and so edge information spreads across all spatial frequencies.

There are actually only two independent equations here, namely,

$$
\begin{bmatrix} H_0'(z) & H_1'(z) \\ H_0'(-z) & H_1'(-z) \end{bmatrix} \begin{bmatrix} H_0(1/z) \\ H_1(1/z) \end{bmatrix} = \begin{bmatrix} 2 \\ 0 \end{bmatrix}, \tag{6.39}
$$

since the other two equations follow from the above by changing $z$ to $-z$. We now choose the relations (5.124) written in terms of $\mathbf{Z}$ transforms[4]

$$
H_1(z) = -z^{-1} H_0'(-1/z), \quad H_1'(z) = -z^{-1} H_0(-1/z). \tag{6.40}
$$

With these choices the second equation in (6.39) (corresponding to the zero on the right hand side) is automatically satisfied, while the first equation relates $H_0$ and $H_0'$,

$$
H_0(z) H_0'(1/z) + H_0(-z) H_0'(-1/z) = 2. \tag{6.41}
$$

Let us assume finite length sequences $h_0[n]$, $N_2 \leq n \leq N_1$ and $h_0'[n]$, $N_2' \leq n \leq N_1'$. Then the corresponding scaling functions, $\phi(t)$ and $\phi'(t)$, defined in (5.108a) and (5.108b), have support in $[N_1, N_2]$ and $[N_1', N_2']$, respectively. In addition, if

$$
h_0[N_1], \ h_0[N_2], \ h_0'[N_1'], \ h_0'[N_2'] \ \neq \ 0, 2^{-1/2},
$$

we have

$$
\phi(N_1) = \phi(N_2)) = \phi'(N_1') = \phi'(N_2')) = 0.
$$

Equation (6.40) shows that the sequence $h_1[n]$ is nonzero for $n \in [1 - N_2', 1 - N_1']$ while $h_1'[n]$ is nonzero for $n \in [1 - N_2, 1 - N_1]$. To determine the support of the wavelet functions we consider equation (5.113a) for $\psi(t)$ which is now

$$
\psi(t) = \sqrt{2} \sum_{n=1-N_2'}^{1-N_1'} h_1[n] \phi(2t - n), \tag{6.42}
$$

which clearly shows that

$$
t \leq \frac{N_1 - N_2' + 1}{2} \quad \text{or} \quad t \geq \frac{N_2 - N_1' + 1}{2} \quad \Rightarrow \quad \psi(t) = 0.
$$

Thus, the support of $\psi(t)$ is the interval $[(N_1 - N_2' + 1)/2, (N_2 - N_1' + 1)/2]$ and the support of $\psi'(t)$ is $[(N_1' - N_2 + 1)/2, (N_2' - N_1 + 1)/2]$. These results [52] are displayed in table 6.1. Furthermore, defining $N \equiv N_2 - N_1$,

---

[4]More generally we could choose $H_1(z) = -z^K H_0'(-1/z)$ and $H_1'(z) = -z^K H_0(-1/z)$. Then the second equation in (6.39) is satisfied provided that $K$ is an odd integer. The usual choice is $K = -1$ [52].

| Function | Support |
|----------|---------|
| $h_0[n]$ | $[N_1,\ N_2]$ |
| $h_0'[n]$ | $[N_1',\ N_2']$ |
| $h_1[n]$ | $[1 - N_2',\ 1 - N_1']$ |
| $h_1'[n]$ | $[1 - N_2,\ 1 - N_1]$ |
| $\phi(t)$ | $[N_1,\ N_2]$ |
| $\psi(t)$ | $[(N_1 - N_2' + 1)/2,\ (N_2 - N_1' + 1)/2]$ |
| $\phi'(t)$ | $[N_1',\ N_2']$ |
| $\psi'(t)$ | $[(N_1' - N_2 + 1)/2,\ (N_2' - N_1 + 1)/2]$ |

Table 6.1: Support of compact biorthogonal scaling and wavelet functions in terms of support of corresponding filter sequences.

$N' \equiv N_2' - N_1'$, and using (6.40) in the first equation of (6.39) we find

$$z^{N_1 - N_2'} \left\{ \sum_{n=0}^{N} h_0\left[N_1 + n\right] z^n \sum_{n'=0}^{N'} h_0'\left[N_2' - n'\right] z^{n'} + \right.$$
$$\left. (-1)^{N_1 - N_2'} \sum_{n=0}^{N} h_0\left[N_1 + n\right] (-z)^n \sum_{n'=0}^{N'} h_0'\left[N_2' - n'\right] (-z)^{n'} \right\} = 2.$$

Comparison of coefficients of $z$ on both sides gives

$$N_1 < N_2', \quad N_1' < N_2, \quad N_2 - N_1' = 2k + 1, \quad N_2' - N_1 = 2k' + 1.$$

Thus,

$$(N_2' - N_1' + 1) - (N_2 - N_1 + 1) \equiv N' - N = 2j, \quad j = 0, 1, 2 \ldots,$$

and since the difference between their lengths is an even number, $h_0$ and $h_0'$ either both have even lengths or both have odd lengths [52].

Matrix implementation of a biorthogonal wavelet pair is analogous to equation (6.23) of section 6.3 with an important difference related to the differing indices of the filter coefficients. For instance, as we shall see later in section 7.6, the 10/6 biorthogonal pair $h_0$ and $h_0'$ have supports in intervals $[-2, 4]$ and $[-5, 3]$, respectively. Using table 6.1 we find the supports of $h_1$ and $h_1'$ to be in intervals $[-2, 3]$ and $[-4, 5]$, respectively, while table 6.1 gives the supports of the filters $h_1$ and $h_1'$ to be $[-4, 5]$ and $-2, 3]$, respectively. Thus, to create the two output sequences that are time-aligned we must use input data $x$ indices $[n - 4, n + 5]$ and $[n - 2, n + 3]$, respectively. The matrix shown in figure 5.8 (with appropriate circular shifts for the end rows to have a square matrix), when applied to input data $x[n]$, corresponds to the following two correlation sums followed by taking the even numbered indices, i.e., the two output sequences of the forward biorthogonal wavelet transform, using the 10/6 filter pair, are

$$\sum_{m=-4}^{5} h'_0[m]\, x\,[2n+m], \qquad \sum_{m=-2}^{3} h'_1[m]\, x\,[2n+m].$$

$$\begin{bmatrix} h'_0[-4] & h'_0[-3] & h'_0[-2] & h'_0[-1] & h'_0[0] & h'_0[1] & h'_0[2] & h'_0[3] & h'_0[4] & h'_0[5] & 0 & \cdots & 0 & 0 \\ 0 & 0 & h'_1[-2] & h'_1[-1] & h'_1[0] & h'_1[1] & h'_1[2] & h'_1[3] & 0 & 0 & \cdots & & 0 & 0 \\ 0 & 0 & h'_0[-4] & h'_0[-3] & h'_0[-2] & h'_0[-1] & h'_0[0] & h'_0[1] & h'_0[2] & h'_0[3] & h'_0[4] & h'_0[5] & \cdots & 0 \\ 0 & 0 & 0 & 0 & h'_1[-2] & h'_1[-1] & h'_1[0] & h'_1[1] & h'_1[2] & h'_1[3] & 0 & 0 & \cdots & 0 \\ 0 & 0 & 0 & 0 & h'_0[-4] & h'_0[-3] & h'_0[-2] & \cdots & & & & & & 0 \\ 0 & 0 & 0 & 0 & 0 & 0 & h'_1[-2] & \cdots & & & & & & 0 \end{bmatrix} \begin{bmatrix} x_{n-4} \\ x_{n-3} \\ x_{n-2} \\ x_{n-1} \\ x_n \\ \vdots \end{bmatrix}$$

Figure 6.9: 10/6 biorthogonal wavelet: forward transform matrix.

The inverse of this matrix is obtained by using the filter pair $\{h_0, h_1\}$ as columns. Positioning of filters in the first two columns may require a linear shift. Once the first two columns are constructed the rest of the columns are found by shifting the first two columns down by two places with appropriate circular shifts for the end columns, as required by the square matrix size. For instance, in the above example, the transpose of the inverse transform matrix is depicted below in figure 6.10. The inverse property is, of course, guaranteed by equation (5.120) and the QMF conditions (5.124).

$$\begin{bmatrix} 0 & 0 & h_0[-2] & h_0[-1] & \cdots & & & \\ h_1[-4] & h_1[-3] & h_1[-2] & h_1[-1] & \cdots & & & \\ 0 & 0 & 0 & 0 & h_0[-2] & h_0[-1] & \cdots \\ 0 & 0 & h_1[-4] & h_1[-3] & h_1[-2] & h_1[-1] & \cdots \end{bmatrix}$$

Figure 6.10: 10/6 biorthogonal wavelet: transpose of inverse transform matrix.

## 6.6 The Lazy Filters

A generalized function solution to the biorthogonal wavelet equations (6.40) and (6.41) is provided by the lazy filters [14]

$$H_0(z) = H_0'(z) = 1, \ H_1(z) = H_1'(z) = -z^{-1}, \tag{6.43}$$

or equivalently

$$H_0(e^{i\omega}) = H_0'(e^{i\omega}) = 1, \ H_1(e^{i\omega}) = H_1'(e^{i\omega}) = -e^{-i\omega}. \tag{6.44}$$

The corresponding time domain filters are $h_0[n] = h_0'[n] = \delta[n]$ and $h_1[n] = h_1'[n] = \delta[n-1]$ which merely separate out the even and odd samples of the input sequence (thus, the name lazy), as shown in figure 6.11, and the associated $\phi(t)$ and $\psi(t)$ are

$$\phi(t) = \phi'(t) = \delta(t), \quad \psi(t) = \psi'(t) = \delta(t - 1/2). \tag{6.45}$$

The lazy functions are not in $L_2(\mathbb{R})$ and $\psi(t)$ is not a wavelet since its integral over the real line does not vanish. However, the method of lifting, described in section 7.8, can be used to build biorthogonal wavelets that are in $L_2(\mathbb{R})$ starting with the lazy filters.

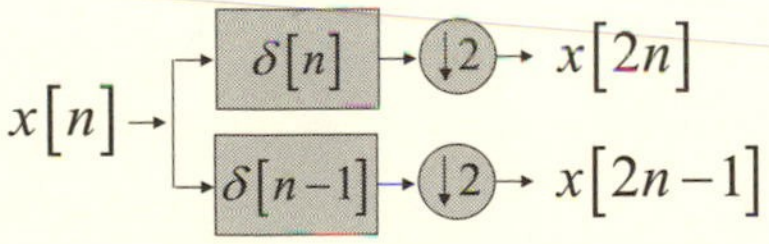

Figure 6.11: The lazy wavelet filter bank.

## 6.7 Exercises

**1** Implement both the matrix and the correlation/convolution forms of the orthogonal wavelet transform. Use the signal $x[n]$ in equation (6.33) and $x[n] + 5$ to reproduce figures 6.2, 6.3, and 6.4.

**2** Verify the equivalence of the matrix and the correlation/convolution forms of the orthogonal wavelet transform (equations (6.34) through (6.37)) using the DAUB-4 wavelet (see table 5.1).

**3** Write a matrix implementation of the UDWT and its inverse using the results of section 5.4. Apply the results to the signal of exercise 1 above.

CHAPTER 7

# Wavelet Regularity and Daubechies Solutions

## 7.1 Introduction

The usefulness of compactly supported wavelet analysis primarily lies in the possibility of representing data with few coefficients, in addition to a reconstruction of the original data that is reasonably smooth. The Haar wavelet is discontinuous in the time domain (or the space domain if used on an image) which allows for excellent localization in that domain but very poor resolution in the Fourier transform domain. Most functions and images of interest, however, consist of smooth sections separated by discontinuities (edges) and a good wavelet would produce coefficients with smaller magnitudes for the smooth parts and larger magnitudes for the edges (discontinuities). For instance, consider the function shown in figure 7.1 with two polynomial sections separated by a very small jump discontinuity at $t = 0$ (not visible). The DWT of this function using the DAUB-16 wavelet

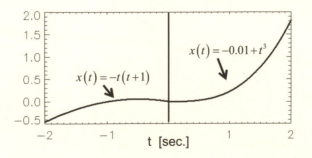

Figure 7.1: A function consisting of a quadratic and a cubic with a very small jump discontinuity at $t = 0$.

for four levels, figure 7.2, shows that at each level the only nonzero coeffi-

193

cients are around the discontinuity at $t = 0$ while they are exactly zero for the rest of the time (except at the discontinuous end points). As we shall see in the next section, the zero response of this wavelet to quadratic and cubic polynomial sections of the data is the result of the mother wavelet's smoothness, which in turn is characterized by the number of zero moments of the mother wavelet function.

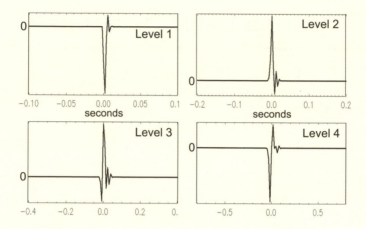

Figure 7.2: Four levels of DWT coefficients using DAUB-16.

## 7.2 Zero Moments of the Mother Wavelet

Given a function $x(t)$ we define its $k$th moment, $k \geq 0$, by

$$\left\langle t^k \right\rangle_x \equiv \int_{-\infty}^{\infty} t^k x(t) \, dt, \tag{7.1}$$

provided that the integral exists. The first moment, $k = 1$, is also denoted by $\bar{t}$, the mean time, as, e.g., in equation (6.13). In what follows we will not include the zeroth moment in our discussion of moments. Thus, a moment will refer to the above definition for $k \geq 1$, and we will explicitly refer to the zeroth moment if the need arises. The reason for this convention is as follows. The zeroth moment of a mother wavelet $\psi(t)$ must always be zero because of the wavelet admissibility condition (5.53) with no restrictions on the higher moments. The zeroth moment of a scaling function $\phi(t)$, however, is always nonzero. Vanishing moment requirements, as additional constraints, always refer to $k$th moments for $k \geq 1$ since the values of the zeroth moments for both the wavelet and the scaling function are fixed by theory. Thus, when we speak of $K$ vanishing moments of $x(t)$ we mean $\langle t^k \rangle_x = 0, 1 \leq k \leq K$. The range for the variable $k$ always includes 0 for vanishing moments of the mother wavelet $\psi(t)$ and never includes 0 for vanishing moments of the scaling function $\phi(t)$.

In order to understand the zero response of the DAUB-16 wavelets to quadratic and cubic portions of the test signal depicted in figure 7.1 consider the general coefficients $d_{mn}$ for a wavelet whose support is limited to the interval $[0, N-1]$ and whose filter coefficients $h_0$ and $h_1$ have $N$ nonzero values,

$$d_{mn} = \langle \psi_{mn} \mid x \rangle = 2^{m/2} \int_0^{(N-1)} x \left( 2^m \left( t + n \right) \right) \psi \left( t \right) dt. \qquad (7.2)$$

Now assuming that $x(t)$ is of the form $t^K$ for some integer $K \geq 1$ we find

$$d_{mn} = \sum_{k=0}^{K} 2^{m(k+1/2)} \binom{K}{k} n^{K-k} \left\langle t^k \right\rangle_\psi, \qquad (7.3)$$

where

$$\binom{K}{k} \equiv \frac{K!}{k! \left( K - k \right)!} \qquad (7.4)$$

is the usual combinatoric factor. The $k = 0$ term vanishes because of the wavelet admissibility condition, equation (5.53), of theorem 5.5. If we further assume that the mother wavelet has exactly $K$ zero moments,

$$\left\langle t^k \right\rangle_\psi = 0, \ 1 \leq k \leq K, \qquad (7.5)$$

then all the wavelet coefficients in (7.3) vanish.

**Theorem 75.** *A mother wavelet $\psi(t)$ with $K \geq 1$ vanishing moments when applied to a polynomial function of degree $\leq K$ will produce wavelet coefficients $d_{mn}$ that are identically zero for all $m, n \in \mathbb{Z}$.*

Vanishing moments of a mother wavelet function also improve the approximation accuracy in a wavelet expansion as stated in the following theorem [15, 14].

**Theorem 76.** *If $x_m(t)$ is the wavelet approximation at level $m$ to a function $x(t)$ in $L_2(\mathbb{R})$, and if the mother wavelet has $K \geq 1$ vanishing moments then the approximation error is bounded as follows*

$$\|x - x_m\|^2 \leq C 2^{-m(K+2)},$$

where $C$ is independent of scale $m$ and the number of zero moments $K$, but depends on the function $x$ and the mother wavelet $\psi$.

The number of zero moments of $\psi(t)$ is related to the number of zero derivatives of $\Psi(\omega)$ at $\omega = 0$. Using the Fourier transform relation (1.62) we have

$$\left\langle t^k \right\rangle_\psi = 2\pi i^k \frac{d^k \Psi(\omega)}{d\omega^k}\bigg|_{\omega=0}, \quad k \geq 1. \tag{7.6}$$

Thus, $K$, the number of zero moments of $\psi$, represents the degree of flatness of the spectrum of the mother wavelet at $\omega = 0$, i.e., flatness of the DC component of the spectrum. It is useful to relate the number of zero derivatives of $H_1(\omega)$ at $\omega = 0$ to the degree of flatness of the DC component of the mother wavelet spectrum, and to the number of zero moments of the mother wavelet. Starting with the frequency domain relation (5.69) and differentiating $k \geq 1$ times we obtain

$$\frac{d^k \Psi(\omega)}{d\omega^k} = 2^{-1/2} \sum_{l=0}^{k} \binom{k}{l} \frac{d^l H_1(\omega/2)}{d\omega^l} \frac{d^{k-l} \Phi(\omega/2)}{d\omega^{k-l}}. \tag{7.7}$$

Setting $\omega = 0$ on both sides, the term $l = 0$ vanishes because $H_1(\omega = 0) = 0$. If, in addition, $H_1(\omega)$ has $k \geq 1$ zero derivatives at $\omega = 0$, we have the result

$$\frac{d^l H_1(\omega)}{d\omega^l}\bigg|_{\omega=0} = 0, \quad 1 \leq l \leq k, \; \Rightarrow \; \frac{d^l \Psi(\omega)}{d\omega^l}\bigg|_{\omega=0} = 0 \; \text{and} \; \left\langle t^l \right\rangle_\psi = 0. \tag{7.8}$$

It is easy to see that the converse holds too. For instance, setting $k = 1$ and $\omega = 0$ in equation (7.7) and using the relations $H_1(0) = 0$ and $\Phi(0) = 1$ will lead to the result

$$\frac{d\Psi(\omega)}{d\omega}\bigg|_{\omega=0} = 0 \Rightarrow \frac{dH_1(\omega)}{d\omega}\bigg|_{\omega=0} = 0 \;.$$

Continuing on to $k = 2$, etc., we have the result that if $\Psi(\omega)$ has $k \geq 1$ zero derivatives at $\omega = 0$, then so does the high-pass filter function $H_1(\omega)$. Next we use the Fourier transform relation

$$\frac{d^k H_1(\omega)}{d\omega^k} = (-i)^k \sum_{n=-\infty}^{\infty} n^k h_1[n] e^{-in\omega}, \; k \geq 1, \tag{7.9}$$

and define the $k$th moment of a discrete time sequence $h[n]$, $k \geq 0$, by

$$\left\langle n^k \right\rangle_h \equiv \sum_{n=-\infty}^{\infty} n^k h[n], \quad k \geq 0. \tag{7.10}$$

Thus, we arrive at the following result.

**Theorem 77.** *For an integer $k \geq 1$ the following conditions are equivalent:*

- *The Fourier transform of the mother wavelet $\Psi(\omega)$ has $k \geq 1$ zero derivatives at $\omega = 0$.*

- *The mother wavelet has $k$ zero moments, $\langle t^l \rangle_\psi = 0$, $1 \leq l \leq k$.*

- *The Fourier transform of $h_1[n]$, i.e., $H_1(\omega)$, has $k$ zero derivatives at $\omega = 0$.*

- *The high-pass filter sequence $h_1[n]$ has $k$ zero moments, $\langle n^l \rangle_{h_1} = 0$, $1 \leq l \leq k$.*

Theorem 77 suggests that the simplest way to ensure a number of vanishing moments on the mother wavelet is to require the same number of zero derivatives of the Fourier transform function $H_1(\omega)$ at $\omega = 0$. It is more useful, however, to find an equivalent requirement on the function $H_0(\omega)$ instead of $H_1(\omega)$. To do this we recall that for an orthogonal wavelet system the two functions $H_1$ and $H_0$ are related through equation (5.81), viz.,

$$H_1(\omega) \equiv -e^{-i\omega} H_0^*(\omega + \pi).$$

We may, therefore, relate the zero derivative requirements on $H_1(\omega)$ at $\omega = 0$ to zero derivatives of $H_0(\omega)$ at $\omega = \pi$. For instance, taking one derivative of the above relation we have

$$\frac{dH_1(\omega)}{d\omega} = -e^{-i\omega}\left(-iH_1^*(\omega + \pi) + \frac{dH_1^*(\omega + \pi)}{d\omega}\right),$$

which upon setting $\omega = 0$ and using the low-pass relation $H_0(\pi) = 0$ gives

$$\frac{dH_1(\omega)}{d\omega}\bigg|_{\omega=0} = -\frac{dH_0^*(\omega + \pi)}{d\omega}\bigg|_{\omega=0} = -\frac{dH_0^*(\omega)}{d\omega}\bigg|_{\omega=\pi}.$$

Thus,

$$\frac{dH_1(\omega)}{d\omega}\bigg|_{\omega=0} = 0 \quad \Leftrightarrow \quad \frac{dH_0(\omega)}{d\omega}\bigg|_{\omega=\pi} = 0.$$

**Theorem 78.** *If the high-pass function $H_1(\omega)$ and the low-pass function $H_0(\omega)$ (Fourier transforms of the high- and low-pass filters) of an orthogonal wavelet system satisfy the relation (5.81), then $H_1(\omega)$ possesses $K \geq 1$ zero derivatives at $\omega = 0$ if, and only if, $H_0(\omega)$ has $K$ zero derivatives at $\omega = \pi$.*

The results of theorems 77 and 78 and the low-pass nature of $H_0(\omega)$, i.e., the condition $H_0(\pi) = 0$, can be used to find the most general form for $H_0(\omega)$ for a compact orthogonal wavelet system whose support is $[0, N-1]$, with $N$ an even integer. To find this general form let us consider the equivalent **Z**-transform function $H_0(z)$. The latter must have a zero of order at least 1 at $z = -1$ since this corresponds to the condition $H_0(\omega = \pi) = 0$. Let us assume that the zero has multiplicity $K \geq 1$, which will ensure that $H_0(\omega)$ has $K-1$ zero derivatives at $\omega = \pi$ and that the mother wavelet has $K-1$ vanishing moments. A further condition is found at $z = 1$ by setting $\omega = 0$ in equation (5.63) to find

$$H_0\left(z = 1\right) = \sqrt{2}.$$

The above together with the $K$ zeros at $z = -1$ restrict $H_0(z)$ to have the form [15]

$$H_0\left(z\right) = \sqrt{2}\left(\frac{1 + z^{-1}}{2}\right)^K P\left(z\right), \qquad (7.11)$$

where $P(z)$ is a polynomial in $z^{-1}$ of degree $N - 1 - K$ (the compact support property ensures that $H_0(z)$ is a polynomial in $z^{-1}$ of degree $N - 1$) and further, $P\left(1\right) = 1$.

A trivial example is provided by the Haar wavelet system for which we found (see section 4.5)

$$H_0\left(z\right) = \left(1 + z^{-1}\right)\big/\sqrt{2},$$

which is clearly a polynomial in $z^{-1}$ of degree 1. Thus, $N = 2$ and $K = 1$, and the Haar $H_0(z)$ has a single zero at $z = -1$, corresponding to $\omega = \pi$, with no zero derivatives at that point. Consequently, the Haar mother wavelet has no vanishing moments.[1]

We will use the above form for $H_0(z)$ and the half-band equation (5.84) (equivalent **Z**-transform form of (5.75a)) to find the coefficients $h_0[n]$, $0 \leq n \leq N - 1$, for Daubechies orthogonal wavelets of compact support for general even $N$. In the next section, however, we explore the effect of the wavelet vanishing moments on the decay rate of the Fourier transform of the scaling function $\Phi(\omega)$. The decay rate of $\Phi(\omega)$, as shown in theorem 29 of the general theory of Fourier transforms, has implications for the differentiability, and hence smoothness, of the scaling function $\phi(t)$.

---

[1]The zeroth moment is, of course, zero for all admissible mother wavelets.

## 7.3 The Form of $H_0(z)$ and the Decay Rate of $\Phi(\omega)$

Evaluating both sides of (7.11) on the unit circle, i.e., $z = e^{i\omega}$, and using the identity

$$1 + e^{-i\omega} \equiv 2e^{-i\omega/2} \cos(\omega/2),$$

we find

$$H_0\left(e^{i\omega}\right) = \sqrt{2}e^{-i\omega K/2} \cos^K(\omega/2) P\left(e^{i\omega}\right), \tag{7.12}$$

which we substitute on the right hand side of the infinite product equation (5.66) in theorem 5.7 to obtain the following expression for the Fourier transform of the scaling function:

$$\Phi(\omega) = \prod_{n=1}^{\infty} \left( e^{-i\omega K/2^{n+1}} \cos^K\left(\omega/2^{n+1}\right) P\left(e^{i\omega/2^n}\right) \right). \tag{7.13}$$

Using the identity

$$\cos\left(\omega/2^{n+1}\right) \equiv \sin\left(\omega/2^n\right)/2\sin\left(\omega/2^{n+1}\right)$$

on the right hand side of (7.13) we obtain

$$\begin{aligned}
\Phi(\omega) &= \lim_{m\to\infty} \left( \frac{\sin(\omega/2)}{2^m \sin\left(\omega/2^{m+1}\right)} \right)^K \prod_{n=1}^{\infty} P\left(e^{i\omega/2^n}\right) \\
&= \left( \frac{\sin(\omega/2)}{\omega/2} \right)^K \prod_{n=1}^{\infty} P\left(e^{i\omega/2^n}\right).
\end{aligned} \tag{7.14}$$

The infinite product on the right hand side of (7.14) is bounded above. Taking absolute values of both sides of (7.14) and using the Cauchy-Schwartz inequality (1.7), without loss of generality, we may write

$$|\Phi(\omega)| \le A |1 + \omega|^{-K}, \tag{7.15}$$

for some positive integer $K$. Theorem 29 then shows that the corresponding scaling function $\phi(t)$ is $K$ times continuously differentiable. The order of the vanishing moments of the mother wavelet $\psi(t)$ is, therefore, equivalent to a differentiability requirement on the associated scaling function $\phi(t)$.

The results of this and the last section can be stated in reverse. The regularity (differentiability) of the scaling function $\phi(t)$ is defined by the decay rate $|\omega|^{-K}$ of its Fourier transform. This in turn is a requirement on the flatness of the function $H_0(z)$ at $z = -1$ ($\omega = \pi$) which translates to a zero of multiplicity $K$ at $z = -1$. The quadrature mirror property, equation (5.88) that relates $h_0$ and $h_1$, then implies that $H_1(z)$ has a zero of multiplicity $K$ at $z = 0$. This in turn implies the vanishing moment condition $\langle n^k \rangle_{h_1} = 0$ on the high-pass sequence $h_1[n]$, which implies the zero-moment condition on the wavelet function $\langle t^k \rangle_\psi = 0$, for $1 \le k \le K-1$, ensuring zero wavelet coefficients for polynomial data sections where the polynomials are of degree $\le K - 1$. The relationship among regularity of wavelets, vanishing moments, and the form of $H_0(z)$ are summarized in the following theorem [15].

**Theorem 79.** *(Daubechies) If the scaling and wavelet function pair $\{\phi(t),$ $\psi(t)\}$ is $K - 1$ times continuously differentiable and produces an orthogonal wavelet system in $L_2(\mathbb{R})$, and it decays according to*

$$|\psi(t)| \le \frac{A_\psi}{|1 + t|^k}, \quad |\phi(t)| \le \frac{A_\phi}{|1 + t|^k}, \quad k > K,$$

*for constants $A_\psi$ and $A_\phi$, then the mother wavelet has $K - 1$ vanishing moments (excluding the 0 moment) and*

$$H_0(z) = \sqrt{2} \left( \frac{1 + z^{-1}}{2} \right)^K P(z),$$

*where $P(e^{i\omega})$ is $2\pi$ periodic.*

## 7.4   Daubechies Orthogonal Wavelets of Compact Support

We assume the support of the scaling and wavelet functions to be $[0, N - 1]$ for even $N$. Thus, the filters $h_0[n]$ and $h_1[n]$ are nonzero for $0 \le n \le N - 1$. We also assume the mother wavelet to have $K - 1$ vanishing moments and use the form (7.11) for $H_0(z)$ to find

$$H_0(z) H_0(1/z) = 2^{-(K-1)} \left( 1 + \frac{z + z^{-1}}{2} \right)^K P(z) P(1/z), \qquad (7.16)$$

where $P(z)$ is a polynomial of degree $N - 1 - K$ in $z^{-1}$ and $P(1) = 1$. Evaluating the above on the unit circle we find

$$\left| H_0(e^{i\omega}) \right|^2 = H_0(e^{i\omega}) H_0(e^{-i\omega}) = 2 \cos^{2K}(\omega/2) \left| P(e^{i\omega}) \right|^2. \qquad (7.17)$$

We now use the results of section 2.6 for spectra of finite length real sequences, equations (2.14) and (2.16), to write

$$\left| P\left(e^{i\omega}\right) \right|^2 = p_0 + 2 \sum_{k=1}^{N-K-1} p_k \cos\left(k\omega\right). \tag{7.18}$$

Defining the variable $s$

$$s \equiv \sin^2\left(\omega/2\right), \tag{7.19}$$

and using the identities

$$\cos\omega \equiv 1 - 2s, \quad \sin^2\omega \equiv 4s(1-s),$$

and

$$\cos\left(k\omega\right) = \sum_{l=0}^{[k/2]} \binom{k}{k-2l} \left(-1\right)^l \cos^{k-2l}\left(\omega\right) \sin^{2l}\left(\omega\right)$$

where $[k/2]$ denotes the largest integer not exceeding $k/2$, we can write (7.17) as a function of $s$,

$$\left| H_0\left(e^{i\omega}\right) \right|^2 = 2(1-s)^K u(s), \tag{7.20}$$

where $u(s)$, given by

$$u\left(s\right) \equiv \left| P\left(e^{i\omega}\right) \right|^2 \equiv P\left(z\right) P\left(1/z\right), \tag{7.21}$$

is constructed by converting all multiple angle cosines in equation (7.18) into powers of $\cos(\omega)$ that are then converted into powers of $s$. Note that the above equation and the fact that $0 \le s \le 1$ mean that $u(s) \ge 0$. In addition, we have the relation

$$s(z) = 1/2 - \left(z + z^{-1}\right)/4, \tag{7.22}$$

which clearly shows that

$$s\left(-z\right) = 1 - s\left(z\right) = 1/2 + \left(z + z^{-1}\right)/4. \tag{7.23}$$

Finally we substitute (7.21) into the perfect reconstruction condition, equation (5.75a) of theorem 71, viz.,

$$\left| H_0\left(e^{i\omega}\right) \right|^2 + \left| H_0\left(e^{i(\omega+\pi)}\right) \right|^2 = 2,$$

and use the relations (7.22) and (7.23) to find the following functional relation for $u(s)$

$$(1 - s)^K u(s) + s^K u(1 - s) = 1. \qquad (7.24)$$

Note that if a solution $u(s)$ exists then the function $u(s) + s^K v(s)$ is also a solution so long as

$$v(s) = -v(1 - s). \qquad (7.25)$$

The specific form of $v(s)$ is undetermined but must be restricted to produce a positive function $u(s)$ for $0 \le s \le 1$.[2] To find a polynomial solution for $u(s)$ we substitute the following form into the functional relation (7.24):

$$u(s) = \sum_{k=0}^{K-1} a_k s^k, \qquad (7.26)$$

in addition to using the binomial expansion

$$(1 - s)^K \equiv \sum_{k=0}^{K} \binom{K}{k} (-s)^k,$$

and determine the coefficients $a_k$ by setting the coefficient of $s^0$ on the left hand side equal to 1, and the coefficients of $s^k$, $1 \le k \le K - 1$, equal to 0. The first four coefficients are shown below:

$$a_0 = 1, \ a_1 = K, \ a_2 = K(K + 1)/2, \ a_3 = K(K + 1)(K + 2)/6,$$

and so the solution to (7.24), including the additional function $v(s)$ that satisfies the condition (7.25), is [15]

$$u(s) = \sum_{k=0}^{K-1} \binom{K + k - 1}{k} s^k + s^K v(s). \qquad (7.27)$$

The Daubechies family of filter coefficients are found by setting $v(s) = 0$ and $K = N/2$. Thus, given even $N$ and $2K = N$, we find $u(s)$ from equation (7.27) and evaluate it as a function $z$ using the relation (7.22), viz., $s = 1/2 - (z + z^{-1})/4$. According to equation (7.21) this is now the function $P(z)P(1/z)$ which we proceed to factor using the spectral factorization theorem 43, to find $P(z)$. The latter must be normalized to have $P(1) = 1$. Next we form $H_0(z)$ using equation (7.11), which when written as a

---

[2]If the function $v(s)$ is written as $v(1/2 - s)$ then the constraint (7.25) is equivalent to $v(1/2 - s) = -v(s - 1/2)$.

polynomial in $z^{-1}$ will finally yield the finite set of coefficients $h_0[n]$, $0 \le n \le N - 1$. The Daubechies solution for a given even integer $N$ corresponds to a magnitude squared low-pass filter $\left|H_0\left(e^{i\omega}\right)\right|^2$ that has equal zero derivatives at $\omega = 0$ and $\omega = \pi$ and is known as the maximally flat solution.[3]

As an example consider the DAUB-4 wavelet defined by $N = 4$ and $K = 2$. This wavelet has one vanishing moment (in addition to the vanishing zeroth moment). We form $u(s)$:

$$u(s) = 1 + 2s = 2 - \frac{z + z^{-1}}{2} = -\frac{1}{2}\left(1 - z_1 z^{-1}\right)\left(z - z_2\right),$$

where the two roots $z_{1,2}$ are $z_1 = 2 - \sqrt{3}$ and $z_2 = 2 + \sqrt{3}$. According to the spectral factorization theorem 43 we keep the root $z_1$ which is inside the unit circle and obtain the suitably normalized function

$$P(z) = \frac{1 + \sqrt{3}}{2}\left(1 - z_1 z^{-1}\right),$$

which when used in equation (7.11) yields $H_0(z)$:

$$H_0(z) = \sqrt{2}\left(\frac{1 + z^{-1}}{2}\right)^2 \frac{1 + \sqrt{3}}{2}\left(1 - z_1 z^{-1}\right)$$

$$= \left\{1 + \sqrt{3} + \left(3 + \sqrt{3}\right)z^{-1} + \left(3 - \sqrt{3}\right)z^{-2} + \left(1 - \sqrt{3}\right)z^{-3}\right\} / 4\sqrt{2}.$$

This corresponds to precisely the same set of four coefficients $h_0[m]$, $0 \le m \le 3$, shown in equation (5.59) of section 5.6 that were found using time domain methods, and whose corresponding scaling and wavelet functions are shown in figures 5.6 and 5.7.

It can be shown that the Daubechies orthogonal wavelets whose support is $[0, N - 1]$ for even integers $N$ attain the maximum number $N/2 - 1$ of vanishing moments of the mother wavelet (excluding the vanishing zeroth moment). The corresponding scaling functions, however, do not have any vanishing moments. In fact, the only moment condition they satisfy is the normalization $\left\langle t^0 \right\rangle_\phi = 1$. As shown in section 6.2, approximation to functions using scaling function expansions of discrete time data requires at least one vanishing moment of the scaling function. Although from the standpoint of orthogonal transform theory this is not necessary, in approximation problems that use scaling function series expansions, it is desirable

---

[3]Note that the maximally flat solutions have no vanishing scaling function moments. Indeed, vanishing scaling function moments would require zero derivatives of the function $H_0(\omega)$, and not its magnitude squared, at $\omega = \pi$, as shown in theorem 80 in the next section.

to have vanishing moments of the scaling function, in addition to the vanishing moments of the mother wavelet. We will look at these requirements in the next section.

## 7.5    Wavelet and Scaling Function Vanishing Moments

We found in section 6.2 that in order to use scaling function approximations with discrete sampled data we must use scaling functions with at least one zero moment. In this section we investigate the Daubechies solutions that have a number of vanishing scaling function moments in addition to vanishing wavelet function moments.

Let us require $K - 1$ vanishing moments for both the wavelet and the scaling functions. If we include the vanishing of the wavelet zero moment, i.e., the mother wavelet admissibility condition, we have the following requirements

$$\left\langle t^k \right\rangle_\psi = 0, \ 0 \le k \le K - 1, \quad \left\langle t^j \right\rangle_\phi = 0, \ 1 \le j \le K - 1, \quad \left\langle t^0 \right\rangle_\phi = 1. \quad (7.28)$$

Now the $k$th zero moment of the scaling function $\phi(t)$ implies that the $k$th derivative of the Fourier transform $\Phi(\omega)$ vanishes at $\omega = 0$. In addition, differentiation of the frequency domain relation (5.63) with respect to $\omega$ and setting $\omega = 0$ gives

$$\frac{d\Phi(\omega)}{d\omega}\bigg|_{\omega=0} = \frac{1}{\sqrt{2}} \left( \frac{dH_0(\omega/2)}{d\omega}\bigg|_{\omega=0} + \sqrt{2}\ \frac{d\Phi(\omega/2)}{d\omega}\bigg|_{\omega=0} \right),$$

which shows that a zero derivative of $\Phi(\omega)$ at $\omega = 0$ implies that $H_0(\omega)$ has a zero derivative there, etc. Thus, we arrive at the following theorem for the scaling function vanishing moments (the scaling function form of theorem 77 for the wavelet function).

**Theorem 80.** *For an integer $k \ge 1$ the following conditions are equivalent:*

- *The Fourier transform of the scaling function $\Phi(\omega)$ has $k \ge 1$ zero derivatives at $\omega = 0$.*

- *The scaling function has $k$ zero moments, $\left\langle t^l \right\rangle_\phi = 0, \ 1 \le l \le k$.*

- *The Fourier transform of $h_0[n]$, i.e., $H_0(\omega)$, has $k$ zero derivatives at $\omega = 0$.*

- *The low-pass filter sequence $h_0[n]$ has $k$ zero moments, $\langle n^l \rangle_{h_0} = 0$, $1 \le l \le k$.*

The $K-1$ vanishing moments of the wavelet require the form (7.11) for $H_0(z)$ while the scaling function vanishing moments require the form [15]

$$H_0(z) = \sqrt{2} + \sqrt{2} \left( \frac{1 - z^{-1}}{2} \right)^K Q(z), \qquad (7.29)$$

where $Q(z)$ is a polynomial in $z^{-1}$, and the sequence $h_0[n]$ is nonzero for $N_1 \le n \le N_2$ where $N_1, N_2 \in \mathbb{Z}$ and must be determined in terms of $K$. Major simplifications occur when $K$ is even and so we set $K = 2L$. Equating the two equivalent expressions for $H_0(z)$, equations (7.11) and (7.29), and using the relation

$$\left( \frac{1 \pm z^{-1}}{2} \right)^{2L} \equiv z^{-1} \left( \frac{1}{2} \mp \frac{z + z^{-1}}{4} \right)^L , \qquad (7.30)$$

and (7.22) we find the equation

$$(1 - s)^L z^{-L} P(z) - s^L (-z)^{-L} Q(z) = 1, \qquad (7.31)$$

which upon introduction of two new functions

$$\hat{P}(z) \equiv z^{-L} P(z), \quad \hat{Q}(z) \equiv -(-z)^{-L} Q(z) \qquad (7.32)$$

reduces to

$$(1 - s)^L \hat{P}(z) + s^L \hat{Q}(z) = 1. \qquad (7.33)$$

Comparing this equation with (7.24) shows that the solutions are of the form

$$\hat{P}(z) = \sum_{l=0}^{L-1} \binom{L + l - 1}{l} s^l + s^L p(z),$$

$$\hat{Q}(z) = \sum_{l=0}^{L-1} \binom{L + l - 1}{l} (1 - s)^l - (1 - s)^L p(z), \qquad (7.34)$$

where $p(z)$ is an arbitrary function of $z$. Next we must find an appropriate polynomial $p(z)$ to ensure that the perfect reconstruction condition, equation (5.75a) of theorem 71, holds. That condition is

$$\cos^{2K}(\omega/2) \left| P(e^{i\omega}) \right|^2 + \sin^{2K}(\omega/2) \left| P(e^{i(\omega + \pi)}) \right|^2 = 1, \qquad (7.35)$$

which is equivalent to

$$(1-s)^{2L}\, P\left(z\right) P\left(1/z\right) + s^{2L} P\left(-z\right) P\left(-1/z\right) = 1. \qquad (7.36)$$

The form

$$p\left(z\right) = \sum_{l=0}^{2L-1} p_l z^{-l} \qquad (7.37)$$

is the choice made by Daubechies [15] when solving for these orthogonal wavelets, that were named Coiflets, after Coifman, who had first suggested the vanishing moments of the scaling function as an additional requirement.

For instance, consider the case $K = 2$, $L = 1$. The corresponding Coiflet is limited to the range $[-2, 3]$ and so the low-pass filter coefficients $h_0[n]$ are nonzero for $N_1 = -2 \leq n \leq N_2 = 3$ and the filter length is 6. There are two coefficients $p_0$ and $p_1$ for $p(z)$; substituting them into (7.36) and equating the coefficients of even powers of $z$ on both sides (the odd power coefficients vanish identically), we obtain the following set of dependent equations:

$$p_0^2 + p_1^2 - 4p_1 = 0,$$
$$p_0^2 + p_1^2 + 4p_0 - 4 = 0, \qquad (7.38)$$
$$3p_0^2 + 3p_1^2 + 4p_1 + 16p_0 + 48 = 64.$$

The solution that leads to an orthonormal wavelet system is given by

$$\{p_0, p_1\} = \left\{\sqrt{7} - 1, 3 - \sqrt{7}\right\}\Big/2 = \{0.82287566,\ 0.17712434\}, \qquad (7.39)$$

and the corresponding low-pass filter $H_0(z)$ is

$$H_0\left(z\right) = -0.0727326\ z^2\ + 0.3378977\ z\ + 0.852572+$$
$$0.3848648\ z^{-1} - 0.072733\ z^{-2} - 0.0156557\ z^{-3}. \qquad (7.40)$$

Figure 7.3 shows this Coiflet's scaling and wavelet functions.

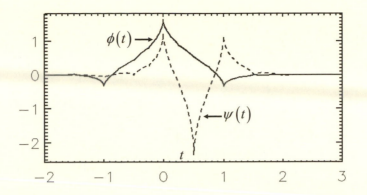

Figure 7.3: Coiflet scaling and wavelet functions of order 6 with two vanishing scaling function moments and one vanishing wavelet function moment.

## 7.6   Biorthogonal Wavelets of Compact Support

A biorthogonal wavelet pair is characterized by filter sequence pairs $\{h_0, h_1\}$ and $\{h_0', h_1'\}$ as shown in figure 5.8 and discussed in section 6.5. The corresponding frequency domain functions satisfy equations (5.117), (5.119), and (5.121). The biorthogonal frame property of finite length sequences (biorthogonal wavelets of compact support) and the required decay properties of the scaling function spectra are summarized in the following theorem [52].

**Theorem 81.** *(Cohen, Daubechies, Feauveau) Consider the finite length real sequences $h_0[n]$, $0 \le n \le N - 1$, and $h_0'[n]$, $0 \le n \le N' - 1$ satisfying the finite length equivalent of equation (5.116), namely,*

$$\sum_{m=m_0}^{m=m_1} h_0\left[2n + m\right] h_0'\left[m\right] = \delta_{0n},$$

*where $m_0 = Max\,[0, -2n]$ and $m_1 = Min\,[N' - 1, N - 1 - 2n]$, define the Fourier transform functions $\Phi(\omega)$ and $\Phi'(\omega)$ through equations (5.127a) and (5.127b), and assume that for some real and positive constants $C$ and $\epsilon$*

$$\left|\Phi\left(\omega\right)\right| \le C\left(1 + |\omega|\right)^{-1/2-\varepsilon}, \quad \left|\Phi'\left(\omega\right)\right| \le C\left(1 + |\omega|\right)^{-1/2-\varepsilon}.$$

*Then the wavelet functions $\psi_{mn}$ and $\psi_{mn}'$ defined through equations (5.113) and (5.125) constitute a frame and its dual in $L_2(\mathbb{R})$. Moreover, they satisfy the orthogonality relation $\langle \psi_{mk} \mid \psi_{ln}' \rangle = \delta_{ml}\delta_{kn}$ if, and only if, $\langle \phi_{00} \mid \phi_{0k}' \rangle = \delta_{0k}$.*

To construct compactly supported wavelets using $\mathbf{Z}$ transforms we use the QMF conditions (6.40), viz.,

$$H_1\left(z\right) = -z^{-1}H_0'\left(-1/z\right), \quad H_1'\left(z\right) = -z^{-1}H_0\left(-1/z\right), \qquad (7.41)$$

and solve the perfect reconstruction (PR) equation (6.41), viz.,

$$H_0\left(z\right)H_0'\left(1/z\right) + H_0\left(-z\right)H_0'\left(-1/z\right) = 2. \qquad (7.42)$$

Using the relation (7.22) for the variable $s$ in terms of $z$, we define the function $g(s)$ by

$$g\left(s\right) \equiv 2^{-1}H_0\left(z\right)H_0'\left(1/z\right). \qquad (7.43)$$

Equation (7.42) together with the relation (7.23) then lead to the following equation for the function $g(s)$:

$$g(s) + g(1 - s) = 1. \qquad (7.44)$$

Polynomial solutions to the above equation upon factorization will produce solutions for the finite filter sequences $h_0$ and $h_0'$ [52]. Now equation (7.44) is formally identical to equation (7.24) if we identify the function $g(s)$ as follows:

$$g(s) \equiv (1 - s)^K u(s), \qquad (7.45)$$

where $u(s)$ is given by equation (7.27). Using the relation (7.22) and setting $v(s) = 0$ in (7.27) we write $g(s)$ in terms of $z$

$$g\left(z\right) = 2^{-2K}z^{-K}\left(1 + z\right)^{2K}\sum_{k=0}^{K-1}\binom{K + k - 1}{k}\left(\frac{1}{2} - \frac{z + z^{-1}}{4}\right)^k. \qquad (7.46)$$

Thus, choosing an integer $K$ we calculate the function $g(z)$ and factorize it to find the desired biorthogonal filter functions $H_0(z)$ and $H_0'\left(1/z\right)$ according to equation (7.43). Factorizations based on the spectral decomposition theorem 2.6 yield the Daubechies orthogonal filters (wavelets) of compact support with $K$ vanishing moments, as was shown earlier in this chapter. Other factorizations will yield biorthogonal filters (wavelets) of compact support.

In this section we illustrate a factorization method based on including zeros of the function $u(z)$ in both $H_0(z)$ and $H_0'\left(1/z\right)$ for the case $K = 4$. The associated $u(s)$ from equation (7.27) with $v(s) = 0$, is

$$u(s) = 1 + 4s + 10s^2 + 20s^3, \qquad (7.47)$$

which together with the relation (7.22), expressing $s$ in terms of $z$, will give

$$g(z) = 2^{-12}z^{-4}(1+z)^8(-5z^{-3} + 40z^{-2} - 131z^{-1} +$$
$$208 - 131z + 40z^2 - 5z^3). \qquad (7.48)$$

The last factor on the right hand side has six roots, namely,

$$0.328876, \quad 3.04066, \quad 0.284096 \pm 0.243228i, \quad 2.03114 \pm 1.73895i,$$

which by theorem 2.6 occur in pairs that are inside and outside the unit circle; complex roots ,of course, occur in conjugate pairs. Thus, if we denote two roots by $z_1 = 0.328876$ and $z_2 = 0.284096 + 0.243228i$, then all six roots are

$$z_1, \ 1/z_1, \ z_2, \ z_2^*, \ 1/z_2, \ 1/z_2^*.$$

Of the six roots three, namely, $z_1$, $z_2$, and $z_2^*$ are inside the unit circle. Writing $g(z)$ in equation (7.47) as the product of its elementary factors and using the definition of $g$ in equation (7.43) we find the equation

$$H_0(z) H_0'(1/z) = -\frac{5}{2^{11}} z^{-7} (1+z)^8 (z - z_1) (z - 1/z_1) \times$$
$$(z - z_2) (z - z_2^*) (z - 1/z_2) (z - 1/z_2^*). \qquad (7.49)$$

There are many ways to factor the right hand side to obtain appropriate functions $H_0(z)$ and $H_0'(1/z)$. In order to obtain symmetric filters as described in exercise 4 of section 2, however, we will choose a factorization that corresponds to functions of the form (2.36). For instance, consider the factorization

$$H_0(z) \propto z^{-4} (1+z)^5 (z - z_1) (z - 1/z_1),$$
$$H_0'(1/z) \propto z^{-3} (1+z)^3 (z - z_2) (z - z_2^*) (z - 1/z_2) (z - 1/z_2^*),$$

corresponding to values $L = -4, K = 5, M = 1$ and $L = -4, K = 5, M = 1$, respectively, in equation (2.36). Choosing normalization constants so that the filter coefficients $h_0$ and $h_0'$ satisfy equation (5.109), or equivalently,

$$H_0(1) = H_0'(1) = \sqrt{2}, \qquad (7.50)$$

we arrive at

$$H_0(z) = -0.032269457 \left(1 + z^{-1}\right)^4 (1+z)(z - z_1)(z - 1/z_1),$$
$$H_0'(1/z) = \quad 0.0756567 \left(1 + z^{-1}\right)^3 (z - z_2)(z - z_2^*) \times$$
$$(z - 1/z_2)(z - 1/z_2^*),$$

where the product of the normalizing factors is precisely equal to $-5/2^{11}$, as in equation (7.49). The five zeros at $z = -1$ for $H_0(z)$ imply that the corresponding mother wavelet function $\psi(t)$ has four vanishing moments. The biorthogonal mother wavelet $\psi'(t)$, on the other hand, has two vanishing moments since $H_0'(z)$ has three zeros at $z = -1$.[4] Expanding the factors we find

$$H_0\left(z\right) = -0.0322695z^3 - 0.0526142z^2 + 0.188702z + 0.603289+$$
$$0.603289z^{-1} + 0.188702z^{-2} - 0.0526142z^{-3} - 0.0322695z^{-4}$$

and

$$H_0'(z) = 0.0756567z^3 - 0.123355z^2 - 0.0978927z + 0.852696+$$
$$0.852696z^{-1} - 0.0978927z^{-2} - 0.123355z^{-3} + 0.0756567z^{-4},$$

which clearly show the symmetry of the corresponding filter coefficients, each of which has eight coefficients and they are known as the 8/8 biorthogonal pair. The support of $h_0$ and $h_0'$ is then $[-3, 4]$, which is also the support of the scaling functions $\phi(t)$ and $\phi'(t)$. Using the results of table 6.1 we find the supports of both $\psi(t)$ and $\psi'(t)$ to be $[-3, 4]$. The scaling and wavelet functions for the 8/8 pair are shown in figure 7.4. The 9/7 and 10/6 filters

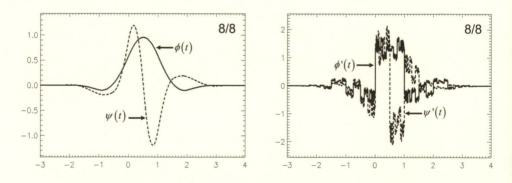

Figure 7.4: The 8/8 biorthogonal pair scaling and wavelet functions.

are based on other factorizations that still satisfy the symmetric form of (2.36) and are discussed in exercises 6 and 7 below.

---

[4]Using the convention adopted earlier in this chapter we are not counting the zeroth moment of the mother wavelet that must vanish because of the wavelet admissibility condition.

## 7.7 Biorthogonal Spline Wavelets

Another method of obtaining symmetric biorthogonal filters is to observe that the $g(z)$ shown in equation (7.46) has a factor of $(1 + z)^{2K}$ [52, 14]. If we now choose a scaling function $\phi(t)$ whose filter sequence $h_0$ has a **Z** transform that is a power of $(1+z)$ while choosing the second scaling function so as to correspond to the other factors in $g(z)$, the associated biorthogonal filters are still symmetric. For instance, consider the case $K = 2$ for which we have

$$u(z) = 1 + 2s = 2 - \left(z + z^{-1}\right)/2 \tag{7.52}$$

and

$$g(z) = -2^{-5}z^{-3}(1 + z)^4(z - z_1)(z - 1/z_1), \quad z_1 = 2 - \sqrt{3}. \tag{7.53}$$

We choose $H_0(z)$ to correspond to the Haar scaling function

$$H_0(z) = 2^{-1/2}\left(1 + z^{-1}\right), \tag{7.54}$$

which will result in the following $H_0'(1/z)$ using the other factors of $g(z)$ (and normalizing):

$$H_0'(z) = -2^{-7/2}\left(1 + z^{-1}\right)^3(1 - zz_1)(1 - z/z_1), \quad z_1 = 2 - \sqrt{3}. \tag{7.55}$$

The resulting symmetric filter $h_0'$, biorthogonal to the Haar $h_0$, has support on $[-2, 3]$ and is given by

$$h_0'[n] = [-0.0883883, \ 0.0883883, \ 0.707107,$$
$$0.707107, \ 0.0883883, \ -0.0883883]. \tag{7.56}$$

The scaling function $\phi(t)$ is, of course the Haar function with support in $[0, 1]$. The corresponding wavelet function now has support in $[-1, 2]$ and is shown in figure 7.5. The biorthogonal pair $\phi'(t)$ and $\psi'(t)$ with supports in $[-2, 3]$ and $[-1, 2]$, respectively, are shown in figure 7.6.

More general spline functions can be defined as follows. Let us denote the (normalized) Haar scaling function by $b_0(t)$, the spline of order 0, defined to be 1 when $0 \le t \le 1$ and 0 otherwise. Then $b_0(t)$ satisfies a two-term scaling equation

$$b_0(t) = b_0(2t) + b_0(2t - 1). \tag{7.57}$$

Higher-order splines are defined recursively,

$$b_{N+1}(t) \equiv b_N * b_0(t) = \int_{-\infty}^{\infty} b_N(\tau)b_0(t - \tau)\,d\tau, \quad N = 0, 1, 2, \dots. \tag{7.58}$$

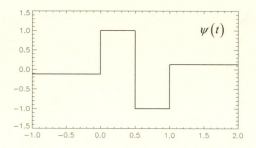

Figure 7.5: The wavelet function corresponding to the Haar scaling function $b_0(t)$.

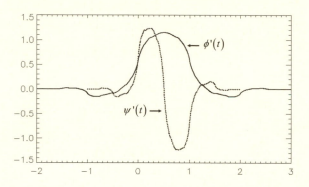

Figure 7.6: The scaling and wavelet functions biorthogonal to $b_0(t)$ and $\psi(t)$ of figure 7.5.

For instance, $b_1(t)$, also defined in equation (5.89), has support in $[-1, 1]$, $b_2(t)$ has support in $[-2, 1]$, and $b_N(t)$ has support in $[-N, 1]$. We can use the following general result to show that all the splines satisfy a finite scaling equation.

**Theorem 82.** *If $x(t)$ and $y(t)$ satisfy the scaling equations*

$$x(t) = \sum_{n=-\infty}^{\infty} a_n\, x(2t - n), \quad y(t) = \sum_{n=-\infty}^{\infty} b_n\, x(2t - n), \qquad (7.59)$$

*then $z(t) = x * y(t)$, the convolution of $x$ and $y$, satisfies a scaling equation*

$$z(t) = \sum_{n=-\infty}^{\infty} c_n\, z(2t - n),$$

*where the coefficients $c_n$ are given by one-half the discrete convolution of $a_n$ and $b_n$,*

$$c_n = 2^{-1} \sum_{m=-\infty}^{\infty} a_m b_{n-m}. \qquad (7.60)$$

For instance, the linear spline $b_1(t)$ satisfies a three-term scaling equation as seen in equation (5.90),

$$b_1(t) = \sum_{n=-1}^{1} \beta_1[n]b_1(2t-n), \quad \beta_1[-1] = \beta_1[1] = 0.5, \ \beta_1[0] = 1. \qquad (7.61)$$

This is, of course, consistent with theorem 82 since $b_1(t)$ is the convolution of $b_0(t)$ with itself and so $b_1(t)$ should satisfy a scaling equation whose coefficients are one half the discrete convolution of the sequence $[1, 1]$ with itself, i.e., $[.5, 1, .5]$. The **Z** transform of these coefficients is $2^{-1}z^{-1}(1+z)^2$.

Theorem 82 shows that the spline of order $N$ satisfies a scaling equation

$$b_N(t) = \sum_{n=-N}^{1} \beta_N[n]b_N(2t-n), \qquad (7.62)$$

where the scaling coefficients are (note the range of the index $n$)

$$\beta_N[n] = 2^{-N} \binom{N+1}{n+N}, \quad -N \le n \le 1, \qquad (7.63)$$

with a **Z** transform equal to $2^{-N}z^{-1}(1+z)^{N+1}$.

Continuing with the same case $K = 2$ instead of the Haar filter (7.54), we may now choose

$$H_0(z) = 2^{-3/2}z^{-1}(1+z)^2 \qquad (7.64)$$

in equation (7.53). This time we find the following form for $H'_0(z)$:

$$H'_0(z) = -0.17677673 \, (1+z^{-1})^2(1-zz_1)(1-z/z_1). \qquad (7.65)$$

Once again both $H_0(z)$ and $H'_0(z)$ are normalized so that $H_0(1) = H'_0(1) = \sqrt{2}$. Finally we have the following biorthogonal filter sequences: $h_0$ with support $[-1, 1]$ and $h'_0$ with support $[-2, 2]$, corresponding to the spline scaling function of order 1,

$$h_0 = 2^{-3/2}[1, 2, 1] = [0.35355339, \ 0.70710678, \ 0.35355339],$$
$$h'_0 = [-0.176777, \ 0.353554, \ 1.06066, \ 0.353554, \ -0.176777]. \qquad (7.66)$$

The scaling function $\phi(t)$ with support in $[-1, 1]$ and wavelet function $\psi(t)$ with support in $[-1.5, 1.5]$ corresponding to the above $h_0$ are shown in figure 7.7. The biorthogonal pair $\phi'(t)$ with support in $[-2, 2]$ and $\psi'(t)$ with support in $[-1.5, 1.5]$ corresponding to the above $h'_0$ are shown in figure 7.8.

Thus, the scaling function $\phi_N(t)$ corresponding to the spline $b_N(t)$ of order $N$, according to theorem 82, has scaling coefficients (normalized so they sum to $\sqrt{2}$)

$$h_0^{(N)}[n] = 2^{-(N+1/2)} \binom{N+1}{n+N}, \quad -N \le n \le 1, \tag{7.67}$$

whose **Z** transform is

$$H_0^{(N)}(z) = 2^{-(N+1/2)} z^{-1} (1+z)^{N+1}, \tag{7.68}$$

which has the correct value $\sqrt{2}$ at $z = 1$ and when used as a factor of $g(z)$ in equation (7.46) will yield the associated biorthogonal function $H'_0(1/z)$ with $g(z) = H_0(z) H'_0(1/z)$.

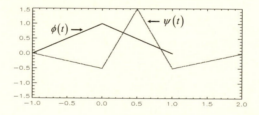

Figure 7.7: The scaling and wavelet functions corresponding to the spline $b_1(t)$.

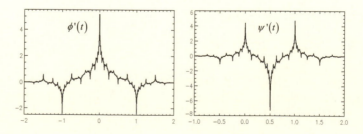

Figure 7.8: The scaling and wavelet functions biorthogonal to the functions in figure 7.7.

## 7.8   The Lifting Scheme

Suppose that a dual pair of compact support filters $h_0$ and $h_0'$, generating a biorthogonal scaling and wavelet function pair, is known. Assume that a second dual filter $h_0''$ also exists that together with $h_0$ generates a second biorthogonal pair. Then the following perfect reconstruction equations (see (6.41)) must hold

$$H_0\left(1/z\right)H_0'\left(z\right) + H_0\left(-1/z\right)H_0'\left(-z\right) = 2, \qquad (7.69a)$$

$$H_0\left(1/z\right)H_0''\left(z\right) + H_0\left(-1/z\right)H_0''\left(-z\right) = 2. \qquad (7.69b)$$

Subtracting these two equations we find

$$H_0\left(1/z\right)\left(H_0'\left(z\right) - H_0''\left(z\right)\right) + H_0\left(-1/z\right)\left(H_0'\left(-z\right) - H_0''\left(-z\right)\right) = 0, \quad (7.70)$$

whose solution is of the form

$$H_0'\left(z\right) - H_0''\left(z\right) = z^{-1}H_0\left(-1/z\right)S'\left(z^2\right). \qquad (7.71)$$

This, together with the relation between $H_1'$ and $H_0$ in (6.40), becomes

$$H_0''\left(z\right) = H_0'\left(z\right) + H_1'\left(z\right)S'\left(z^2\right). \qquad (7.72)$$

Thus, a new filter $h_0''$ dual to $h_0$ can be found through "lifting" of $h_0'$ by the above equation [57, 14]. This also replaces the filter $h_1$ by $\tilde{h}_1$ using the equation (see (6.40))

$$\tilde{H}_1(z) = -z^{-1}H_0''\left(-1/z\right). \qquad (7.73)$$

The original biorthogonal pair characterized by $\{h_0, h_1, h_0', h_1'\}$ is then replaced by a biorthogonal pair $\left\{h_0, \tilde{h}_1, h_0'', h_1'\right\}$ in which the two filters $\tilde{h}_1$ and $h_0''$ have replaced the originals $h_1$ and $h_0'$.

Lifting increases the support of the original biorthogonal wavelet pair $\psi(t)$ and $\psi'(t)$ by the support of the filter $s'[n]$ corresponding to the function $S'(z^2)$ [14]. The function $S'(z^2)$ is designed so as to increase the regularity of $\phi'(t)$ and the number of zero moments of $\psi(t)$ (see chapter 7 for a description of regularity and vanishing moments). To increase the regularity of $\phi(t)$ and the number of zero moments of $\psi'(t)$ we would lift the filter $h_0$ by the equation

$$H_0''\left(z\right) = H_0\left(z\right) + H_1\left(z\right)S\left(z^2\right), \qquad (7.74)$$

and replace $h_0$ and $h_1'$ in the original pair to obtain $\left\{h_0'', h_1, h_0', \tilde{h}_1\right\}$ where $\tilde{H}_1(z)$ is found just as before from $H''$.

We now apply the lifting scheme to the lazy filters of equations (6.43) and (6.44) of section 6.6 using

$$S(z^2) = 2^{-4}\left(z^4 - 9z^2 - 9 + z^{-2}\right),\qquad(7.75)$$

which will ensure three vanishing moments for $\phi(t)$ and $\psi'(t)$. Equation (7.74) together with (7.73) then give

$$H_0''(z) = 2^{-4}(-z^3 + 9z + 16 + 9z^{-1} - z^{-3}),\qquad(7.76a)$$

$$\tilde{H}_1(z) = 2^{-4}(-z^2 + 9 - 16z^{-1} + 9z^{-2} - z^{-4}).\qquad(7.76b)$$

Thus, the newly lifted filter $h_0[n]$, $-3 \le n \le 3$, satisfying the normalization (7.50) is

$$h_0 = [-0.0441942, 0., 0.397748, 0.707107, 0.397748, 0., -0.0441942],$$

and has an associated scaling function $\phi(t)$, known as the Deslauriers-Dubuc interpolating function [14], that has support in $[-3, 3]$. The new filter $h_1'[n]$, $-2 \le n \le 4$ is found from (7.73):

$$h_1' = [-0.0441942, 0., 0.397748, -0.707107, 0.397748, 0., -0.0441942],$$

and when used to calculate $\psi'(t)$ leads to a sum of delta distributions at points $\{-1, -1/2, 0, 1/2, 1, 3/2, 2\}$. The other functions are $\psi(t) = \sqrt{2}\,\phi(2t - 1)$, an invalid wavelet, and $\phi'(t) = \delta(t)$. To create a valid biorthogonal set a lifting dual to the one shown here must be performed to replace $h_0'$ and $h_1$. Requiring three vanishing moments for $\phi'(t)$ and $\psi(t)$ leads to the following newly lifted filter coefficients [14]

$$\begin{aligned}
h_0' = [&-0.00276214, \ 0., \ 0.0497184, \ -0.0441942, \ -0.174015, \ 0.397748, \\
&+0.961223, \ 0.397748, \ -0.174015, \ -0.0441942, \ 0.0497184, \ 0., \\
&-0.002762142],
\end{aligned}$$

and

$$\begin{aligned}
h_1 = [&-0.0027621359, \ 0., \ 0.049718444, \ 0.044194174, \ -0.17401456, \\
&-0.39774755, \ +0.96122326, \ -0.39774755, \ -0.17401456, \\
&+0.044194174, \ 0.049718444, \ 0., -0.0027621359].
\end{aligned}$$

The biorthogonal scaling and wavelet functions associated with the above filters are shown in figure 7.9.

Lifting of lazy filters is a general construction method for biorthogonal wavelet systems as stated in the following theorem [58].

**Theorem 83.** *(Daubechies, Sweldens) Any biorthogonal set $\{h_0, h_1, h_0', h_1'\}$ can be constructed using lifting and dual lifting of lazy filters, up to shifts and a multiplicative constant.*

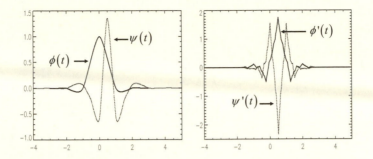

Figure 7.9: The Deslauriers-Dubuc interpolating function $\phi(t)$ and the associated biorthogonal wavelet system.

## 7.9 Exercises

**1** Use spectral factorization and the Daubechies solution shown in (7.27) to construct the Daubechies wavelet of order 6. Construct an approximation to the mother wavelet function using the cascade algorithm (section 5.9) and verify that it has the right number of vanishing moments.

**2** Verify that the functions $\hat{P}(z)$ and $\hat{Q}(z)$ of (7.34) satisfy equation (7.33).

**3** Derive the set of equations (7.38) for the coefficients of the function $p(z) = p_0 + p_1 z^{-1}$ for the $K = 2$ Coiflet using equation (7.36). Verify that the coefficients of the even powers of $z$ are identically zero. Verify that (7.39) is a solution and use this solution to derive the coefficients of $H_0(z)$ given in (7.40). Then use the ascade algorithm (section 5.9) to construct approximations to the scaling and the wavelet functions and verify that they have the right number of vanishing moments (two for the scaling function and one for the wavelet function, excluding the vanishing zeroth moment of the mother wavelet), and that they form an orthonormal pair.

**4** Derive the equations for the coefficients of the function $p(z) = p_0 + p_1 z^{-1} + p_2 z^{-2} + p_3 z^{-3}$ for the $K = 4$ Coiflet using equation (7.36). Verify that $p_0 = 2.96642, p_1 = 0.494045, p_2 = -0.330036, p_3 = -0.130433$ is a solution. Calculate the function $H_0(z)$, determine its coefficients $h_0[n]$, and show that the range of the index $n$ is $[-4, 7]$ (as is expected from the formula $[-2K, 4K - 1]$). Construct approximations to $\psi(t)$ and $\phi(t)$ and verify that they are an orthonormal pair and that they have the expected number of vanishing moments. Figure 7.10 shows this Coiflet's scaling and wavelet functions.

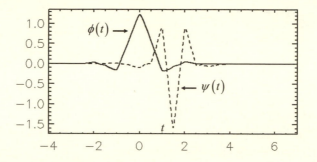

Figure 7.10: Coiflet scaling and wavelet functions of order 12 with four vanishing scaling function moments and three vanishing wavelet function moments (excluding the zeroth moment).

**5**   The $K = 6$ coiflet has the following sequence $h_0[n]$, $-6 \leq n \leq 11$:

$$h_0 = [-0.00379352, \ 0.00778260, \ 0.0220695, \ -0.0657719,$$
$$-0.0528240, \ 0.405177, \ 0.773029, \ 0.428484,$$
$$-0.0441353, \ -0.0823019, \ 0.0138065, \ 0.0158805,$$
$$-0.000708563, \ -0.00257452, \ -0.000265718, \ 0.000466218,$$
$$-7.09834 \times 10^{-5}, \ -3.45999 \times 10^{-5}].$$

Verify that the function

$$p(z) = 10.9872 + 1.45919z^{-1} - 2.00314z^{-2} - 0.749039z^{-3} +$$
$$0.20559z^{-4} + 0.100212z^{-5}$$

reproduces the given $h_0$ (exercise 3 of chapter 2 is useful). Figure 7.11 shows this Coiflet's scaling and wavelet functions.

**6**   Referring to equation (7.49) and the six roots following that equation we find the 9/7 biorthogonal pair by the factorizations

$$H_0(z) = -0.064538913 \ z^{-4}(1+z)^4(z-z_1)(z-1/z_1),$$
$$H_0'(1/z) = 0.0378283 \ z^{-3}(1+z)^4(z-z_2)(z-z_2^*)(z-1/z_2)(z-1/z_2^*),$$

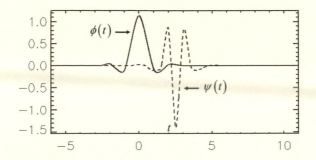

Figure 7.11: Coiflet scaling and wavelet functions of order 18 with six vanishing scaling function moments and five vanishing wavelet function moments (excluding the zeroth moment).

corresponding to the values $L = -4, K = 4, M = 1$ and $L = -3, K = 4, M = 2$, respectively, in the symmetric form (2.36). Compute the associated symmetric filters $h_0$ and $h'_0$ and show that their supports are $[-2, 4]$ and $[-3, 5]$, respectively. Verify that the corresponding scaling and wavelet function pairs are those shown in figure 7.12. The supports of $\phi(t)$ and $\phi'(t)$ are the same as their corresponding filters, i.e., $[-2, 4]$ and $[-3, 5]$, while the both wavelet functions have support in $[-3, 4]$.

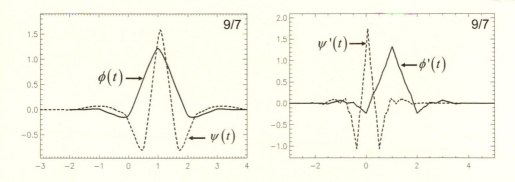

Figure 7.12: The 9/7 biorthogonal pair scaling and wavelet functions.

**7**  The 10/6 biorthogonal pair is found by the factorizations

$$H_0(z) = -0.129078\, z^{-3} (1 + z)^3 (z - z_1)(z - 1/z_1),$$

$$H'_0(1/z) = 0.0189142\, z^{-4} (1 + z)^5 (z - z_2)(z - z_2^*)(z - 1/z_2)(z - 1/z_2^*),$$

corresponding to the values $L = -3, K = 3, M = 1$ and $L = -4, K = 5, M = 2$, respectively, in the symmetric form (2.36). Compute the associated filters $h_0$ and $h_0'$ and show that their supports are $[-2, 3]$ and $[-4, 5]$, respectively. Verify that the corresponding scaling and wavelet function pairs are those shown in figure 7.13. The supports of $\phi(t)$ and $\phi'(t)$ are $[-2, 3]$ and $[-4, 5]$ while the both wavelet functions have support in $[-3, 4]$.

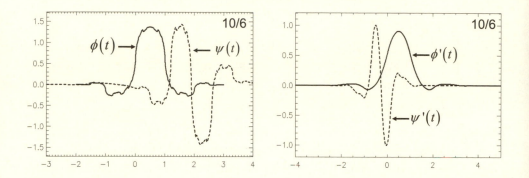

Figure 7.13: The 10/6 biorthogonal pair scaling and wavelet functions.

# CHAPTER 8

# Orthogonal Wavelet Packets

## 8.1 Introduction

Daubechies orthogonal wavelets are but one of many other related orthogonal bases for analysis and synthesis of $L_2(\mathbb{R})$. Although they form an excellent basis for representation of images they are by no means the optimal basis for representation of time series such as music. The main issue here is that orthogonal wavelets are based on the union of orthogonal bases of the subspaces $\mathscr{V}_m^\perp$ and it is the approximation subspaces $\mathscr{V}_m$ that are always sub divided. Orthogonal wavelet packets are based on sub dividing the detail as well as the approximation subspaces. Thus, they provide an overcomplete set of functions for signal representation. The subdivision of detail as well as approximation subspaces provides enormous flexibility in applications such as data compression: methods that rely on neglecting subspaces with small coefficient magnitudes would now have the chance of further refining (localizing) the coefficient subspaces.[1] The optimal basis, in the sense of finding a minimal set of transform coefficients with large magnitudes while maintaining orthogonality, is found by the best basis algorithm described in section 8.5.

## 8.2 Review of the Orthogonal Wavelet Transform

The finite resolution orthogonal wavelet transform, applied to finite length discrete time data, can be summarized as the computation of a series of detail (wavelet) coefficients $d_{mn}$, $m = 1, \ldots, M$, together with the final set of scaling coefficients $c_{Mn}$ starting from the data $c_{0n}$. For an initial data of length $2^N$, there are precisely $2^N$ coefficients in an $M$-level orthogonal

---

[1]Pathological signals that defy compression have been constructed and are known as noiselets [59]: they exhibit a noiselike property that all their packet transform coefficients are of equal magnitude.

wavelet decomposition, namely, $2^{N-m}$ coefficients $d_{mn}$ for each $m$, $m = 1, \ldots, M$, and the final $2^{N-M}$ coefficients $c_{Mn}$. At every level the coefficients are given by

$$c_{mn}^{(0)} = \langle \phi_{mn}, x \rangle, \quad d_{mn}^{(0)} = \langle \psi_{mn}, x \rangle. \tag{8.1}$$

This process is illustrated in figure 8.1 in which three levels of wavelet coefficients denoted by $d_{1n}^{(0)}$, $d_{2n}^{(0)}$, $d_{3n}^{(0)}$ and the final scaling coefficients $c_{3n}^{(0)}$ are calculated. Although redundant in the present example, in anticipation of calculating new sets of packet coefficients we have introduced the superscript (0). The solid arrows indicate the usual operations of correlating with the filters $h_0[m]$ and $h_1[m]$ (or equivalently, convolving with the time-reversed sequences $h_0[-m]$ and $h_1[-m]$) followed by taking the even sampled coefficients (decimation by a factor of 2).

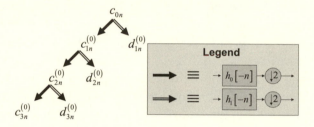

Figure 8.1: Three-level orthogonal wavelet transform.

Following the frequency domain result shown in figure 4.8 for the Haar wavelet transform, if the initial discrete time data (assumed to be real valued) is sampled at a rate denoted by $f_s$, then the coefficients $d_{1n}^{(0)}$ correspond to that portion of the data whose frequencies are limited to the normalized upper half-band, i.e., $[-0.5, -0.25] \cup [0.25, 0.5]$, since $h_1$ (and its time-reversed form) is an (upper) half-band filter. Similarly, the lower half-band nature of the spectrum of $h_0$ (and its time-reversed form) show that coefficients $c_{1n}^{(0)}$ correspond to that portion of the data whose frequencies are limited to the normalized lower half-band, i.e., $[-0.25, 0.25]$. The first two decomposition levels, and the corresponding effects in the frequency domain are illustrated in figure 8.2. Note that only the positive frequencies are displayed since, for real data, the negative frequency values of the spectra are mirror images in the vertical axis of the positive frequency values.

Each set of $d$ coefficients corresponds to taking the upper half of the signal band at the previous level, while retaining the even samples corresponds to reducing the time resolution of the previous level by a factor of 2. Thus, the division of the time and frequency plane of the signal for this three-level

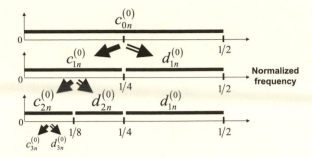

Figure 8.2: Two decomposition levels in the orthogonal wavelet transform in the frequency domain (positive frequencies are shown).

decomposition is as illustrated in figure 8.3. The initial real data set $c_{0n}$ is depicted on the left hand side. The initial data band in normalized linear frequencies is in the interval $[-0.5, 0.5]$, but we show only the positive half of the frequencies since the data is real and the negative frequency band is a mirror image (about a vertical axis at 0) of the positive frequencies. The first column of the wavelet coefficients, $d_{1n}^{(0)}$, shows the eight coefficients occupying the band $[2^{-2}, 2^{-1}]$, and sampling in the time dimension that is twice the original. The next set of wavelet coefficients $d_{2n}^{(0)}$ comprise four numbers that have four times the original sampling time and occupy the band $[2^{-3}, 2^{-2}]$. The last set of wavelet coefficients $d_{3n}^{(0)}$ comprise two numbers that have eight times the original sampling time and occupy the band $[2^{-4}, 2^{-3}]$. Finally, the scaling coefficients, $c_{3n}^{(0)}$ comprise two numbers that have eight times the original sampling time and occupy the band $[0, 2^{-4}]$.

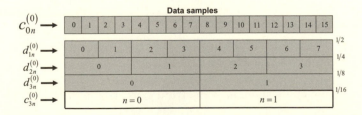

Figure 8.3: Time and frequency plane division of a three-level orthogonal wavelet transform of a 16-point data set.

## 8.3    Packet Functions for Orthonormal Wavelets

Whereas in the orthogonal DWT both correlation operations are performed on the $c_{mn}$ coefficients, as shown in figure 8.1, we now generalize this operation to include the $d_{mn}$ coefficients. This generalization is based on the following theorem [60].

**Theorem 84.** *(Coifman, Meyer, Wickerhauser) Let $\psi_{0n}(t)$ be an orthonormal (wavelet) basis for the subspace $\mathscr{V}_0^\perp$ where $\psi_{0n}(t) \equiv \psi(t-n)$. If $h_0[n]$ and $h_1[n]$ are a pair of conjugate mirror filters as described in theorem 71 and equation (5.82), and defining the pair of functions*

$$\phi^{(1)}(t) = \sum_{k=-\infty}^{\infty} h_0\,[k]\,\psi_{0k}\,(t), \quad \psi^{(1)}(t) = \sum_{k=-\infty}^{\infty} h_1\,[k]\,\psi_{0k}\,(t),$$

*the translated and dilated functions $\mathscr{D}_2\mathscr{T}_n\phi^{(1)}(t)$ and $\mathscr{D}_2\mathscr{T}_n\psi^{(1)}(t)$ form an orthonormal basis for $\mathscr{V}_0^\perp$.*

Returning to the three-level example of the previous section we now have the packet decomposition shown in figure 8.4. We observe that coefficients with the 0 superscript, namely, $c_{mn}^{(0)}$ and $d_{mn}^{(0)}$, are the usual orthogonal DWT coefficients, while those with superscripts $1, 2, \ldots$ are new additional coefficients. Thus, at stage $m+1$, $m = 0, 1, \ldots$, there are $2^{m+1}$ coefficients that

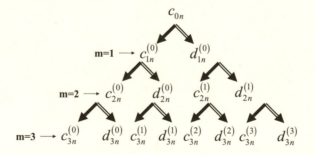

Figure 8.4: Orthogonal wavelet packet three-level decomposition (see figure 8.1 for the meaning of the arrows).

we divide into c-type , those arising from correlation with $h_0$, and d-type , those arising from correlation with $h_1$. We further divide the two sets into those with even and those with odd superscripts:

- The c-type: at level $m+1$, even superscripts $2j$, $c_{m+1,n}^{(2j)}$, come from the previous level $m$ and superscript $j$ coefficients, $c_{mn}^{(j)}$, and odd superscripts $2j+1$, $c_{m+1,n}^{(2j+1)}$, come from previous $d_{mn}^{(j)}$ coefficients.

- The d-type: at level $m+1$, even superscripts $2j$, $d_{m+1,n}^{(2j)}$, come from the previous level $m$ and superscript $j$ coefficients, $c_{mn}^{(j)}$, and odd superscripts $2j+1$, $d_{m+1,n}^{(2j+1)}$, come from previous $d_{mn}^{(j)}$ coefficients.

Figure 8.5 shows the above notation for decomposition at two consecutive levels. Thus, coefficients with even superscripts come from previous c-type coefficients, while those with odd superscripts come from d-type coefficients.

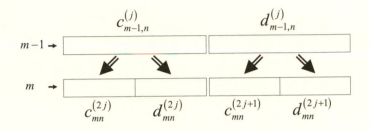

Figure 8.5: General structure of packet coefficients: stage $m-1$ to stage $m$ (see figure 8.1 for the meaning of the arrows).

The coefficients $c_{mn}^{(0)}$ are found by taking the inner product of $x(t)$ with the scaling functions $\phi_{mn}(t)$, while $d_{mn}^{(0)}$ are found by taking the inner product of $x(t)$ with the wavelet functions $\psi_{mn}(t)$. If we call these coefficients the packet coefficients of order 0, they are determined in terms of the translated and scaled versions of the scaling function $\phi(t)$ and the wavelet function $\psi(t)$, which we also call the packet functions of order 0. We express the relations for all the coefficients with 0 superscripts in terms of the newly defined packet functions of order 0, i.e., $p^{(0)}(t)$ and $q^{(0)}(t)$, and their translated and scaled versions $p_{mn}^{(0)}(t)$ and $q_{mn}^{(0)}(t)$ (see definitions 47 and 48 for the translation and dilation operators $\mathscr{T}$ and $\mathscr{D}$),

$$p_{mn}^{(0)}(t) \equiv \phi_{mn}(t) = \mathscr{D}_{2m}\mathscr{T}_n\phi(t) \equiv \mathscr{D}_{2m}\mathscr{T}_n p^{(0)}(t), \tag{8.2a}$$

$$q_{mn}^{(0)}(t) \equiv \psi_{mn}(t) = \mathscr{D}_{2m}\mathscr{T}_n\psi(t) \equiv \mathscr{D}_{2m}\mathscr{T}_n q^{(0)}(t), \quad m,n \in \mathbb{Z}. \tag{8.2b}$$

Using the Dirac notation, the initial time series $c_{0n}$ is, of course, given by the inner product $\langle \phi_{0n} \mid x \rangle$ while the packet coefficients of order 0 are given by the inner products between $x(t)$ and the packet functions of order 0, namely,

$$c_{mn}^{(0)} = \left\langle p_{mn}^{(0)} \mid x \right\rangle, \quad d_{mn}^{(0)} = \left\langle q_{mn}^{(0)} \mid x \right\rangle, \quad m,n \in \mathbb{Z}. \tag{8.3}$$

The issue before us is the construction of packet functions of higher order, $j \geq 1$, namely, $p^{(j)}(t)$ and $q^{(j)}(t)$, whose translated and scaled versions

$p_{mn}^{(j)}(t)$ and $q_{mn}^{(j)}(t)$ would be used to calculate all the packet coefficients of order $j \geq 1$, as depicted in figure 8.6.

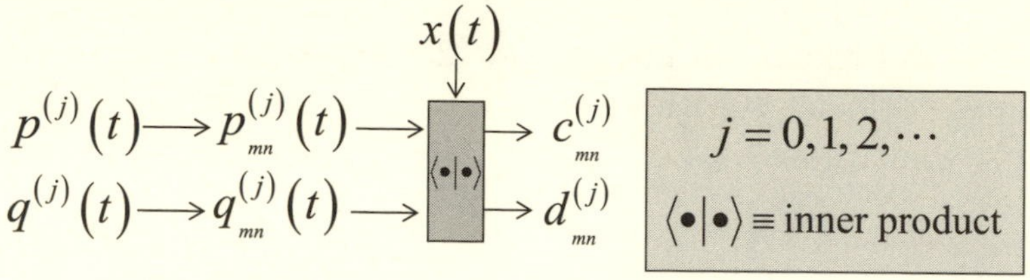

$$x(t)$$

$$p^{(j)}(t) \longrightarrow p_{mn}^{(j)}(t) \longrightarrow \boxed{\langle \bullet | \bullet \rangle} \longrightarrow c_{mn}^{(j)} \qquad \boxed{\begin{array}{l} j = 0, 1, 2, \cdots \\ \\ \langle \bullet | \bullet \rangle \equiv \text{inner product} \end{array}}$$

$$q^{(j)}(t) \Longrightarrow q_{mn}^{(j)}(t) \longrightarrow \qquad \longrightarrow d_{mn}^{(j)}$$

Figure 8.6: Packet functions, their translated and scaled versions, and calculation of packet coefficients.

We illustrate the construction by calculating some of the coefficients using figure 8.4 beginning with $c_{2,n}^{(1)}$ and $d_{2,n}^{(1)}$,[2] assuming throughout this chapter that the sequences $h_0$ and $h_1$ are real quantities. Using the definition of the arrows shown in figure 8.1 we have

$$c_{2,n}^{(1)} = \sum_{k=-\infty}^{\infty} d_{1k}^{(0)} h_0 \left[k - 2n\right], \quad d_{2,n}^{(1)} = \sum_{k=-\infty}^{\infty} d_{1k}^{(0)} h_1 \left[k - 2n\right], \qquad (8.4)$$

which upon substitution from equation (8.3) lead to

$$c_{2,n}^{(1)} = \left\langle \sum_{k=-\infty}^{\infty} h_0 \left[k - 2n\right] q_{1k}^{(0)} \,\middle|\, x \right\rangle \equiv \left\langle p_{2,n}^{(1)} \,\middle|\, x \right\rangle, \qquad (8.5a)$$

$$d_{2,n}^{(1)} = \left\langle \sum_{k=-\infty}^{\infty} h_1 \left[k - 2n\right] q_{1k}^{(0)} \,\middle|\, x \right\rangle \equiv \left\langle q_{2,n}^{(1)} \,\middle|\, x \right\rangle, \qquad (8.5b)$$

defining the two functions $p_{2,n}^{(1)}(t)$ and $q_{2,n}^{(1)}(t)$ whose inner products with $x(t)$ yield the sets of coefficients $c_{2,n}^{(1)}$ and $d_{2,n}^{(1)}$ (recall that we assume that $h_0[k]$ and $h_1[k]$ are real). We proceed to rewrite the above in a form that will allow us to derive the functions $p^{(1)}(t)$ and $q^{(1)}(t)$ whose translation by $n$ and dilation by $2^2$ will produce the functions $p_{2,n}^{(1)}(t)$ and $q_{2,n}^{(1)}(t)$ whose inner product with $x(t)$ will result in the packet coefficients $c_{2,n}^{(1)}$ and $d_{2,n}^{(1)}$.

---

[2]Note that we are using a slightly altered notation by separating two subscripts with a comma to avoid possible confusion. For instance, $c_{2,n}^{(1)}$, instead of $c_{2n}^{(1)}$, refers to the coefficient $c_{mn}^{(1)}$ for $m = 2$.

Changing the summation index $k$ in equation (8.5a) we have

$$p_{2,n}^{(1)}(t) = \sum_{k=-\infty}^{\infty} h_0\,[k]\, q_{1,k+2n}^{(0)}\,(t). \tag{8.6}$$

Now using definitions 47 and 48 for the translation and the dilation operators we have

$$
\begin{aligned}
q_{1,k+2n}^{(0)}(t) &= \mathscr{D}_{2^1}\,\mathscr{T}_{2n}\,\psi_{0k}(t) = \mathscr{D}_{2^2}\,\mathscr{D}_{2^{-1}}\,\mathscr{T}_{2n}\,\psi_{0k}(t) \\
&= \mathscr{D}_{2^2}\,\mathscr{T}_n\,\mathscr{D}_{2^{-1}}\,\psi_{0k}(t) = \mathscr{D}_{2^2}\,\mathscr{T}_n\,q_{-1,k}^{(0)}(t),
\end{aligned}
\tag{8.7}
$$

where theorem 49 was used to write $\mathscr{D}_{2^{-1}}\,\mathscr{T}_{2n} = \mathscr{T}_n\,\mathscr{D}_{2^{-1}}$. Note that

$$q_{-1,k}^{(0)}(t) = \mathscr{D}_{2^{-1}}\,\psi_{0k}(t) = 2^{1/2}\psi\,(2t-k)\,.$$

Thus, equation (8.6) becomes

$$p_{2,n}^{(1)}(t) = \mathscr{D}_{2^2}\,\mathscr{T}_n \sum_{k=-\infty}^{\infty} h_0\,[k]\, q_{-1,k}^{(0)}(t) \equiv \mathscr{D}_{2^2}\,\mathscr{T}_n p^{(1)}(t), \tag{8.8}$$

and so

$$p^{(1)}(t) \equiv \sum_{k=-\infty}^{\infty} h_0\,[k]\, q_{-1,k}^{(0)}(t). \tag{8.9}$$

Similarly,

$$q_{2,n}^{(1)}(t) = \mathscr{D}_{2^2}\,\mathscr{T}_n \sum_{k=-\infty}^{\infty} h_1\,[k]\, q_{-1,k}^{(0)}(t) \equiv \mathscr{D}_{2^2}\,\mathscr{T}_n q^{(1)}(t), \tag{8.10}$$

and

$$q^{(1)}(t) \equiv \sum_{k=-\infty}^{\infty} h_1\,[k]\, q_{-1,k}^{(0)}(t). \tag{8.11}$$

Finally, we have the relations

$$p^{(1)}(t) \equiv \sum_{k=-\infty}^{\infty} h_0\,[k]\, q_{-1,k}^{(0)}(t) = \sum_{k=-\infty}^{\infty} h_0\,[k]\, 2^{1/2}\psi\,(2t-k), \tag{8.12a}$$

$$q^{(1)}(t) \equiv \sum_{k=-\infty}^{\infty} h_1\,[k]\, q_{-1,k}^{(0)}(t) = \sum_{k=-\infty}^{\infty} h_1\,[k]\, 2^{1/2}\psi\,(2t-k). \tag{8.12b}$$

We maintain that all packet coefficients with superscript 1 at all levels, not just $m = 2$, are found by taking the inner product of appropriately translated and dilated versions of the above two functions with $x(t)$, i.e.,

$$c_{mn}^{(1)} = \left\langle \mathscr{D}_{2^m} \mathscr{T}_n p^{(1)}(t), x(t) \right\rangle, \quad d_{mn}^{(1)} = \left\langle \mathscr{D}_{2^m} \mathscr{T}_n q^{(1)}(t), x(t) \right\rangle. \qquad (8.13)$$

For instance, consider $m = 3$ in figure 8.4. We have

$$c_{3,n}^{(1)} = \sum_{k=-\infty}^{\infty} h_0[k - 2n] d_{2,k}^{(0)} = \left\langle \sum_{k=-\infty}^{\infty} h_0[k] q_{2,k+2n}^{(0)} \,\middle|\, x \right\rangle \equiv \left\langle p_{3,n}^{(1)} \,\middle|\, x \right\rangle.$$

Using theorem 49 to write $\mathscr{D}_{2^2} \mathscr{T}_{2n} \equiv \mathscr{D}_{2^3} \mathscr{D}_{2^{-1}} \mathscr{T}_{2n} = \mathscr{D}_{2^3} \mathscr{T}_n \mathscr{D}_{2^{-1}}$ we find

$$p_{3,n}^{(1)}(t) = \sum_{k=-\infty}^{\infty} h_0[k] \mathscr{D}_{2^2} \mathscr{T}_{2n} q_{0,k}^{(0)} = \mathscr{D}_{2^3} \mathscr{T}_n \sum_{k=-\infty}^{\infty} h_0[k] q_{-1,k}^{(0)}. \qquad (8.14)$$

Thus, equations (8.12a) and (8.12b) and (8.13) allow us to compute all the packet coefficients with superscript 1.

Next we construct $p^{(2)}(t)$ and $q^{(2)}(t)$ whose translated and dilated versions will, on taking inner products with $x(t)$, produce all the coefficients with superscript 2. To this end we calculate $c_{3,n}^{(2)}$ and $d_{3,n}^{(2)}$ from figure 8.4. We find

$$c_{3,n}^{(2)} = \sum_{k=-\infty}^{\infty} h_0[k - 2n] c_{2,k}^{(1)} = \left\langle \sum_{k=-\infty}^{\infty} h_0[k] \mathscr{D}_{2^3} \mathscr{T}_n p_{-1,k}^{(1)}(t), x(t) \right\rangle,$$

$$d_{3,n}^{(2)} = \sum_{k=-\infty}^{\infty} h_1[k - 2n] c_{2,k}^{(1)} = \left\langle \sum_{k=-\infty}^{\infty} h_1[k] \mathscr{D}_{2^3} \mathscr{T}_n p_{-1,k}^{(1)}(t), x(t) \right\rangle.$$

and

$$p^{(2)}(t) \equiv \sum_{k=-\infty}^{\infty} h_0[k] p_{-1,k}^{(1)}, \quad q^{(2)}(t) \equiv \sum_{k=-\infty}^{\infty} h_1[k] p_{-1,k}^{(1)}. \qquad (8.15)$$

Thus, all packet coefficients with superscript 2 are calculated by taking inner products of appropriately shifted and scaled versions of $p^{(2)}(t)$ and $p^{(2)}(t)$ with $x(t)$. Continuing on to the superscript 3 packet coefficients we find

$$p^{(3)}(t) \equiv \sum_{k=-\infty}^{\infty} h_0[k] q_{-1,k}^{(1)}, \quad q^{(3)}(t) \equiv \sum_{k=-\infty}^{\infty} h_1[k] q_{-1,k}^{(1)}. \qquad (8.16)$$

The pattern for the calculation of the packet functions is now easy to see by inspecting equations (8.12), (8.15), and (8.16), and is stated in the following theorem.

**Theorem 85.** *The packet functions associated with orthogonal wavelets satisfy the following recursive relations,*

$$p^{(2j)}(t) = \sum_{k=-\infty}^{\infty} h_0[k]\, p_{-1,k}^{(j)}, \quad 1 \leq j \leq 2^m - 1, \tag{8.17a}$$

$$q^{(2j)}(t) = \sum_{k=-\infty}^{\infty} h_1[k]\, p_{-1,k}^{(j)}, \quad 1 \leq j \leq 2^m - 1, \tag{8.17b}$$

$$p^{(2j+1)}(t) = \sum_{k=-\infty}^{\infty} h_0[k]\, q_{-1,k}^{(j)}, \quad 0 \leq j \leq 2^m - 1, \tag{8.17c}$$

$$q^{(2j+1)}(t) = \sum_{k=-\infty}^{\infty} h_1[k]\, q_{-1,k}^{(j)}, \quad 0 \leq j \leq 2^m - 1. \tag{8.17d}$$

At each level $m$, $2^{m-1}$ pairs of the packet functions $\{p^{(j)}(t),\, q^{(j)}(t)\}$, $0 \leq j \leq 2^{m-1} - 1$ can be computed. To compute packet functions with higher superscripts we must go to the next level $m + 1$.

A scheme to compute the packet coefficients begins with the initial pair $p^{(0)}(t) = \phi(t)$ and $q^{(0)}(t) = \psi(t)$ at level $m = 1$. At level $m = 2$ we may compute the next pair $p^{(1)}(t)$ and $q^{(1)}(t)$. The next two pairs, superscripts 2 and 3, are found at level $m = 3$, while the next four pairs, superscripts 5 through 7, are found at level $m = 4$, and so on. Thus, when all $2^{m-1}$ pairs at level $m$ are found, the next $2^{m-1}$ pairs found at level $m + 1$ complete the set for that level $m+1$. Figure 8.7 shows the four pairs $\{p^{(j)}, q^{(j)}\}$, $j = 1, \ldots, 4$, for Daubechies orthogonal wavelet system of order 4, namely, DAUB-4. The pair $\{p^{(0)}, q^{(0)}\}$ is, of course, the initial scaling and wavelet function pair shown previously in figure 5.6. All packet functions have the same support as the superscript 0 packets, i.e., the scaling and the wavelet functions, and have unit norm, i.e., $\|p^{(j)}\| = \|q^{(j)}\| = 1$. In general, all packets resulting from Daubechies orthogonal wavelets DAUB-$2N$ (whose filter coefficients have even length $2N$) have support in the interval $[0, 2N - 1]$. As is evident from figure 8.7, packet functions of increasing superscript $j$ have increasing number of oscillations. This can be seen most easily in the Haar (DAUB-2) packet functions, also known as Walsh functions, displayed in figure 8.8. Clearly, packet functions with increasing index $j$ are appropriate for the representation of functions with oscillatory behavior.

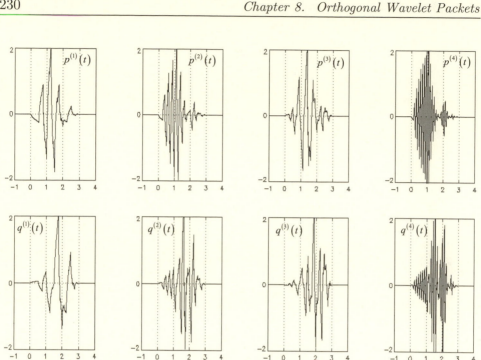

Figure 8.7: DAUB-4 packet function pairs 1 through 4.

Equations (8.17) lead to the following Fourier transform relations:

$$P^{(2j)}(\omega) = 2^{-1/2} H_0(\omega/2) P^{(j)}(\omega/2), \tag{8.18a}$$

$$P^{(2j+1)}(\omega) = 2^{-1/2} H_1(\omega/2) P^{(j)}(\omega/2), \tag{8.18b}$$

$$Q^{(2j)}(\omega) = 2^{-1/2} H_0(\omega/2) Q^{(j)}(\omega/2), \tag{8.18c}$$

$$Q^{(2j+1)}(\omega) = 2^{-1/2} H_1(\omega/2) Q^{(j)}(\omega/2). \tag{8.18d}$$

The above recursions, in the frequency domain, have a simple closed form solution found by inspection and shown in the following theorem.

**Theorem 86.** *Let* $[b_{J-1}\cdots b_1 b_0]$ *denote the binary representation of the superscript* $r$, *where* $r = 2j$ *or* $r = 2j + 1$, *i.e.,* $r = b_{J-1}2^{J-1} + \cdots + b_0 2^0$, *and the bits* $b_k$, $0 \le k \le J - 1$, *are either* 0 *or* 1. *Then,*

$$P^{(r)}(\omega) = 2^{-J/2}\Phi\left(\omega/2^J\right) \prod_{k=0}^{J-1} H_{b_k}\left(\omega\big/2^{k+1}\right), \tag{8.19a}$$

$$Q^{(r)}(\omega) = 2^{-J/2}\Psi\left(\omega/2^J\right) \prod_{k=0}^{J-1} H_{b_k}\left(\omega\big/2^{k+1}\right), \tag{8.19b}$$

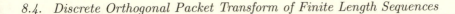

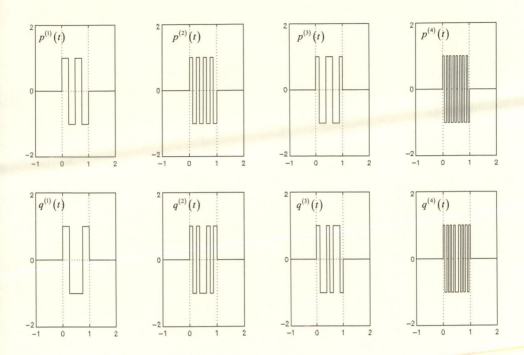

Figure 8.8: DAUB-2 (Haar) packet function pairs — Walsh functions.

*where $P^{(0)}(\omega) \equiv \Phi(\omega)$ and $Q^{(0)}(\omega) \equiv \Psi(\omega)$ are the Fourier transforms of the scaling and the wavelet functions.*

If all packet functions $\{p^{(j)}(t), q^{(j)}(t)\}$, $j = 0, 1, 2, \ldots$, are known, then the set of all time-shifted and scaled packet functions $\{p^{(j)}_{mn}(t),\ q^{(j)}_{mn}(t)\}$, $j = 0, 1, 2, \ldots$ and $m, n \in \mathbb{Z}$, is referred to as the packet library or the packet table. A packet basis is any orthogonal basis chosen from the packet library [61].

## 8.4  Discrete Orthogonal Packet Transform of Finite Length Sequences

The implementation of the discrete orthogonal packet transform for finite length sequences is based on figure 8.4. An example of a two-stage transform for a data set with 16 points is illustrated in figure 8.9. Starting with the data $c_{0n}$, $0 \leq n \leq 15$, at level $m = 0$ we compute the level $m = 1$ coefficients $c^{(0)}_{1n}$ and $d^{(0)}_{1n}$, $0 \leq n \leq 7$, just as we would for a one-stage DWT. Next we perform one-stage discrete wavelet transforms on both sets of coefficients instead of just the $c_{1n}^{(0)}$. The output sequences are $c_{2n}^{(0)}$ and $d_{2n}^{(0)}$ (which would be normally computed for a DWT), and the other two sets of packet coefficients $c_{2n}^{(1)}$ and $d_{2n}^{(1)}$. The next stage (not shown in the figure) corresponding to the level $m = 3$ would produce four pairs of packet coefficients, namely, $c_{3n}^{(j)}$, $d_{3n}^{(j)}$, for $0 \leq j \leq 3$ and $0 \leq n \leq 1$.

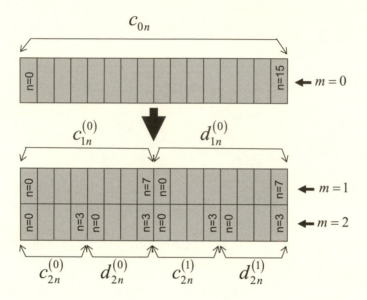

Figure 8.9: Two-stage orthogonal packet transform of 16-point data set.

The reconstruction of the initial data is carried out based on the inversion result for orthogonal wavelets depicted in figure 5.1. Thus, starting at level $m$, each pair of sets of coefficients $c_{mn}^{(2j)}$ and $d_{mn}^{(2j)}$, is used to calculate $c_{m-1,n}^{(j)}$, while $c_{mn}^{(2j+1)}$ and $d_{mn}^{(2j+1)}$ are used to calculate $d_{m-1,n}^{(j)}$, as shown in figure 8.10.

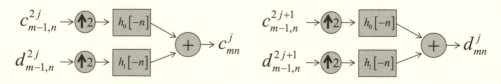

Figure 8.10: Signal reconstruction from packet coefficients: level $m$ from level $m-1$, $0 \leq j \leq m-1$.

Figure 8.11 shows the DAUB-4 discrete wavelet and packet transforms of the linearly chirped signal of figure 6.2 for four levels. Columns 1 through 4 of the DWT image (shown horizontally) correspond to the coefficients $d_{mn}^{(0)}$, $1 \leq m \leq 4$, while the last column contains the coefficients $c_{4n}^{(0)}$. Pixel sizes for each coefficient have been appropriately expanded in the horizontal direction to display a full matrix. Note, however, that each column of the DWT contains half as many coefficients as the previous column, except for the last column that has the same number of coefficients as the one before it. The packet transform image has no pixel expansion and follows the example shown in figure 8.9: boundaries between the c-type and d-type coefficients are shown by horizontal lines on the image. The coefficients with superscript 0, i.e., those of the standard DWT, as well as two arbitrarily chosen sets of packet coefficients $c_{4n}^{(7)}$ and $d_{4n}^{(7)}$, are explicitly marked. It is clear from

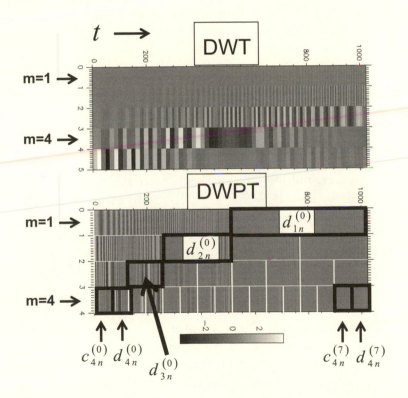

Figure 8.11: Four-level discrete wavelet and packet transforms for a linearly chirped signal using DAUB-4 orthogonal wavelets of compact support.

figure 8.4 that an $M$-level orthogonal packet transform results in a total of $2^{M+1} - 2$ sets of coefficients of varying sizes. At any level $m$ there are

$2 \times 2^{m-1}$ sets of coefficients each of length $N/2^m$, namely, $c_{mn}^{(j)}$ and $d_{mn}^{(j)}$, $0 \le j \le 2^{m-1} - 1$. Thus, there are a total of $N \times M$ numbers specifying all possible packet coefficients in an $M$-level orthogonal packet decomposition of an $N$-point data set which, without loss of generality, we assume to be real. The orthogonal matrix implementation of section 6.3 and equations (6.27)–(6.30) can be extended to compute all the packet coefficients in the following way. Starting with the real data vector $\mathbf{x}$ of length $N$ (a power of 2) the first-level packet coefficients are given by $\mathbf{x}_1 = \mathbf{R}_N \mathbf{H}_N \mathbf{x}$ where the filter matrix $\mathbf{H}_N$ and the reordering matrix $\mathbf{R}_N$ are the first level transformation matrices $\mathbf{H}_N^0$ and $\mathbf{R}_N^0$ defined in equation (6.28). Now $\mathbf{x}_1 = \left[ [\mathbf{d}_1^{(0)}]^T, [\mathbf{c}_1^{(0)}]^T \right]^T$ where $\mathbf{c}_1^{(0)}$ and $\mathbf{d}_1^{(0)}$ are packet coefficient vectors of length $2^{-1}N$. The second-level packet coefficients are given by $\mathbf{x}_2 = \mathbf{R}_{2^{-1}N}^b \mathbf{H}_{2^{-1}N}^b \mathbf{x}_1$ where $\mathbf{H}_{2^{-1}N}^b$ is an orthogonal block matrix (the superscript $b$ denotes the block property) of the form

$$\mathbf{H}_{2^{-1}N}^b = \begin{bmatrix} \mathbf{H}_{2^{-1}N} & \mathbf{0} \\ \mathbf{0} & \mathbf{H}_{2^{-1}N} \end{bmatrix}, \tag{8.20}$$

while $\mathbf{R}_{2^{-1}N}^b$ is an orthogonal block matrix for reordering the output rows so that the second-level transform is of the form

$$\mathbf{x}_2 = \left[ [\mathbf{d}_2^{(0)}]^T, [\mathbf{d}_2^{(1)}]^T, [\mathbf{c}_2^{(0)}]^T, [\mathbf{c}_2^{(1)}]^T \right]^T, \tag{8.21}$$

with the four vectors on the right hand side each of length $2^{-2}N$ (the matrices $\mathbf{H}_{2^{-m}N}$ and $\mathbf{R}_{2^{-m}N}$ are $2^{-m}N \times 2^{-m}N$ matrices $\mathbf{H}_{2^{-m}N}^0$ and $\mathbf{R}_{2^{-m}N}^0$). Since the product of any number of orthogonal matrices is orthogonal, the packet transform vector $\mathbf{x}_m$ at any level $m$ is the result of multiplying the initial data vector $\mathbf{x}$ with a single orthogonal matrix $\mathbf{W}_m$, $1 \le m \le M$. We denote the rows of the latter by $\mathbf{p}_{mn}^{(j)}$ if the corresponding element in $\mathbf{x}_m$ is $c_{mn}^{(j)}$, or by $\mathbf{q}_{mn}^{(j)}$ if the corresponding element in $\mathbf{x}_m$ is $d_{mn}^{(j)}$, where $0 \le j \le 2^{m-1} - 1$ and $0 \le n \le N - 1$. Thus, for each level $m$ there are precisely $N$ orthogonal vectors whose dot product with the initial data vector $\mathbf{c}^{(0)}$ produce the packet transform coefficients at that level and, therefore, there are $N \times M$ vectors altogether whose inner products with the initial data vector produce all the packet transform coefficients,

$$c_{mn}^{(j)} \equiv \left\langle \mathbf{p}_{mn}^{(j)}, \mathbf{c}^{(0)} \right\rangle, \quad d_{mn}^{(j)} \equiv \left\langle \mathbf{q}_{mn}^{(j)}, \mathbf{c}^{(0)} \right\rangle. \tag{8.22}$$

The dimension of the space spanned by these vectors is $N$ and so they cannot all be independent. There are many ways to choose $N$ orthogonal vectors from this set. For instance, one orthogonal basis in the set of all packet vectors consists of the vectors $\mathbf{q}_{mn}^{(0)}$, $1 \leq m \leq M$, $0 \leq n \leq N-1$, and the vectors $\mathbf{p}_{Mn}^{(0)}$, $0 \leq n \leq N-1$, which comprise exactly $N$ orthogonal vectors $\mathbf{w}_n$ of equation (6.30) for the orthogonal wavelet transform. This orthogonal basis produces all the superscript-0 packet coefficients, i.e., the usual orthogonal wavelet coefficients, shown in figure 8.4.

A graphical method is used to establish the orthogonal basis property of any collection of vectors in the packet table. Suppose that we have a packet table comprising four levels, $m = 1, 2, 3, 4$, as shown in figure 8.12. Each rectangular portion corresponding to a set of packet coefficients $c_{mn}^{(j)}$ or $d_{mn}^{(j)}$ is marked by their associated vectors $\mathbf{p}_{mn}^{(j)}$ or $\mathbf{q}_{mn}^{(j)}$, $n = 0, 2^{-m}N - 1$, i.e., the vectors whose inner product with the initial discrete data $\mathbf{c}^0$ of length $N$, equations (8.22), would produce the corresponding coefficients. Given a rectangular region $R_{mj}$ specified by the integer pair $\{m, j\}$ corresponding to a set of $2^{-m}N$ orthogonal vectors $\mathbf{p}_{mn}^{(j)}$ or $\mathbf{q}_{mn}^{(j)}$, then all vectors contained in regions below whose vertical boundaries are within the vertical boundaries of $R_{mj}$ and regions above whose vertical boundaries contain the vertical boundaries of $R_{mj}$ are not orthogonal to all vectors of $R_{mj}$. To form an orthogonal basis, all chosen regions satisfying the previous requirements for orthogonality must span the width of the full rectangle containing all of the regions (the Tetris property). For instance, the set $\left\{\mathbf{q}_{1n}^{(0)}, \mathbf{q}_{2n}^{(0)}, \mathbf{q}_{3n}^{(0)}, \mathbf{q}_{4n}^{(0)}\right\}$

| $p_{1n}^{(0)}$ | | | | | | | | $q_{1n}^{(0)}$ | | | | | | | | ← $m=1$ |
|---|---|---|---|---|---|---|---|---|---|---|---|---|---|---|---|---|
| $p_{2n}^{(0)}$ | | | | $q_{2n}^{(0)}$ | | | | $p_{2n}^{(1)}$ | | | | $q_{2n}^{(1)}$ | | | | ← $m=2$ |
| $p_{3n}^{(0)}$ | | $q_{3n}^{(0)}$ | | $p_{3n}^{(1)}$ | | $q_{3n}^{(1)}$ | | $p_{3n}^{(2)}$ | | $q_{3n}^{(2)}$ | | $p_{3n}^{(3)}$ | | $q_{3n}^{(3)}$ | | ← $m=3$ |
| $p_{4n}^{(0)}$ | $q_{4n}^{(0)}$ | $p_{4n}^{(1)}$ | $q_{4n}^{(1)}$ | $p_{4n}^{(2)}$ | $q_{4n}^{(2)}$ | $p_{4n}^{(3)}$ | $q_{4n}^{(3)}$ | $p_{4n}^{(4)}$ | $q_{4n}^{(4)}$ | $p_{4n}^{(5)}$ | $q_{4n}^{(5)}$ | $p_{4n}^{(6)}$ | $q_{4n}^{(6)}$ | $p_{4n}^{(7)}$ | $q_{4n}^{(7)}$ | ← $m=4$ |

Figure 8.12: The orthogonal wavelet packet table for four levels.

consists of orthogonal vectors but it is not complete. $\mathbf{p}_{4n}^{(0)}$ is the only vector satisfying the graphical rule that when appended to the set would produce an orthogonal basis $\left\{p_{2n}^{(0)}, p_{3n}^{(1)}, p_{4n}^{(3)}, q_{4n}^{(3)}, p_{4n}^{(3)}\right\}$. The latter is, of course, the familiar orthogonal wavelet basis and it is depicted in figure 8.13. Note that the number of basis vectors is $2^{-1}N + 2^{-2}N + 2^{-3}N + 2^{-4}N + 2^{-4}N = N$, as expected. A second example of an orthogonal basis is the set shown in figure 8.14 consisting of the vectors $\left\{q_{2n}^{(1)}, p_{2n}^{(0)}, p_{3n}^{(2)}, p_{3n}^{(1)}, q_{4n}^{(5)}, p_{4n}^{(5)}, q_{4n}^{(3)}, p_{4n}^{(3)}\right\}$. Again, the basis contains $2 \times 2^{-2}N + 2 \times 2^{-3}N + 4 \times 2^{-4}N = N$ orthogonal vectors.

| | | | | | | | | | | | | | | | | |
|---|---|---|---|---|---|---|---|---|---|---|---|---|---|---|---|---|
| | | | | | | | $q_{1n}^{(0)}$ | | | | | | | | | ← $m=1$ |
| | | | $q_{2n}^{(0)}$ | | | | | | | | | | | | | ← $m=2$ |
| | $q_{3n}^{(0)}$ | | | | | | | | | | | | | | | ← $m=3$ |
| $p_{4n}^{(0)}$ | $q_{4n}^{(0)}$ | | | | | | | | | | | | | | | ← $m=4$ |

Figure 8.13:  The orthogonal wavelet basis location within the orthogonal wavelet packet table.

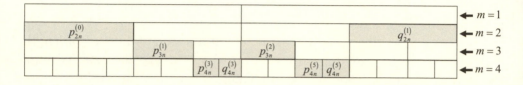

Figure 8.14:  An orthogonal basis, different from the wavelet basis, chosen from the orthogonal wavelet packet table.

## 8.5  The Best Basis Algorithm

The method illustrated in figures 8.13 and 8.14 allows for the choice of many orthogonal bases for the representation of any discrete sequence of length $N$ in an $M$ level orthogonal wavelet packet decomposition. The best basis among all the possibilities should have the property that a given signal's representation in that basis has the fewest coefficients with large magnitude. To illustrate this concept consider the three-dimensional space $\mathbb{R}^3$ with the orthonormal axes $\hat{\mathbf{x}}$, $\hat{\mathbf{y}}$, and $\hat{\mathbf{z}}$. Given a unit vector $\hat{\mathbf{r}}$ whose spherical polar coordinates are denoted by $\theta$ and $\phi$ and which is restricted to the octant defined by $0 \le \theta$, $\phi \le \pi/2$, the components of $\hat{\mathbf{r}}$ in this orthogonal system are

$$(\hat{r}_x, \hat{r}_y, \hat{r}_z) = (\sin\theta\cos\phi, \ \sin\theta\sin\phi, \ \cos\theta) \, .$$

Using these components we form the quantity known as the entropy[3] of the magnitude squared coefficients $\mathscr{E}(\theta, \phi)$

$$\mathscr{E}(\theta, \phi) = -\hat{r}_x^2 \log \hat{r}_x^2 - \hat{r}_y^2 \log \hat{r}_y^2 - \hat{r}_z^2 \log \hat{r}_z^2 \, .$$

---

[3]Our definition here closely resembles the information theoretic definition of entropy. For a discrete random variable $X_n$ which takes values $x_n$ with probability $p_n > 0$, where $\sum_n p_n = 1$, we define the entropy as $-\sum_n p_n \log_2 (p_n)$. The base of the logarithm is 2 in information theory applications.

We use base 10 for the logarithm and show a plot of this quantity, as a function of $\theta$ and $\phi$, in figure 8.15. Clearly, the magnitude squared coefficient

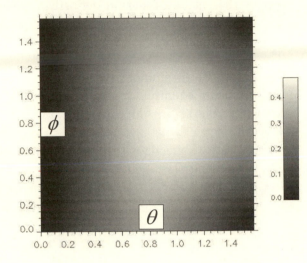

Figure 8.15: Magnitude squared coefficient entropy function of a unit vector in $\mathbb{R}^3$.

entropy is small on the $z$ axis, $\theta = 0$, and on the $x$ and $y$ axes, defined by the points $(\theta = \pi/2, \phi = 0, \pi/2)$, i.e., when one of the components is nearly 1 and the other two components are nearly 0. Thus, if we rotate our axes so that one axis lies along our unit vector we will be able to represent our vector by the single large component and neglect the other two, a storage reduction of 70%. This result can be generalized to a unit vector $\hat{\mathbf{x}}$ in $\mathbb{R}^N$ where $N > 3$ in which case the magnitude coefficient entropy is defined by

$$\mathscr{E}_{\hat{\mathbf{x}}} = - \sum_{n=0}^{N-1} |\langle \mathbf{p}_n, \hat{\mathbf{x}} \rangle|^2 \log |\langle \mathbf{p}_n, \hat{\mathbf{x}} \rangle|^2 , \qquad (8.23)$$

and the vectors $\mathbf{p}_n$ are an orthonormal basis of $\mathbb{R}^N$. A criterion for the best basis of orthogonal packets is then the requirement that the magnitude squared coefficient entropy function attains its minimum value for that basis.[4] Since all orthogonal transformations of the basis vectors that will be used in the minimization process keep the length of the vector unchanged, there is no loss of generality in assuming a unit norm vector $\hat{\mathbf{x}}$. We expect

---

[4] The entropy function defined in equation (8.23) is one example of an additive cost function that can be used in selecting the best basis. Other additive cost functions exist but the entropy has shown excellent results in most applications.

that in such a basis the fewest number of large coefficients would be sufficient to represent the signal of interest.

The key property of the entropy function defined in equation (8.23) is that it is additive: if $\left\{p_{i_1}, \ldots, p_{i_{N_0}}\right\}$ is an orthonormal basis of $\mathbb{R}^{N_0}$ and $\left\{p_{i_{N_0+1}}, \ldots, p_{i_N}\right\}$ is an orthonormal basis of the orthogonal complement of $\mathbb{R}^{N_0}$ in $\mathbb{R}^N$, $N_0 < N$, then $\mathscr{E}_{\hat{\mathbf{x}}}$ is the sum of two entropy functions defined in each subspace separately using equation (8.23). Additivity of the entropy function is used in an iterative algorithm to compute the best orthogonal packet basis from a given packet table: the best basis algorithm [61].

We use figure 8.5 to illustrate the algorithm to compute the best orthogonal packet basis by minimizing the magnitude entropy function for data of length $N$ (a power of 2). At some level $m - 1$ two sets of packet coefficients $c_{m-1,n}^{(j)}$ and $d_{m-1,n}^{(j)}$ are used to obtain the coefficients of the next level $m$. Consider first the vertical block associated with $c_{m-1,n}^{(j)}$ for which we calculate the following entropy functions:

$$\mathscr{E}_{m-1}^j = - \sum_{n=0}^{2^{-(m-1)}N-1} \left|c_{m-1,n}^{(j)}\right|^2 \log \left|c_{m-1,n}^{(j)}\right|^2, \tag{8.24a}$$

$$\mathscr{E}_m^{2j} = - \sum_{n=0}^{2^{-m}N-1} \left|c_{mn}^{(2j)}\right|^2 \log \left|c_{mn}^{(2j)}\right|^2 - \sum_{n=0}^{2^{-m}N-1} \left|d_{mn}^{(2j)}\right|^2 \log \left|d_{mn}^{(2j)}\right|^2. \tag{8.24b}$$

The best basis algorithm consists of choosing the basis vectors $\mathbf{p}_{m-1,n}^{(j)}$ if $\mathscr{E}_{m-1}^j \leq \mathscr{E}_m^{2j}$, or the basis vectors $\left\{\mathbf{p}_{mn}^{(2j)}, \mathbf{q}_{mn}^{(2j)}\right\}$ if $\mathscr{E}_{m-1}^j > \mathscr{E}_m^{2j}$. The algorithm begins with all the orthogonal basis vectors at the last level $M$ and works its way up to level 1. A similar selection rule applies to the vertical block associated with $d_{m-1,n}^{(j)}$. Once the selection of orthogonal basis vectors has been made, the entropy values at level $m$ are updated to include those of the selected coefficients. For instance, if in the above example the set $\left\{\mathbf{p}_{mn}^{(2j)}, \mathbf{q}_{mn}^{(2j)}\right\}$ were selected then the entropy value associated with it, (8.24b), will be used in the comparison at the next stage between levels $m - 1$ and $m - 2$.

Figure 8.16 shows the best packet basis (black rectangles) in a six-level DAUB-8 orthogonal wavelet packet table for the bowhead whale sound of figure 3.1; the DAUB-8 orthogonal wavelet basis appears on the left.

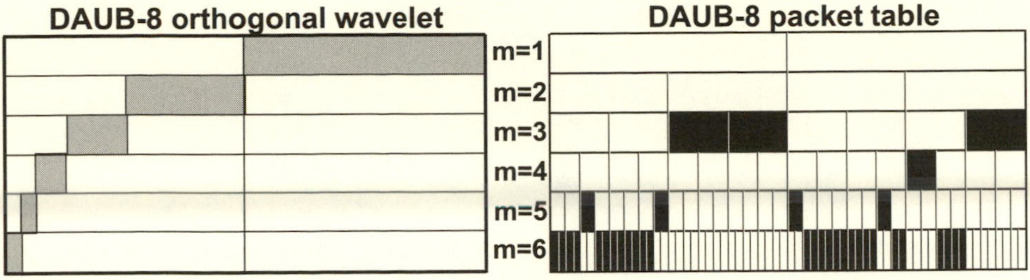

Figure 8.16: Bowhead whale sound and its best packet basis (black) using a six-level DAUB-8 wavelet packet table (the orthogonal wavelet basis is shown in gray).

## 8.6   Exercises

**1**  Use the results of theorem 85, equations (8.17), to compute the packet functions for DAUB-4 orthogonal wavelet of compact support (figure 8.7).

**2**  Show that the packet functions (8.19a) and (8.19b) of theorem 86 satisfy the frequency domain recursions (8.18).

**3**  Create a 512-point signal consisting of a linearly chirped time series, sampled at 1 Hz, whose maximum frequency is 0.25 Hz, together with a burst of a sinusoid at 0.3 Hz starting at around 4.4 minutes and lasting for about 2 minutes. Use the best basis algorithm described in section 8.5 to find the best packet basis (DAUB-4) for this signal and compare with the DAUB-4 orthogonal wavelet basis.

# CHAPTER 9

# Wavelet Transform in Two Dimensions

## 9.1 Introduction

We define an image as a real function $f(x, y)$ of two real variables, $x, y \in \mathbb{R}$, relaxing the requirement that image values, being intensities, must be positive quantities. Assuming that we have an orthonormal wavelet system based on a scaling and a wavelet function pair $\phi(x)$ and $\psi(x)$, we wish to construct a multi-resolution analysis subspace structure for the Hilbert space of all square integrable images, namely, $L_2(\mathbb{R}^2)$, using the tensor product of two identical pairs: $\{\phi(x), \psi(x)\}$ and $\{\phi(y), \psi(y)\}$.[1] If $\mathscr{V}_m^{(2)}$ denotes the approximation subspace at resolution $m$, the projection $\mathbf{P}_m f(x, y)$ onto this subspace

$$f_m(x, y) \equiv \mathbf{P}_m f(x, y) \tag{9.1}$$

can be written as the product of two one-dimensional projections $\mathbf{P}_m^x$ and $\mathbf{P}_m^y$ where

$$
\begin{aligned}
\mathbf{P}_m^x f(x, y) &= \sum_{n=-\infty}^{\infty} \langle \phi_{mn}(x), f(x, y) \rangle \phi_{mn}(x) \\
&\equiv \sum_{n=-\infty}^{\infty} c_{mn}^x(y) \phi_{mn}(x),
\end{aligned}
\tag{9.2}
$$

---

[1]The tensor product of orthogonal wavelets is not the only method of constructing two-dimensional wavelets but it is the simplest [14].

and

$$f_m\left(x,y\right) = P_m^y\left(P_m^x f\left(x,y\right)\right) = \sum_{k=-\infty}^{\infty} \left\langle \phi_{mk}\left(y\right), P_m^x f\left(x,y\right)\right\rangle \phi_{mk}\left(y\right)$$

$$\equiv \sum_{k=-\infty}^{\infty}\sum_{n=-\infty}^{\infty} \left\langle \phi_{mk}\left(y\right), c_{mn}^x\left(y\right)\right\rangle \phi_{mk}\left(y\right)\phi_{mn}\left(x\right)$$

$$\equiv \sum_{k=-\infty}^{\infty}\sum_{n=-\infty}^{\infty} \mathbf{cc}_{mnk}\,\phi_{mn}(x)\phi_{mk}(y). \tag{9.3}$$

The final set of coefficients $\mathbf{cc}_{mnk}$, defined above, are easily seen to be given by[2]

$$\mathbf{cc}_{mnk} \equiv \int_{-\infty}^{\infty}\int_{-\infty}^{\infty} \phi_{mk}\left(y\right)\phi_{mn}\left(x\right)f\left(x,y\right)dxdy. \tag{9.4}$$

Thus, we define the subspace $\mathscr{V}_m^{(2)}$ to be spanned by the functions

$$\xi_{mnk}^{LL}(x,y) \equiv \phi_{mn}(x)\phi_{mk}(y) \equiv 2^{-m}\phi(2^{-m}x - n)\phi(2^{-m}y - k), \tag{9.5}$$

where the superscript $LL$ refers to the fact that both projections were made using the scaling functions $\phi_{mn}(x)$ and $\phi_{mk}(y)$ and so will be related to the application of the low-pass filters $h_0$ in both the $x$ and the $y$ directions, respectively. In the next two sections we will investigate the decomposition rule for $\mathscr{V}_m^{(2)}$, i.e., the analog of the one-dimensional result

$$\mathscr{V}_m = \mathscr{V}_{m+1} \oplus \mathscr{V}^{\perp}{}_{m+1}.$$

## 9.2   The Forward Transform

Equation (9.3) defines the relation

$$\mathbf{cc}_{mnk} \equiv \left\langle \phi_{mk}\left(y\right), c_{mn}^x\left(y\right)\right\rangle. \tag{9.6}$$

The coefficients $c_{mn}^x(y)$, i.e., the inner product of $\phi_{mn}(x)$ with $f(x,y)$ for a fixed $y$, satisfy the general recursion (5.13), i.e.,

$$c_{mn}^x(y) = \sum_{j} h_0\left[j - 2n\right] c_{m-1,j}^x(y), \tag{9.7}$$

---

[2]All scaling and wavelet functions and associated filters are assumed to be real.

which together with the scaling equation (5.3) written in the following equivalent form (see exercise 1 in chapter 6),

$$\phi_{mk}(x) = \sum_{l=-\infty}^{\infty} h_0 [l - 2k] \phi_{m-1,l}(x), \tag{9.8}$$

will give

$$\mathbf{cc}_{mnk} = \left\langle \sum_{l=-\infty}^{\infty} h_0 [l - 2k] \phi_{m-1,l}(y), \sum_{j=-\infty}^{\infty} h_0 [j - 2k] c_{m-1,j}^x(y) \right\rangle$$

$$= \sum_{l=-\infty}^{\infty} \sum_{j=-\infty}^{\infty} h_0 [l - 2k] h_0 [j - 2n] \, \mathbf{cc}_{m-1,jl}. \tag{9.9}$$

The interpretation of equation (9.9) is as follows. The summation over the index $l$, i.e., the columns of the matrix $\mathbf{cc}_{m-1,jl}$, indicates a correlation of each row $j$ of $\mathbf{cc}_{m-1,jl}$ with $h_0$, taking the even numbered indices. The summation over the index $j$ is then the correlation of the result of the last set of operations with $h_1$, taking the even numbered indices. Thus, the $\mathbf{cc}_{mnk}$ coefficients at stage $m$ are related to the those at the previous stage $m - 1$ as shown in figure 9.1. In addition to (9.7), the coefficients $c_{mn}^x(y)$

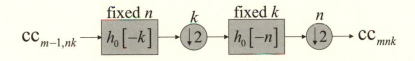

Figure 9.1: Recursive computation of the $\mathbf{cc}_m$ coefficients (the letter above the operation of taking the even numbered samples indicates the index over which the operation is performed).

also satisfy the relation (5.25) for scaling function coefficients, namely,

$$c_{mn}^x(y) = \sum_{j=-\infty}^{\infty} c_{m+1,j}^x(y) h_0 [n - 2j] + \sum_{j=-\infty}^{\infty} d_{m+1,j}^x(y) h_1 [n - 2j], \tag{9.10}$$

where the coefficients $d_{mn}^x(y)$ are equal to the inner products of $\psi_{mn}(x)$ with $f(x, y)$ for a fixed $y$,

$$d_{mn}^x(y) \equiv \langle \psi_{mn}(x), f(x, y) \rangle. \tag{9.11}$$

In addition to the $\mathbf{cc}_{mnk}$, and in analogy to equation (9.6), we now define three more sets of coefficients,

$$\mathbf{dc}_{mnk} \equiv \langle \psi_{mk}(y), c_{mn}^x(y) \rangle = \int\limits_{-\infty}^{\infty} \int\limits_{-\infty}^{\infty} \psi_{mk}(y)\phi_{mn}(x)f(x,y)dxdy, \quad (9.12a)$$

$$\mathbf{cd}_{mnk} \equiv \langle \phi_{mk}(y), d_{mn}^x(y) \rangle = \int\limits_{-\infty}^{\infty} \int\limits_{-\infty}^{\infty} \phi_{mk}(y)\psi_{mn}(x)f(x,y)dxdy, \quad (9.12b)$$

$$\mathbf{dd}_{mnk} \equiv \langle \psi_{mk}(y), d_{mn}^x(y) \rangle = \int\limits_{-\infty}^{\infty} \int\limits_{-\infty}^{\infty} \psi_{mk}(y)\psi_{mn}(x)f(x,y)dxdy. \quad (9.12c)$$

The corresponding bases, i.e.,

$$\xi_{mnk}^{HL}(x,y) \equiv \psi_{mk}(y)\phi_{mn}(x), \quad (9.13a)$$

$$\xi_{mnk}^{LH}(x,y) \equiv \phi_{mk}(y)\psi_{mn}(x), \quad (9.13b)$$

$$\xi_{mnk}^{HH}(x,y) \equiv \psi_{mk}(y)\psi_{mn}(x), \quad (9.13c)$$

form subspaces of $L_2(\mathbb{R}^2)$ that we denote by $\mathbf{DC}_m$, $\mathbf{CD}_m$ and $\mathbf{DD}_m$, respectively. Together with $\xi_{mnk}^{HL}(x,y)$ defined in equation (9.5), we have (all functions are assumed real)

$$\left\langle \xi_{mnk}^{ab}, \xi_{mpq}^{cd} \right\rangle = \int\limits_{-\infty}^{\infty} \int\limits_{-\infty}^{\infty} \left( \xi_{mnk}^{ab}(x,y) \right) \xi_{mpq}^{cd}(x,y)\, dxdy$$

$$= \delta^{ac}\delta^{bd}\delta_{np}\delta_{kq}, \quad \{a,b,c,d\} = \{L,H\}, \quad (9.14)$$

which implies mutual orthogonality of all the subspaces $\mathbf{CC}_m$, $\mathbf{DC}_m$, $\mathbf{CD}_m$, and $\mathbf{DD}_m$.

The coefficients $d_{mn}^x(y)$ satisfy the general recursion relation (5.14), i.e.,

$$d_{mn}^x(y) = \sum_j h_1[j-2n]c_{m-1,j}^x(y), \quad (9.15)$$

which together with the wavelet equation (5.8), written in the equivalent form (see exercise 1 in chapter 6)

$$\psi_{mk}(t) = \sum_{l=-\infty}^{\infty} h_1[l-2k]\phi_{m-1,l}(t), \quad (9.16)$$

and using equations (9.7) and (9.8) will give

$$\mathbf{dc}_{mnk} = \sum_{l=-\infty}^{\infty} \sum_{j=-\infty}^{\infty} h_1 \left[ l - 2k \right] h_0 \left[ j - 2n \right] \mathbf{cc}_{m-1,jl}, \qquad (9.17a)$$

$$\mathbf{cd}_{mnk} = \sum_{l=-\infty}^{\infty} \sum_{j=-\infty}^{\infty} h_0 \left[ l - 2k \right] h_1 \left[ j - 2n \right] \mathbf{cc}_{m-1,jl}, \qquad (9.17b)$$

$$\mathbf{dd}_{mnk} = \sum_{l=-\infty}^{\infty} \sum_{j=-\infty}^{\infty} h_1 \left[ l - 2k \right] h_1 \left[ j - 2n \right] \mathbf{cc}_{m-1,jl}. \qquad (9.17c)$$

These equations have interpretations similar to that of the $\mathbf{cc}_{mnk}$ coefficients, and together with equation (9.9) define the forward two-dimensional orthogonal discrete wavelet transform coefficients, summarized in figure 9.2,[3] which clearly shows the relation

$$\mathscr{V}_{m-1}^{(2)} \subseteq \mathbf{CC}_m \oplus \mathbf{CD}_m \oplus \mathbf{DC}_m \oplus \mathbf{DD}_m. \qquad (9.18)$$

An example of a one-stage transform using orthogonal DAUB-4 coefficients

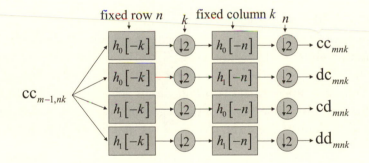

Figure 9.2: The forward two-dimensional orthogonal wavelet transform coefficients.

is shown in figure 9.3.[4] The three detail subimages labeled LH, HL, and HH

---

[3]The notation in the definition of the transform coefficients is consistent with figure 9.2 in which earlier operations appear to the right of later ones. For instance, the coefficients $\mathbf{dc}$ correspond to the application of the low-pass filter $h_0$ to each row (hence the rightmost letter $\mathbf{c}$) followed by the application of the high-pass filter $h_1$ to the columns (hence the leftmost letter $\mathbf{d}$).

[4]XDWT2D, an IDL widget program to compute and display two dimensional orthogonal and biorthogonal discrete wavelet transforms, is available from the author: najmi@jhuapl.edu.

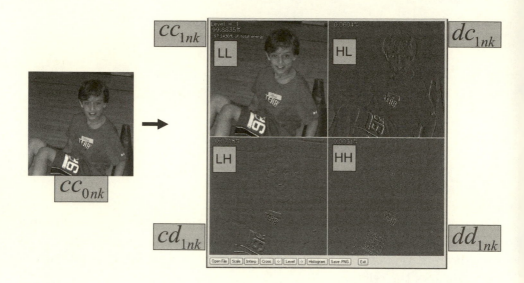

Figure 9.3: Example of a one-stage forward two-dimensional orthogonal wavelet transform: $\mathbf{cc}_{onk}$ is a $512 \times 512$ image decomposed into four orthogonal components of size $256 \times 256$ each.

tend to enhance edges in the two horizontal and vertical directions. This can be seen more clearly in the transform of the image shown in figure 9.4 consisting of a circle and straight lines. The HH detail image enhances lines in the cross directions, while LH enhances lines in the horizontal direction and HL enhances lines in the vertical, at all scales. Note the that the top and bottom of the circle (horizontal segments) are also enhanced in LH, while left and right of the circle (vertical segments) are enhanced in HL.

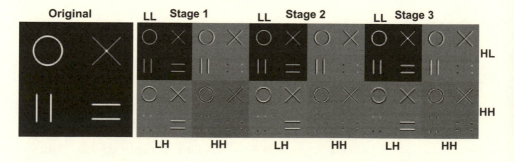

Figure 9.4: Three-stage forward Haar transform of the leftmost image.

## 9.3  The Inverse Transform

To find the inverse relations we use equation (9.10) in the definition of $\mathbf{cc}_{mnk}$, contained in equation (9.3), to obtain

$$\mathbf{cc}_{mnk} \equiv \langle \phi_{mk}(y), \langle \phi_{mn}(x), f(x,y) \rangle \rangle$$

$$= \sum_{j=-\infty}^{\infty} h_0[n-2j] \langle \phi_{mk}(y), c_{m+1,j}^x(y) \rangle +$$

$$\sum_{j=-\infty}^{\infty} h_0[n-2j] \langle \phi_{mk}(y), d_{m+1,j}^x(y) \rangle.$$

We now substitute equation (5.37), i.e.,

$$\phi_{mk}(y) = \sum_{j=-\infty}^{\infty} h_0[k-2j] \phi_{m+1,j}(y) + \sum_{j=-\infty}^{\infty} h_1[k-2j] \psi_{m+1,j}(y),$$

and finally obtain

$$\mathbf{cc}_{mnk} = \sum_{j=-\infty}^{\infty} \sum_{l=-\infty}^{\infty} h_0[n-2j] h_0[k-2l] \mathbf{cc}_{m+1,jl} +$$

$$\sum_{j=-\infty}^{\infty} \sum_{l=-\infty}^{\infty} h_0[n-2j] h_1[k-2l] \mathbf{dc}_{m+1,jl} +$$

$$\sum_{j=-\infty}^{\infty} \sum_{l=-\infty}^{\infty} h_1[n-2j] h_0[k-2l] \mathbf{cd}_{m+1,jl} +$$

$$\sum_{j=-\infty}^{\infty} \sum_{l=-\infty}^{\infty} h_1[n-2j] h_1[k-2l] \mathbf{dd}_{m+1,jl}.$$

$$(9.19)$$

The interpretation of this equation is depicted in figure 9.5. For instance, consider the first double summation on the right hand side. The sum over the index $l$ is a convolution of the low-pass filter $h_0$ with the up-sampled sequence (fixed $n$)

$$\hat{\mathbf{cc}}_{m+1,nk} = \begin{cases} \mathbf{cc}_{m+1,nk}, & k \text{ even,} \\ 0, & k \text{ odd,} \end{cases}$$

for every row, index $n$. The second sum over the index $j$ is then a similar convolution with the up-sampled rows of the resulting matrix. Similar operations are then performed on the coefficients $\mathbf{cd}_{mnk}$, $\mathbf{dc}_{mnk}$, and $\mathbf{dd}_{mnk}$, after which all results are summed to produce the coefficients $\mathbf{cc}_{m-1,nk}$.

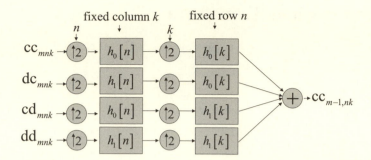

Figure 9.5: The forward two-dimensional orthogonal wavelet transform coefficients.

Together with equation (9.18) we arrive at the following orthogonal decomposition of $\mathscr{V}_{m-1}^{(2)}$:

$$\mathscr{V}_{m-1}^{(2)} = \mathbf{CC}_m \oplus \mathbf{CD}_m \oplus \mathbf{DC}_m \oplus \mathbf{DD}_m. \tag{9.20}$$

The generalization to biorthogonal pairs is now a simple exercise: the forward transform uses the pair $\{\phi_{mn}, \psi_{mk}\}$ while the inverse uses the pair $\{\phi'_{mn}, \psi'_{mk}\}$.

## 9.4   Implementing the Two-Dimensional Wavelet Transform

As in the one-dimensional case, a given image matrix $\mathbf{F}$ with $2^K$ rows and columns is taken as the starting set of the coefficients $\mathbf{cc}_{0nk}$. The operations of correlation and taking the even numbered sampled for $h_0$ and $h_1$ are accomplished using the matrix formulation of section 6.3. Given the coefficients $h_0$ of length $N$ and $h_1$ of length $M$,[5] the matrix $\mathbf{H}$ is constructed as follows. The first two rows are

$$\begin{bmatrix} h_0\,[0] & \cdots & h_0\,[N-1] & 0 & \cdots & 0 \\ h_1\,[0] & \cdots & h_1\,[M-1] & 0 & \cdots & 0 \end{bmatrix}. \tag{9.21}$$

The other rows are constructed by circular shifts of two places of the first two rows until the size of $\mathbf{H}$ matches the image size. For instance, rows 3 and 4 are

$$\begin{bmatrix} 0 & 0 & h_0\,[0] & h_0\,[1] & \cdots & 0 \\ 0 & 0 & h_1\,[0] & h_1\,[1] & \cdots & 0 \end{bmatrix}. \tag{9.22}$$

---

[5]In general and for biorthogonal wavelets $N$ and $M$ are not equal.

The correlations with $h_0$ and $h_1$ and the even sample selection can now be performed by a single multiplication by the transpose of $\mathbf{H}$ on the right, namely,

$$\mathbf{FH}^T \to \begin{bmatrix} c & d & \cdots & c & d \\ \vdots & \vdots & \vdots & \vdots & \vdots \\ c & d & \cdots & c & d \end{bmatrix}, \tag{9.23}$$

where the right hand side shows the structure of the ensuing matrix in terms of c- and d-type coefficients. A permutation of columns is then performed to put all the c-type coefficients to the left of all the d-type coefficients. Thus,

$$\mathbf{FH}^T \underset{\text{columns}}{\to} \begin{bmatrix} c & c & \cdots & d & d \\ \vdots & \vdots & \vdots & \vdots & \vdots \\ c & c & \cdots & d & d \end{bmatrix}. \tag{9.24}$$

A further multiplication on the left by $\mathbf{H}$ and row permutations will result in

$$\mathbf{H} \begin{bmatrix} c & c & \cdots & d & d \\ \vdots & \vdots & \vdots & \vdots & \vdots \\ c & c & \cdots & d & d \end{bmatrix} \to \begin{bmatrix} cc & cc & \cdots & cd & cd \\ dc & dc & \cdots & dd & dd \\ \vdots & \vdots & \vdots & \vdots & \vdots \\ cc & cc & \cdots & cd & cd \\ dc & dc & \cdots & dd & dd \end{bmatrix} \underset{\text{rows}}{\to} \begin{bmatrix} CC & CD \\ DC & DD \end{bmatrix}$$
$$\tag{9.25}$$

where all the $\mathbf{cc}$ coefficients are in the top left quadrant, the $\mathbf{cd}$ coefficients are in the top right quadrant, the $\mathbf{dc}$ coefficients are in the bottom left quadrant, and the $\mathbf{dd}$ coefficients are in the bottom right quadrant. If the original matrix is of size $2^K$, then each quadrant is now of size $2^{K-1}$. This completes the first stage of the transform, i.e.,

$$[\mathbf{cc}_0] \to \begin{bmatrix} [\mathbf{cc}_1] & [\mathbf{cd}_1] \\ [\mathbf{dc}_1] & [\mathbf{dd}_1] \end{bmatrix}. \tag{9.26}$$

For the next stage the same process is repeated on the $[\mathbf{cc}_1]$ sub-matrix of size $2^{K-1}$ with a correspondingly resized matrix $\mathbf{H}$. A description of the inverse transform in terms of orthogonal matrices is left as an exercise.

## 9.5   Application to Image Compression

The most widespread application of discrete wavelets is to compression of image data. All image compression schemes can be broadly described as

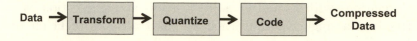

Figure 9.6: Data compression using transform coding.

transform coding which consists of three generic stages shown in figure 9.6: data is first transformed, then quantized, and finally coded.

The two broad objectives of transforming data are to decorrelate and to concentrate energy, i.e., the transform coefficients should show little statistical correlation among themselves and most of the energy in the data should be concentrated in a few coefficients.

The ideal decorrelating transform is the Karhunen-Loeve expansion [31]. Consider a zero-mean real random vector $\mathbf{x}$ with correlation matrix $\mathbf{R}_{xx} = E[\mathbf{x}\mathbf{x}^T]$ which is symmetric and has a decomposition in the form $\mathbf{R}_{xx} = \mathbf{U}\boldsymbol{\Lambda}\mathbf{U}^T$.[6] The columns of the orthogonal matrix $\mathbf{U}$ denoted by $\mathbf{u}_n$, $0 \leq n \leq N - 1$, are the normalized eigenvectors of the correlation matrix and $\boldsymbol{\Lambda}$ is a diagonal matrix of eigenvalues. The transformed vector $\mathbf{y} = \mathbf{U}^T\mathbf{x}$ is now a zero mean random vector with uncorrelated components. The Karhunen-Loeve expansion is then given by

$$\mathbf{x} = \mathbf{U}\mathbf{y} = \sum_{n=0}^{N-1} y_n \mathbf{u}_n. \tag{9.27}$$

Lossy compressions based on this transform rely on retaining only those eigenvectors corresponding to the largest eigenvalues of the correlation matrix. The Karhunen-Loeve transform, however, is impractical because the transform matrix $U$ is data dependent: it has to be computed for each data set separately. The most widely used transform before the new JPEG2000 image compression standard was the discrete cosine transform (DCT) in the form of the JPEG standard. The one-dimensional DCT for a discrete sequence $x[n]$, $0 \leq n \leq N - 1$, is defined by [62]

$$X_k = \sum_{n=0}^{N-1} x_n \cos\left[\frac{\pi k}{2N}(2n + 1)\right], \tag{9.28}$$

---

[6]A scalar random variable $\mathbf{x}$ has an associated probability density function $f_\mathbf{x}(x)$ which is positive and whose integral over the real line is equal to 1. The expectation operation is then defined with respect to this density function; for instance, $E[\mathbf{x}] = \int_{-\infty}^{\infty} x f_\mathbf{x}(x) dx$. A random vector is composed of a finite number of scalar random variables each with its own density function.

and its inverse is

$$x_n = \frac{X_0}{2} + \sum_{k=0}^{N-1} X_k \cos\left[\frac{\pi n}{2N}(2k+1)\right]. \tag{9.29}$$

The two-dimensional DCT is performed, like the two-dimensional DWT, first on rows and then on columns.

Before a more thorough discussion of the transform stage we will briefly describe the other two stages of transform coding that were shown in figure 9.6: quantization and coding. Consider an image $f(i,j)$ as the input data. If we denote the output of the transform stage by an identically sized matrix $F(i,j)$, then the quantizer is defined by

$$Q[F(i,j)] = v_m, \quad u_m \le F(i,j) < u_{m+1}. \tag{9.30}$$

For instance, figure 9.7 shows a four-level quantizer which maps all values in the range $[-u_2, u_1]$ to four values $v_n$, $1 \le n \le 4$. Transforms with good energy concentration properties (such as the DCT and the DWT) are preferred because they allow for the concentration of almost all the energy into a small subset (2%) of transform coefficients. Thus, coefficients with large magnitudes can be quantized with more precision while the others can be quantized coarsely. The last stage in figure 9.6 is the bit allocation, or coding, stage at which binary code words are assigned to each quantized value. In the most elementary coding scheme in the example shown in figure 9.7 one would assign four binary codewords $\{00, 01, 10, 11\}$ to the four available levels. More sophisticated entropy coding takes advantage of the frequency of occurrence of quantized values to assign longer codewords to values that occur least and shorter codewords to those occurring most in order to get as close as possible to limits derived from information theoretic arguments.

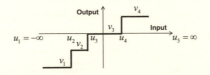

Figure 9.7: A simple four-level quantizer.

The JPEG standard[7] is a lossy compression scheme based on the DCT applied to $8 \times 8$ blocks of an image. The block application is necessary so

---

[7]www.jpeg.org.

that the coefficients have local spatial information since the DCT basis vectors are discrete spatial cosine functions, and the $8 \times 8$ block size is a good compromise for efficiency defined as the ratio of the memoryless entropy of the original signal to the mean entropy of the transform coefficients. Each block is then quantized with different precisions: higher for lower frequencies and lower for higher frequencies. The quantized values are next encoded. The blocks do not overlap and so JPEG encoded images show distinct blocking artifacts at higher compression ratios. On the other hand, the spatial frequency information provided by the DCT transformation basis vectors is related to the sensitivity of human vision to spatial frequencies. In addition, the DCT is closely related to the Karhunen-Loeve expansion when the source image has a spectrum that decays rapidly as a function of spatial frequency: a property shared by most images.

The newer JPEG2000 system, depicted in figure 9.8, was developed with a desire to provide not only higher compression efficiency but also a multi-resolution representation of an image that would be far more versatile than the prevailing DCT-based JPEG system in a communications environment with differing resolution requirements: from high-resolution monitors to much lower-resolution handheld devices [63]. The wavelet transform, of course, provides the multi-resolution image representation for free, whereas the DCT-based JPEG encoded images would have to decompress the image first before recompressing at other resolutions. A second advantage of the wavelet transform is that it can be applied to the entire image, retaining local spatial information, since the basis functions are localized in space. Thus, the full frame applicability of the DWT, as opposed to the $8 \times 8$ blocks of the DCT, allows for reducing artifacts at high compression ratios. In addition, the multi-resolution property of the DWT is in conformity with models of early human vision that show that human perception of objects (edges) relies on their detectability at multiple scales. As we have seen, sufficiently regular wavelets produce large coefficients at signal discontinuities on all scales while smooth backgrounds have negligibly small coefficient magnitudes [64].

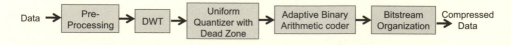

Figure 9.8: JPEG2000 fundamental building blocks.

To compare the energy concentration property between the discrete cosine transform and the discrete wavelet transform we use the image shown in figure 9.3 and perform a two-dimensional DCT based on $8 \times 8$ blocks, as well as a discrete biorthogonal wavelet transform using the 9/7 wavelet filter pair (see exercise 6 of chapter 7). We normalize the output coefficients in both cases so as to have unit total energy (the sum of the squares of the coefficients is 1), then sort the square of the coefficients by their magnitudes in descending order, and finally calculate partial sums of the sorted coefficients.

Figure 9.9 shows the energy concentration curve for the DCT: almost all the energy is packed into the top $1 - 2\%$ of the transform coefficients.

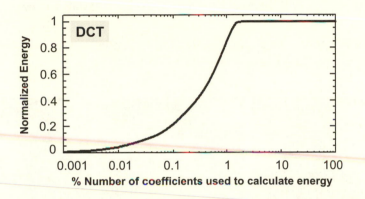

Figure 9.9: DCT: energy versus % number of coefficients sorted on magnitude (decreasing left to right).

Figure 9.10 shows two similar curves for the 9/7 biorthogonal wavelet transform of the same image: one for a three-level transform and one for a five-level transform: the three-level transform curve is nearly identical to the corresponding DCT curve while the five-level transform curve indicates better performance.

Energy concentration performance, however, must be measured in combination with the quality of the reconstructed image. Here we reconstruct the image based on retaining the top 0.1%, 1%, and 2% of the coefficients (in terms of magnitude) and setting the rest of the coefficients to 0. Figure 9.11 shows three reconstructed images for the DCT performed on $8 \times 8$ blocks. Clearly, $1 - 2\%$ of the largest magnitude coefficients are required for a reasonable quality reconstruction. At these small percentages, however, reconstructed images suffer from "blockiness": artifacts due to the $8 \times 8$ block application.

Similar reconstructed images for both the three-level and the five-level biorthogonal DWT are shown in figures 9.12 and 9.13. A map of the square

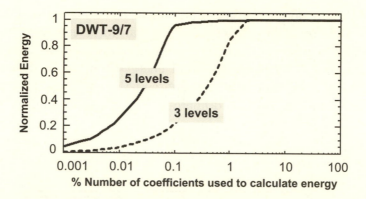

Figure 9.10: 9/7 biorthogonal DWT: energy versus % number of coefficients sorted on magnitude (decreasing left to right).

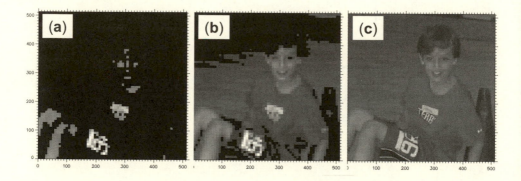

Figure 9.11: DCT reconstructed images using: (a) top 0.1%, (b) 1%, and (c) 2% of the coefficients with largest magnitude.

of the magnitude of the retained coefficients (in dB scale with a 60 dB range) is displayed above each image and clearly shows that much of the energy is concentrated in the final LL portion of the DWT plane: for the three-level example the final LL portion of the transform accounts for 98.47% of the total energy of the transform.

Corresponding reconstructions and map of coefficients for the five-level transform confirm that most of the energy is concentrated in the final LL portion of the DWT plane which now accounts for 95.65% of the total transform energy. Table 9.1 summarizes the percentage of energy, relative to the total transform energy, in the final LL portion of the 9/7 biorthogonal wavelet transform and as a function of the number of transform levels 1–5.

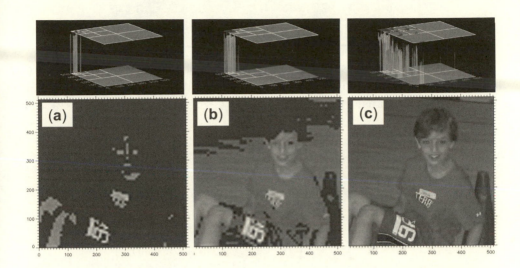

Figure 9.12: 9/7 biorthogonal 3-level DWT reconstructed images using: **(a)** top 0.1%, **(b)** 1%, and **(c)** 2% of the coefficients with largest magnitude — map of used coefficients, with a 60 dB range, is shown above each image.

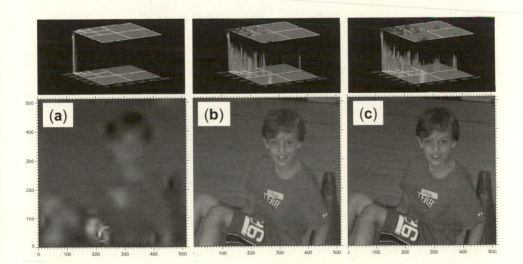

Figure 9.13: 9/7 biorthogonal 5-level DWT reconstructed images using: **(a)** top 0.1%, **(b)** 1%, and **(c)** 2% of the coefficients with largest magnitude — map of used coefficients, with a 60 dB range, is shown above each image.

| 1 | 2 | 3 | 4 | 5 |
|---|---|---|---|---|
| 99.85% | 99.34% | 98.47% | 97.17% | 95.65% |

Table 9.1: 9/7 biorthogonal wavelet transform of $512 \times 512$ image (figure 9.3): fraction of total transform energy for the final LL portion as a function of the number of transform levels 1–5.

A quantity that is often used in comparing image reconstructions is the Peak Signal to Noise Ratio[8] (PSNR) defined by

$$
\text{PSNR} = 10 \, \text{Log}_{10} \left( \frac{Max\left(i^2\right)}{N^{-2} \sum_{n=0}^{N-1} \sum_{m=0}^{N-1} \left(i_{mn} - r_{mn}\right)^2} \right)
$$

where $i$ denotes the original $N \times N$ image and $r$ is the reconstructed image. Figure 9.14 shows a comparison of the PSNR values for the DCT and the 9/7 biorthogonal DWT on a $1024 \times 1024$ version of the same image we have used above. At very low compressions (using $\leq 1.5\%$ of the top magnitude transform coefficients) the PSNR value for the DWT reconstruction is at least 7 dB above that of the DCT in this particular case.

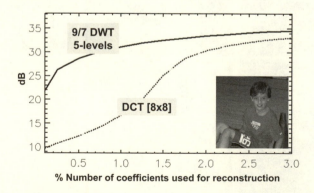

Figure 9.14: PSNR values for 5-level 9/7 biorthogonal DWT and DCT— $1024 \times 1024$ version of image shown in figure 9.3.

---

[8]We make no association between this quantity and a meaningful (in the sense of detection theory) signal-to-noise-ratio (SNR) here. For purposes of comparison, however, this is a useful quantity whose difference between two reconstruction methods is a reasonable measure of their relative fidelity.

The three images shown in figure 9.15 from the JPEG2000 test suite seem to confirm the general conclusion that for reconstructions based on less than 3% of the top magnitude transform coefficients (low compression ratios) the PSNR curves for DWT and DCT retain the relative performance characteristics illustrated in figure 9.15.

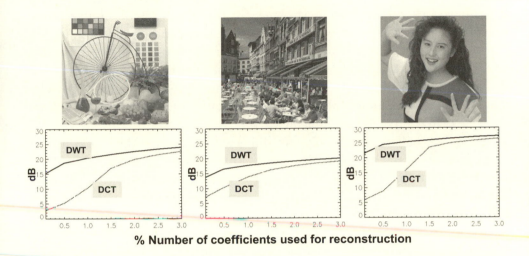

Figure 9.15: PSNR curves for JPEG2000 test images: bike, cafe, and woman.

## 9.6 Image Fusion

In section 6.4 we discussed applications of thresholding the wavelet coefficients to detect edges in imagery. Another useful application of wavelets is to fusion of imagery collected on multiple sensors, e.g., infrared and visible cameras, or MRI and PET (positron emission tomography) scans, to produce a single image that combines all useful information from those images. A simple fusion scheme is based on the following: normalize all images to have the same total energy, perform a wavelet transform of each image, and finally compare the wavelet coefficients in each scale across sensors and add those coefficients exceeding calculated thresholds in each scale. The resulting wavelet transform coefficients are then inverted to obtain the fused image. More sophisticated selection and combining methods are required if the same object presents opposite contrasts in different sensors [65].

## 9.7   Wavelet Descendants

Wavelets have led to new developments in the mathematical theory of multi-resolution signal representation to include new families of analyzing functions in two dimensions. As we have seen, one-dimensional wavelets with sufficient regularity can detect discontinuities along a curve at all scales. In two dimensions, however, an image generally consists of discontinuities along boundary curves or edges while regions on either side of those boundary curves are smooth. The usual construction of the two-dimensional wavelet transform using the product of two one-dimensional wavelets can detect the discontinuity across an edge but not along the edge. Ridgelets, wedgelets, and curvelets are among the family of analyzing functions that have been developed to overcome this difficulty.

Ridgelets [66] are wavelet-like two-dimensional surfaces that are constant along lines of a given direction. Ridgelets can be defined as a one-dimensional wavelet transform of slices of the Radon transform of an image. They have shown to substantially reduce ringing around discontinuities along straight lines, i.e., along straight edges, in an image that result from DWT reconstructions [67].

Wedgelets [68] are functions on a square that have constant profiles on either side of a given line (edge). A line in a square is defined by two parameters: a direction and a distance from the center of the square. Wedgelets are more suitable than wavelets at capturing geometrical structures (edges) in images. Multiscale wedgelet decompositions have been used to eliminate ringing around edges that result using DWT reconstructions.

Curvelets [69] are analyzing functions that are produced by applying translations, rotations, and parabolic dilations to a mother wavelet function. These analyzing functions are highly anisotropic at finer scales with support that follows the parabolic scaling with a fixed ratio of the square of the length to the width: at scale $m$ each is an oriented needle shape with effective support over a rectangle of size $2^{-m} \times 2^{-m/2}$. In spatial frequency, it is a wedge with support on a rectangle of size $2^m \times 2^{m/2}$. Thus, curvelets are localized in space, scale, and orientation. Their localization in orientation endows a curvelet frame with a sparsity unmatched by any other frame, including wavelet frames. If an image consists of smooth curves with discontinuities along each curve ($C^2$ curves with isolated singularities) then a curvelet basis is optimal in representing the image in the sense that the curvelet coefficients decay the fastest in this frame, i.e., a curvelet frame will have a better energy concentration curve than any wavelet frame for such images.

Thus, the wavelet story is still unfolding and there is little doubt that rapid developments in digital technology will find ever increasing applications of wavelets and their multiplying descendants.

## 9.8 Exercises

**1** The forward transform for orthogonal wavelets in terms of orthogonal matrices was described in detail in section 9.4. Complete the description for the inverse transform. In addition, extend the forward and inverse matrix implementation to biorthogonal wavelets of compact support.

**2** Use equations (9.28) and (9.29) to implement the discrete cosine transform. Repeat problem 2 with the DCT and compare results at high compressions, e.g., top 0.1%, 1% and 2% the number of (sorted) coefficients used in reconstruction (see figures 9.11).

**3** Using a gray scale $512 \times 512$ image of your choice produce results similar to figures 9.12 and 9.13 for DAUB-12 and the 10/6 biorthogonal wavelet system (see exercise 7 of chapter 7).

# Bibliography

[1] A. Grossman and J. Morlet, "Decomposition of functions into wavelets of constant shape and related transforms," in *Mathematics and Physics, Lectures on Recent Results* (L. Streit, ed.), New Jersey: World Scientific, 1985.

[2] N. Ricker, "Wavelet contraction, wavelet expansion, and the control of seismic resolution," *Geophysics*, vol. 18, no. 4, pp. 769–792, 1953.

[3] J. D. McEwen *et al.*, "Cosmological applications of wavelet analysis on the sphere," *J. Fourier Anal. and Appl.*, vol. 13, pp. 495–510, 2007.

[4] J. C. Burkill, *The Lebesgue Integral*. Cambridge: Cambridge University Press, 1971.

[5] P. R. Halmos, *Finite-Dimensional Vector Spaces*. New York: Springer-Verlag, 1974.

[6] G. Bachman and L. Narici, *Functional Analysis*. New York: Academic Press, 1966.

[7] J. Mathews and R. L. Walker, *Mathematical Methods of Physics*. Menlo Park: Benjamin/Cummings, 1964.

[8] N. Aronszajn, "Theory of reproducing kernels," *Trans. Amer. Math. Soc.*, vol. 68, pp. 337–404, 1950.

[9] I. M. Gel'fand and G. E. Shilov, *Generalized Functions - Volume 1*. New York: Academic Press, 1964.

[10] H. J. Bremermann, *Distributions, Complex Variables and Fourier Transforms*. Reading, MA: Addison-Wesley, 1965.

[11] D. G. Luenberger, *Optimization by Vector Space Methods*. New York: Wiley, 1969.

[12] J. P. Keener, *Principles of Applied Mathematics: Transformation and Approximation*. Reading, MA: Addison-Wesley, 1988.

[13] S. Mallat, "Multiresolution approximations and wavelet orthonormal bases of $\mathbf{L}_2(\mathbb{R})$," *Trans. Amer. Math. Soc.*, vol. 315, pp. 69–87, 1989.

[14] S. Mallat, *A wavelet tour of signal processing*. San Diego: Academic Press, 1999.

[15] I. Daubechies, *Ten Lectures on Wavelets*. Philadelphia: SIAM, 1992.

[16] P. J. Davis, *Interpolation and Approximation*. New York: Dover, 1975.

[17] B. Friedman, *Principles and Techniques of Applied Mathematics*. New York: Wiley and Sons, 1956.

[18] P. A. M. Dirac, *Principles of Quantum Mechanics*. Oxford: Oxford University Press, 1974.

[19] T. W. Körner, *Fourier Analysis*. Cambridge: Cambridge University Press, 1988.

[20] H. F. Davis, *Fourier Series and Othogonal Functions*. Boston: Allyn & Bacon, 1963.

[21] R. Beals, *Analysis: An Introduction*. Cambridge: Cambridge University Press, 2004.

[22] A. V. Oppenheim and R. W. Schafer, *Digital Signal Processing*. Englewood Cliffs, NJ: Prentice-Hall, 1975.

[23] C. E. Shannon, "Communication in the presence of noise," *Proc. Institute of Radio Engineers*, vol. 37, no. 1, pp. 10–21, 1949.

[24] R. Duffin and A. Schaeffer, "A class of nonharmonic fourier series," *Trans. Amer. Math. Soc.*, vol. 72, pp. 341–366, 1952.

[25] I. Daubechies, A. Grossman, and Y. Meyer, "Painless nonorthogonal expansions," *J. Math. Phys.*, vol. 27, no. 5, pp. 1271–1283, 1986.

[26] O. Christensen, *An Introduction to Frames and Riesz Bases*. Boston: Birkhäuser, 2003.

[27] R. K. Young, *An Introduction to Nonharmonic Fourier Series*. New York: Academic Press, 1980.

[28] G. Strang, *Linear Algebra and Its Applications*. Fort Worth, TX: Harcourt Brace Jovanovich, 1988.

[29] S. L. Campbell and C. D. Meyer, *Generalized Inverses of Linear Transformations*. New York: Dover, 1979.

[30] G. H. Golub and F. V. Loan, *Matrix Computations*. Baltimore: Johns Hopkins University Press, 1996.

[31] T. K. Moon and W. C. Stirling, *Mathematical Methods and Algorithms for Signal Processing*. Englewood Cliffs, NJ: Prentice-Hall, 2000.

[32] A. Berlinet and C. T. Agnan, *Reproducing Kernel Hilbert Spaces in Probability and Statistics*. Norwell, MA: Kluwer Academic Publishers, 2004.

[33] A. V. Oppenheim, S. Willsky, and I. T. Young, *Signals and Systems*. Englewood Cliffs, NJ: Prentice-Hall, 1983.

[34] F. Riesz and B. Sz.-Nagy, *Functional Analysis*. New York: Frederick Ungar Publ. Co., 1955.

[35] A. Croisier, D. Esteban, and C. Galand, "Perfect channel splitting by use of interpolation/decimation/tree decomposition techniques," in *Proceedings of the International Conference on Information Sciences and Systems*, Patras, Greece, 1976.

[36] G. Strang and T. Nguyen, *Wavelets and Filter Banks*. Wellesley, MA: Wellesley-Cambridge Press, 1997.

[37] L. Cohen, *Time-Frequency Analysis*. Englewood Cliffs, NJ: Prentice-Hall, 1995.

[38] D. Gabor, "Theory of communication," *Proc. IEE (London)*, vol. 93, no. 3, pp. 429–457, 1946.

[39] P. Goupillaud, A. Grossman, and J. Morlet, "Cycle-octave and related transforms in seismic signal analysis," *Geoexploration*, vol. 23, pp. 85–102, 1985.

[40] D. A. Swick, "A review of wideband ambiguity functions," *Naval Research Laboratory interim report*, 1969.

[41] A. H. Najmi and J. Sadowsky, "The continuous wavelet transform and variable resolution time-frequency analysis," *Johns Hopkins APL Technical Digest*, vol. 18, no. 1, pp. 134–140, 1997.

[42] A. Haar, "Zur theorie der orthogonalen funktionensysteme," *Mathematische Annalen*, vol. 69, no. 3, pp. 331–371, 1910.

[43] M. Holschneider, R. Kronland-Martinet, J. Morlet, and P. Tchamitchian, "A real-time algorithm for signal analysis with the help of the wavelet transform," in *Wavelets: Time-Frequency Methods and Phase Space* (L. Streit, ed.), Berlin: Springer-Verlag, 1989.

[44] M. J. Shensa, "Wedding the a' trous and mallat algorithms," *IEEE Trans. Signal Processing*, vol. 40, no. 10, pp. 2464–2482, 1992.

[45] T. Lui and A. H. Najmi, "Time-frequency decomposition of signals in a current disruption event," *Geophysical Research Letters*, vol. 24, no. 24, pp. 3157–3160, 1997.

[46] S. Mallat, "A theory of multiresolution signal decomposition," *IEEE Trans. Pattern Analysis & Machine Intelligence*, vol. 11, pp. 674–693, 1989.

[47] Y. Meyer, "Principe d'incertitude, bases hilbertiennes et algebres d'operateurs," *Seminaire Bourbaki*, vol. 662, 1985-86.

[48] G. G. Walter, *Wavelets and Other Orthogonal Systems with Applications*. Boca Raton, FL: CRC Press, 1994.

[49] B. Vidakovic, *Statistical Modeling by Wavelets*. New York: Wiley, 1999.

[50] J. E. Fowler, "The redundant discrete wavelet transform," *IEEE Signal Proc. Lett.*, vol. 12, no. 9, pp. 629–632, 2005.

[51] D. Pollen, "Daubechies scaling function on $[0, 3]$," in *Wavelets: A Tutorial in Theory and Applications* (C. K. Chui, ed.), Boston: Academic Press, 1982.

[52] A. Cohen, I. Daubechies, and J. C. Feauveau, "Biorthogonal bases of compactly supported wavelets," *Comm. Pure Applied Mathematics*, vol. 45, pp. 485–560, 1992.

[53] W. H. Press, S. A. Teukolsky, W. T. Vetterling, and B. P. Flannery, *Numerical Recipes: The Art of Scientific Computing*. Cambridge: Cambridge University Press, 1992.

[54] C. S. Burrus, R. A. Gopinath, and H. Guo, *Introduction to Wavelets and Wavelet Transforms*. Englewood Cliffs, NJ: Prentice-Hall, 1998.

[55] D. L. Donoho and I. M. Johnstone, "Ideal spatial adaptation by wavelet shrinkage," *Biometrika*, vol. 81, no. 3, pp. 425–455, 1994.

[56] D. L. Donoho, "De-noising by soft thresholding," *IEEE Trans. Information Theory*, vol. 38, no. 2, pp. 613–627, 1995.

[57] W. Sweldens, "The lifting scheme: A custom design construction of biorthogonal wavelets," *J. Appl. Comp. Harm. Anal.*, vol. 3, no. 2, pp. 186–200, 1996.

[58] I. Daubechies and W. Sweldens, "Factoring wavelet transforms into lifting steps," *J. Fourier Analysis & Applications*, vol. 4, no. 3, pp. 245–267, 1998.

[59] R. Coifman, F. Geshwind, and Y. Meyer, "Noiselets," *Applied and Computational Harmonic Analysis*, vol. 10, pp. 27–44, 2001.

[60] R. R. Coifman, Y. Meyer, and M. V. Wickerhauser, "Adapted waveform analysis, wavelet packets and applications," in *Proceedings of ICIAM '91 Washington DC*, pp. 41–50, Philadelphia: SIAM Press, 1992.

[61] M. V. Wickerhauser, *Adapted Wavelet Analysis from Theory to Software*. New York: IEEE Press, 1996.

[62] N. Ahmed, T. Natarajan, and K. R. Rao, "Discrete cosine transform," *IEEE Trans. Computers*, vol. 23, pp. 88–93, 1974.

[63] M. Rabbani and R. Joshi, "An overview of the jpeg2000 still image compression standard," *Signal Processing: Image Communication*, vol. 17, pp. 3–48, 2002.

[64] S. G. Mallat and W. L. Hwang, "Singularity detection and processing with wavelets," *IEEE Trans. Information Theory*, vol. 37, pp. 617–643, 1992.

[65] G. Pajaras and J. M. de la Cruz, "A wavelet-based image fusion tutorial," *Pattern Recognition*, vol. 37, pp. 1855–1872, 2004.

[66] E. J. Candès and D. L. Donoho, "Ridgelets: a key to higher-dimensional intermittency?," *Phil. Trans. R. Soc. Lond. A.*, pp. 2495–2509, 1999.

[67] M. H. Do and M. Vetterli, "The finite ridgelet transform for image representation," *IEEE Trans. Image Proc.*, 2000.

[68] D. L. Donoho, "Wedgelets: Nearly-minimax estimation of edges," *Annals of Stat.*, vol. 27, pp. 659–897, 19995.

[69] E. J. Candès and D. L. Donoho, "New tight frames of curvelets and optimal representations of objects with piecewise $C^2$ singularities," *Comm. Pure Appl. Math.*, vol. 57, no. 2, pp. 219–266, 2004.

# Index

A' Trous, 104
    Lagrange interpolator, 106
alternating flip, 71, 150, 156, 181
ambiguity function
    narrow band, 79
    wide band, 79
Aronszajn, 17

Balian-Low theorem, 87
Banach, 12
band-limited functions, 20, 38
basis
    biorthogonal, 43
    complete, 26
    complete and orthonormal, 25
    exact, 45
    finite, 7
    infinite, 25
    orthonormal, 51
    Riesz, 43, 50
    Schauder, 50
    unconditional, 50
biorthogonal
    MRA, 167
    symmetric wavelets, 209
    wavelets, 45, 163, 189, 207

Cantor, 3
cascade algorithm, 159
Cauchy sequence, 9
Cauchy-Schwartz inequality, 11
Coiflet, 206
column rank, 45
completeness, 12
condition number, 42, 49, 52

continuous wavelet transform, 90
    admissibility, 93
    completeness, 94
    dual frame, 103
    dual wavelet functions, 94
    frame, 103
    Haar, 100
    inverse, 94
    irregular grid, 101
    Mexican hat, 98
    Morlet, 96, 107
    Morlet scalogram, 108
    Parseval's relation, 94
    range space, 96
    regular grid, 102
    reproducing kernel, 104
    resolution, 92
    scale, 101
    translation invariance, 102
convolution
    continuous, 59
    discrete, 60
    finite length, 61
    theorem, 60
correlation, 65
countable, 3
curvelet, 258

Daubechies, 72
decimation, 69
denoising, 185
dimension, 7
Dirac
    delta, 18
    notation, 28

direct sum, 15
discrete cosine transform, 250
    energy concentration, 252
discrete wavelet transform, 104, 139,
        140, 142, 153, 173, 179, 231,
        245
    energy concentration, 252
    image fusion, 257
    redundant, 107, 143
    shift invariant, 107, 143
    undecimated, 107, 144
down-sampling, 69

eigenvalue, 15
entropy, 237

filter
    banks, 68
    finite impulse response, 63
    infinite impulse response, 63, 131
Fourier
    basis functions, 31
    series expansion, 34
    transform, 31
frame, 45, 50
    bounds, 50–52
    dual, 52, 54
    dual reconstruction, 52
    exact, 50
    inequalities, 50, 87
    operator, 51
    redundancy ratio, 52
    tight, 50, 51
functional, 15
    bounded, 16
    continuous, 16
    evaluation, 15

Gabor, 77, 88
Gibbs phenomenon, 27
Gramm-Schmidt, 20

Haar, 72
    MRA, 24, 112
    orthogonal projection, 115
    orthonormality, 120
    packet functions, 229
    resolution, 114
    scaling equation, 113
    scaling function, 113
    spectra, 122
    wavelet equation, 118
    wavelet function, 117
Hilbert space
    separable, 12

inner product, 9

JPEG, 251
JPEG2000, 252

Karhunen-Loeve expansion, 250
Kroenecker delta, 20

Laplace transform, 64
Laurent series, 64
lazy filters, 191, 216
Lebesgue integral, 3
lifting, 191, 215, 216
linear time-invariant system, 63

Meyer
    scaling function, 132
    wavelet function, 132
MRA, 23, 135

nested subspaces, 22
    multi-resolution, 23
    successive approximation, 23
Neumann expansion, 14, 55
norm, 10

operator
    adjoint, 14

basis, 42
bounded, 13
dilation, 90
linear, 13
modulation, 80
norm, 13
null space, 14
range space, 14
translation, 80
unitary, 15
orthonormal vectors, 20

Parseval's relation, 26, 33, 35, 37, 42, 82
partition of unity, 148
perfect reconstruction, 68, 70
Poisson summation formula, 36, 151
PR-QMF, 69, 71, 179
projection, 21
    orthogonal, 21
pseudoinverse, 47
PSNR, 256

quadrature mirror filter, 68, 70
quantizer, 251

reproducing kernel, 17, 27, 39, 54, 57
    Hilbert space, 15, 39, 54, 82, 95
resolution of identity, 19, 29, 31, 32, 96
ridgelet, 258
Riemann, 3
Riesz-Fischer theorem, 35
row rank, 46

sampling theorem, 40, 57
scaling equation, 136
Shannon
    MRA, 125
    scaling equation, 127
    scaling function, 173

spectra, 130
    wavelet equation, 128
singular values, 49, 50
spectral factorization, 67, 124, 202
spectral radius, 15
spectrogram, 80
spectrum, 66
spline
    function, 157, 171, 211
    wavelet, 214

thresholding, 187
triangle inequality, 8

uncountable, 3
up-sampling, 69

vector space, 5

Walsh functions, 229
wavelet
    admissibility, 148, 195
    compression, 249
    Daubechies family, 202
    packet basis, 231
    packet best basis, 238
    packet table, 231, 235
    packets, 221, 225, 228
    regularity, 199
    transform coding, 250
    two-dimensional, 241
    two-dimensional biorthogonal, 248
    vanishing moments, 194, 204
wavelet equation, 137
wedgelet, 258
windowed Fourier transform, 79
    compact window frame, 87
    completeness, 81
    discretized, 83
    dual window, 85
    inverse, 81

Parseval's relation, 82
range space, 81, 82
reproducing kernel, 83
resolution, 88

Z transform, 64